INFLUENCE OF CLIMATE

IN

NORTH AND SOUTH AMERICA:

SHOWING THE

VARIED CLIMATIC INFLUENCES

OPERATING IN THE

EQUATORIAL, TROPICAL, SUB-TROPICAL, TEMPERATE, COLD AND FRIGID REGIONS,

EXTENDING FROM THE

ARCTIC TO THE ANTARCTIC CIRCLE.

ACCOMPANIED BY AN

AGRICULTURAL, AND ISOTHERMAL MAP OF NORTH AMERICA.

COMPILED BY

J. DISTURNELL,

AUTHOR OF "INFLUENCE OF CLIMATE, RELATING TO THE WORLD," ETC.

NEW YORK:
PUBLISHED BY D. VAN NOSTRAND,
No. 192 BROADWAY.
AND FOR SALE BY BOOKSELLERS GENERALLY.

1867.

Stereotyped by SMITH & McDOUGAL, 82 and 84 Beekman St., N. Y.
Printed by JOHN J. REED, 43 Center St.

TO

PETER COOPER ESQ.

MY DEAR SIR :—

By your munificent endowment of "THE COOPER UNION FOR THE ADVANCEMENT OF SCIENCE AND ART," *in the City of New York, you have conferred on the rising and future generations one of the most princely gifts that has ever been bestowed on the American public. Owing to this consideration, and the liberal manner in which you and the Trustees of the Cooper Union have favored the* INSTITUTION *with which I am connected, and in grateful remembrance of the personal kindness I have uniformly received at your hands, allow me to dedicate this Volume to you.*

With feelings of great regard, I remain,

Your Obedient Servant,

JOHN DISTURNELL.

ROOMS OF THE "ASSOCIATION FOR THE ADVANCEMENT OF SCIENCE AND ART," COOPER INSTITUTE, NEW YORK, *January*, 1867.

ILLUSTRATION AND MAPS.

CONTENTS.

PART I.

PART II.

PART III.

PART IV.

PART V.

PART VI.

PART VII.

PART VIII.

PART IX.

PART X.

PART XI.

PART XII.

PART XIII.

PART XIV.

PART XV.

INTRODUCTION.

Climate, the most important and least understood of all the physical elements, has been most strangely neglected by the scientific writers of the New World. This can only be reasonably accounted for from the fact that the meteorological observations on the American Continent have, until recently, been confined to a few military stations and trading posts, extending over a wide range of country.

The early French discoverers in North America have left on record valuable observations in Lower Canada, commencing with Jacques Cartier, during the winter of 1535–36, which he spent near the mouth of the river St. Charles, in the vicinity of the present site of Quebec. These observations were continued by the Jesuit Fathers, with more or less regularity, for a long period in the ancient capital of the French possessions.

The employees of the Hudson Bay Company, in connection with researches of the English navigators in the northern portion of the continent, have, also, added much valuable information on the subject of Climate—all going to show a great increase of cold on the same parallels of latitude in Northeastern America, from that which exist in Northwestern Europe.

Monthly records of the weather have been compiled and published by a careful observer in Philadelphia [Charles Peirce], commencing in 1790, and running through a period of fifty-seven years, furnishing much valuable information in regard to climate, during the different seasons of the year. For the above period to the present date, there has been a gradual increase of heat, as shown by the mean annual temperatures, which may be ascribed to the clearing of the land, and the mild winters of late years.

Previous to the year 1818, however, we possess no record of meteorological observations, taken in the United States, on a reliable and extensive scale. In that year, Congress, acting upon the recommendation of the then Secretary of War, created the office of Surgeon-General of the Army. The Medical Department was not, however, fully organized upon its present basis, until the year 1821, when the surgeons at the different military posts were required "to keep a diary of the weather, and to note everything of importance relating to the medical topography of his station—the climate, diseases prevalent in the vicinity," &c., and transmit it quarterly to the Medical Bureau at Washington. The earliest registers thus forwarded and on file in the Surgeon-General's Office, are dated Jan., 1819.

Thus did the Medical Department, at its very origin, as a distinct branch of the staff of the Army, enter upon a system of meteorological observations, in which it has since been followed by the Topographical Bureau,* the States of New York, Pennsylvania, Ohio, &c., also by the Smithsonian Institution, and numerous Academies and Colleges located in different sections of the country.

"THE ARMY METEOROLOGICAL REGISTER," as published at different periods by order of the Government, embraces observations made by the surgeons of the army at all the military posts of the United States, from 1822 to 1860, which, in connection with the "MEDICAL STATISTICS OF THE ARMY," brought down to the same time, includes the full Meterological Reports, and a Statistical Report on the Sickness and Mortality in the Regular Army for a period of thirty-eight years.

The "RESULTS OF METEOROLOGICAL OBSERVATIONS" from 1854 to 1859, published in 1861, under the direction of the Smithsonian Institution, together with the "METEOROLOGICAL ABSTRACTS," contained in the Report of the Topographical Bureau relating to the Survey of the Great Lakes, on our northern frontier, all go to furnish additional information on this important subject, which, together with recent published observations along the Rocky Mountain range, and on the Pacific coast, altogether furnish facts of the most interesting and varied character, which are now being justly appreciated by the intelligent porton of the community.

These combined results, in connection with the United States Census of 1860, and the Census of Canada for 1861, furnish data by which the *Influence of Climate* on the vital subjects of population, health, and agricultural productions, can be favorably compared over the greater and most important portions of North America.

* This duty has been transferred to the Corps of Engineers of the U. S. Army.

"The term CLIMATE," says Dr. Forry, "which is limited, in its rigorous acceptation, to a mere geographical division, and in ordinary parlance to the temperature of a region, possesses, in medical science, a wider signification. It embraces not only the temperature of the atmosphere, but all those modifications of it which produce a sensible effect on our organs, such as its serenity and humidity, changes of electric tension, variation of barometric pressure, the admixture of terrestrial emanations dissolved in its moisture, and its tranquillity as respects both horizontal and vertical currents. *Climate*, in a word, as already defined, constitutes the aggregate of all the external physical circumstances appertaining to each locality in its relation to organic nature." 'To observe,' says Professor Rostan, 'the simultaneous effects of light, heat, electricity, of the winds, &c., on the organic productions of the different zones of the earth, to explore the nature of this earth, to deduce from this knowledge the influence which they exercise on the physical and moral state of man, such is the wide field which climates present to our investigation.'

"The little knowledge that we possess upon these various points, is far from being precise. On the one hand, we are ignorant of what constitutes the real elements of climate; and, on the other hand, these complex agents act upon living organs still more complex in their functions. Our knowledge heretofore has consisted mainly of the unexplained results of experience. As the subject does not admit of the precision of the exact sciences, the aid of induction and analogy must be invoked. Having once acquired a knowledge of the distinctive characters of different systems of climate, and of their effects upon the animal economy, both in health and disease, the general laws regulating such influences may be readily ascertained. In regard to the remaining elements of climate, such as the admixture of terrestrial emanations dissolved in atmospheric moisture, our positive knowledge is still more limited. That mysterious agent—Ma-la-ri-a—though too well recognised in its deleterious effects on the human frame, has hitherto remained inscrutable in its nature.

"It is thus seen that there are many circumstances besides mere temperature, which enter into the constitution of climate.

Amongst these, as influencing organized beings, one of the most important is the nature of the *soil*, the formation of which has apparently been the result of the gradual attrition of the solid materials composing the crust of the globe. As all animals and vegetables, at least all animals, are dependent for existence on this stratum of comminuted mineral substances and organic remains, its influence in regard not only to mere health, but the organic modifications which the human frame experiences, constitutes an interesting subject of inquiry."

The same author, in the valuable work, entitled "*The Climate of the United States and its Endemic Influences*," published in New York, 1842, remarks: "In regard to the climate of our own country, we possess no treatise founded on facts. Indeed, so little effort has been made to keep pace with the progress of kindred branches of science, that the work of M. Volney, written more than forty years ago, is still quoted by every writer on the subject. In relation to climate, nearly all facts stand isolated, and inasmuch as to render such data valuable, it is necessary that they be collated, thus determining their relations to one another and to general laws, the attempt has been made to present a systematic arrangement, so far as the facts collected will warrant, leaving the further prosecution of the subject to a period when new data shall have accumulated."

Professor Rogers says: "This continent, being more restricted in its dimensions, especially in its east and west diameter, than that of Europe and Asia, yet comprises almost as wide a range of heat and cold, and of dryness and humidity. Here these extremes are brought together within narrower limits; there they are expanded both in latitude and longitude, over wider zones. Here, geographically, their transitions are more abrupt; there their gradations are more gentle.

"This close packing together of the various belts of temperature and humidity, the result of closer proximity of the earth's tropical and polar currents, oceanic and atmospheric, occasions the several climates of the continent to act and react upon each other with greater potency. Hence the whole region is one of marked climatic contrasts, as striking, when we compare different districts, as when we regard the different seasons."

The Pacific or Western coast of America, extending eastward

to the summit of the Rocky Mountains, including the great basin of Salt Lake in Utah, possesses in many respects a marked difference of climate from the eastern portion of the continent facing the Atlantic. The Temperate Zone or climate takes a much wider range in the former, extending from Sitka, in Russian America, 57° north latitude, to near Monterey, in California, 36° north, running through upwards of twenty degrees of latitude, where the mean annual temperature ranges from 40° to 60° Fahrenheit. The more southern portion of the continent, including Lower California, assumes the tropical character of the Gulf coast of Mexico, embracing Central America and the Isthmus of Panama.

To the north of Mount St. Elias, 60° north latitude, the temperature is cold and forbidding, partaking of the character of the more eastern portion of the continent along the coast of Labrador. Westward, toward Alaska, the climate is modified by the warm waters of the Pacific Ocean, here flowing eastward toward the northwest coast of America. The cold influence setting down from the Arctic Sea, through Behring's Strait, being more perceptibly felt on the opposite side of Asia, along the coast of Kamschatka.

North America, on the Pacific side, may thus be said to be divided into three grand climatic divisions—giving from the Equator, northward, 36 degrees to the Tropical and Semi-tropical climate; 22 degrees to the Temperate climate, where the white race may live in safety and comfort; and 32 degrees to the cold and Frigid climate of the Arctic region.

The volume on the INFLUENCE OF CLIMATE now offered to the Public, is, so far as I am aware, the only American or English one containing, within a moderate compass, so large an amount of reliable information concerning the most important subject of *weather*, or climate, and presenting at the same time correct statistics as to population and agricultural products, all of which are closely allied to the subject under consideration. It is earnestly and respectfully hoped that it will prove useful, and that the further investigation be followed by more competent minds.

J. DISTURNELL.

NEW YORK, *January*, 1867.

CLIMATE.

PHYSICAL CLIMATE IS CHIEFLY DETERMINED by—

1. The Temperature of a country divided into Months and Seasons.
2. The Elevation of the land above the level of the sea.
3. The nature of the Soil; also, whether cleared or woodland.
4. The prevalent Winds.
5. The annual quantity of Rain or Snow that falls.
6. The great Oceanic Currents.

ATMOSPHERE.

Measure and Weight of the Atmosphere, according to Dr. Murray.

Constituent Parts.	By Measure.	By Weight.
Nitrogen gas, or impure air,..................	77.50	75.55
Oxygen gas, or pure air,.....................	21.00	23.32
Aqueous vapor,............................	1.42	1.03
Carbonic acid gas,........................	.08	.10
Total, ...	100	100

Mean Normal Temperature of the Northern Hemisphere of the Earth.

Latitude.	January.	July.	Yearly Mean.	Difference of the Hottest and Coldest Months.
°	° ′	° ′	° ′	° ′
90 *a*	—26 6	+30 6	+2 0	57 0
80 *a*	—20 5	34 1	6 8	54 6
70 *a*	—11 9	45 1	16 0	57 0
65 *b*	— 6 1	51 6	22 6	57 7
60 *b*	+ 3 6	56 4	30 2	52 8
50 *c*	19 8	62 6	41 7	42 8
40 *c*	40 4	72 3	56 5	32 2
30 *d*	58 6	78 4	69 8	22 0
20 *e*	70 1	81 8	77 5	11 7
10 *e*	77 3	80 8	79 9	3 6
0 *e*	79 4	78 6	79 7	2 7
Decrease from Equator to the Pole,........	106°	48°	77° 4′	

The mean Annual Temperature of the whole earth, at the level of the sea, is 50° Fahrenheit.

a. Frigid Zone, mostly within the Arctic Circle.
b. Cold Zone.
c. Temperate Zone.
d. Sub-Tropical Zone.
e. Tropical, or Equatorial Zone.

NOTE.—The *Arctic Circle* of the sphere, parallel to the equator, and distant 23° 28′ from the North Pole, from whence its name. This and its opposite, the *Ant-Arctic,* are called the two *Polar Circles.*

ZONES, OR BELTS OF TEMPERATURE.

THE Tropics, and Polar or Arctic Circles divide the surface of the Earth into *five* great climatic zones or belts, viz.:

1. One *Torrid Zone*, 47° in breadth, or 23½° on each side of the Equator, and bounded by the Tropics of Cancer and Capricorn. Every place in this wide region has the sun vertical to it twice a year; and as the sun's rays never fall very obliquely on any part of it, the temperature at the surface of the earth is here always very high, averaging from 78° to 84° mean annual temperature near the level of the ocean. Here the growth of vegetation is luxuriant and *man* indolent.

2. Two *Temperate Zones*, one northern and the other southern, each 43° in breadth, lying between the Tropics and the Polar Circles. This belt is properly divided into *three* parts, viz.: the Sub-tropical, Temperate, and Cold Zones. The *first* division has a mean annual temperature ranging from 78° to 60° Fahr.; here snow seldom falls nor is ice formed, while fevers of different types are prevalent. On the Atlantic coast of America, it extends from 23½° to 37° north latitude. The *second* division has a mean annual temperature ranging from 60° to 40° Fahr., extending from 37° to 47° north latitude, in the United States and Canada. Here snow and ice are found, particularly in the middle and northern portions. The middle of this zone or belt, 50° mean annual temperature, is the most favored climate on the earth's surface. The *third* division has a mean annual temperature ranging from 40° to 20° Fahr., and may be said to extend in British America from 47° to 66½° north latitude.

These zones, lying between the Tropics and the Arctic Circle, never having the sun vertical, are characterized by a lower temperature than tropical regions; the vegetation and fruits of the earth are less luxuriant and spontaneous; and *man*, compelled to exercise his corporeal and thinking powers, attains to a higher degree of intelligence and civilization than in those regions where his wants are supplied without any exertion on his part.

3. The two *Frigid Zones*, each 23½° in radius, are included within the Polar Circles, called the Arctic and Ant-arctic Circles, by way of distinction. They are deprived of the influence of of the sun for long intervals in winter, and have a correspondingly greater length of day in summer, when his rays fall very obliquely on the surface. These conditions, coupled with the extreme cold of the long winters, are so unfavorable to human culture and human happiness, that the tribes who inhabit the

Northern Frigid Zone have not been able to attain to any considerable degree of civilization—the Southern Frigid Zone being entirely destitute of settlement by the human family. The mean annual temperature in this frigid inhospitable climate varies from 20° to 0°, and even below zero.

It thus appears that about one-third of the earth's surface is too hot, and one-third too cold for the advancement of the human species, the remaining third embraced within the Temperate Zones alone, being fitted for progress and the full development of the human race.

Isothermal Lines and Climatic Zones.

"As the temperature of any place depends on a multitude of causes besides latitude, it is obvious that the old designations of *Torrid, Temperate, and Frigid Zones,* bounded by the Tropics and Polar Circles, do not adequately express the temperature, and far less the general climatic character, of the different parts of the earth's surface. Humboldt and others have accordingly substituted other lines instead of the parallels, as the true boundaries of climatic zones; viz., *Isothermal, Isocheimenal* and *Isotheral lines.* The mean annual temperature of any given place may be readily ascertained by means of the thermometer; and imaginary lines connecting together all the places in the same hemisphere, having the same *mean annual temperature,* are called Isotherms. The Isocheimenals are similar lines connecting places that have the same *winter temperature,* and the Isotheral lines are drawn between places having the same *summer temperature.* These lines of equal temperature approximate more or less to the direction of the equator, though they are nowhere parallel to it. They diverge more from it in the northern than in the southern hemisphere, and greatly more in high than in low latitudes. The hottest portion of the earth's surface is an oval shaped tract in East Africa, extending from Lake Tchad to Mecca, and the Strait of Babelmandeb, having a mean annual temperature of 83°; and the coldest, so far as yet ascertained, is a long narrow belt in the Arctic Ocean, midway between Behring Strait and the North Pole, and extending from Melville Island, in the direction of New Siberia, with an average temperature of 0° Fahr. It appears, therefore, that the hottest region is not under the Equator, nor the coldest under the Pole; and that all the lines of equal temperature in the northern hemisphere attain their highest latitude in the eastern side of the Atlantic Ocean—owing, no doubt, to the high temperature of the Gulf Stream, which flows northward along the western shore of Europe. By means of these Isotherms each hemishphere is divided by the meteorologist into *five* climatic zones, named respectively, the meteorologist into *five* climatic zones, named respectively, the *hot*, or equatorial, the *warm*, *temperate*, *cold*, and *frigid*, or polar zone."

Natural Influences that Produce Rain, &c.

From the facts established in regard to the great Ocean Currents, sweeping across the Atlantic and Pacific Oceans, as well as across the Indian Ocean, it seems evident that the abundance of moisture in the shape of rain which falls in the West Indies, and along the Gulf coast of Mexico (90 to 110 inches), as well as the southern coast of the United States, including Texas, Louisiana, Alabama, Florida, and the Carolinas, is due to the influence of the great Equatorial Current and Gulf Stream.

This benign influence, no doubt, extends far inland, until it is met by other currents of air, produced by similar causes in nature. It penetrates the valley of the Mississippi far inland, being met by a counter descending current of air, sweeping across the Rocky Mountains, north of the 49th parallel of latitude. This counter current of air, charged with a less amount of moisture, no doubt, proceeds from the influence of the great Pacific current first passing over Russian and British America, in its course toward Dakota and Minnesota.

The annual amount of rain falling in the Southern States is fifty-one inches, the quantity decreasing as you ascend northeastward; the Middle States and New England States having a fall of about forty inches annually, while the annual fall of rain and snow in the Northwestern States is only about thirty inches; the average quantity falling in the United States being thirty-six inches.

Sweeping across the Atlantic Ocean in a northeast direction, the Gulf Stream carries its benign influence to Ireland and the west coast of England, continuing still northward to the coast of Norway, and within the Arctic Circle, carrying in its extended course warmth and moisture.

The Pacific current alike performs a beneficent part in distributing heat and moisture along the extended western coast of North America, from the Aleutian Islands to Lower California. The amount of rain decreasing as you proceed southward.

The temperature of this extended region seems to be influenced differently from any other portion of the globe, there being nearly an uniform temperature of many hundred miles running from north to south, along the coast and valleys of Oregon and California. The same mean annual temperature, about 52° Fahr., being found to exist at Astoria, Oregon, as at Fort Humboldt, Cal., situated near Cape Mendocino, the two posts being separated by about five and a half degrees of latitude.

Rain is, in general, most abundant about the Equator, and the quantity decreases in a direction toward the Poles; because heat, which is the origin of vapor and the cause of rain, decreases in the same direction.

This decrease occurs in the following order:

The annual amount of rain under the Tropics of the New World is	112 inches.
The annual amount of rain under the Tropics of the Old World, ...	80 "
Within the Tropics generally, ...	96 "
In the Temperate Zone of the New World (United States), ...	38 "
Temperate Zone of the Old World (Europe), .	34 "
Average within the Temperate Zone, . .	36 "

Annual Mean Temperature.

Starting from the Equator, and going north, along the meridian of the city of Washington, we find the following singular results in regard to annual mean temperature:

STATIONS.			DEG. FAHR.	STATIONS.			DEG. FAHR.
Quayaquil, near the Equator,			85°	New York, .	. 40° 42′	N. Lat.	51°
Panama, . .	8° 56′	N. Lat.	80°	West Point, N. Y.,	41° 23′	"	50°
Kingston, Jamaica,	18°	"	78°	Albany, N. Y.,	. 42° 37′	"	48°
Havana, Cuba, .	23° 9′	"	75°	Montreal, Can.,	. 45° 30′	"	45°
New Orleans, .	29° 57′	"	70°	Quebec, Can.,	. 46° 49′	"	41°
Charleston, S. C.,	32° 45′	"	67°	Nain, Labrador,	. 56° 10′	"	28°
Norfolk, Va., .	. 36° 50′	"	60°	Hudson Bay, .	. 59°	"	20°
Washington, D. C.,	38° 53′	"	56°	Fox Channel,	. 65°	"	10°
Philadelphia, .	. 39° 57′	"	52°	Within the Arctic Circle, .	. 70°	"	0°

From this exhibit, it seems conclusive that the mean annual temperature on the Continent of America, by Fahrenheit's scale, corresponds to the degrees of latitude—thus there being 90 degrees of latitude from the Equator to the Pole, the mean annual temperature varies from *zero*, and below, to 85° Fahr. on the hottest portions of the earth's surface. Allowing for an increase of cold as you approach north, toward the Pole, of five degrees less annual temperature, and you have the result verified. The further fact, that Montreal, Canada, situated in 45° 30′ north latitude, about midway between the Equator and North Pole, having a mean annual temperature of 45° Fahr., is further convincing proof of the above statement.

PART I.

CLIMATE OF NORTH AMERICA.

The Continent of North America, extending from the Isthmus of Panama, near the 9th parallel, to within the Arctic Circle, possesses every variety of climate from the tropical heat of 80° Fahr. to the frigid cold of Zero, mean annual temperature. "It differs from the eastern hemisphere by a greater predominance of cold; it being calculated that the heat is at least ten degrees less upon an average on the American Continent, than under the same parallels in Europe. Thus, while Denmark and the southern part of Sweden enjoy comparatively temperate seasons and mild air, Labrador, and the countries inhabited by the Esquimaux, though lying in about the same parallel, are extremely cold—even the Torrid Zone of America, contiguous to the Caribbean Sea, knows none of those intense heats which are experienced in Asia and Africa." These remarks apply more particularly to the *eastern* than the western side of the Continent of America.

"In North America, cultivation is extended on the eastern coast, and near Hudson Bay, to the parallel of 54° north; in Asia, it has reached 57°; but in Europe it has been carried as high as 70°, to near North Cape, in Lapland."

On the west coast of America, in the Russian Possessions, and on Mackenzie's River, in British America, cultivation extends to 60° north, and upwards; being, no doubt, favorably influenced by warm currents of air from off the Pacific Ocean.

"The great cause of the cold in North America has been attributed to the quantity of land stretching towards the North Pole, a proportion of which is involved in perpetual winter. The wind, in passing over this snow and ice-clad region, brings, it is said, a severity of cold which nothing resists in sweeping over Baffins' and Hudson Bay. The whole of the central and eastern portion of America, north of the 55th parallel, may be considered a frozen inhospitable region. In Greenland, Labrador, and around Hudson Bay, brandy freezes during winter, which begins about the 1st of September, and continues for nine months. In summer the heat is often as great as in New England; but continues for too short a period to bring grass or vegetables to maturity. Between 55° and 47° north, the climate is still severe; the cold of winter is steady and intense, and the snow, which begins to fall in November, remains till May. Here summer advances with such rapidity, that the season of spring is scarcely known." The above described regions includes the *Cold and Frigid Zones* of America.

The *southern* limit of this great belt, extending to 47° north latitude on the eastern portion of the continent, rises as you approach Lake Winnipeg to about 50°, rising to 57° as you approach the North Pacific coast in the Russian Possessions.

The stretch of country along the line of temperature that divides the cold and temperate zones (40° Fahr.) is so distinctly marked, as regards cultivation and settlement, that an observing traveller passing west from near Quebec to the head of navigation on the Ottawa River, and thence to the north shore of Lake Superior, passing up the St. Mary's River, can perceive the last vestige of civilized habitation. This marked and singular circumstance can only be attributed to the liability of killing frosts all along this extended line during the summer months—potatoes, wheat, and other hardy vegetables, and cereals, being destroyed for the want of a season of uninterrupted heat sufficient to bring them to perfection.

The *temperate* portion of America, on the Atlantic coast, may be considered as extending from 47° to 37° north latitude, (Quebec, Can., to Norfolk, Va.) The southern limit of this zone (60° Fahr.) is also strongly marked by different vegetable productions, and a different class of diseases peculiar to the human family. Here commences the cotton region, and the prevalence of malignant fevers; there being no severe frosts, and little or no snow during the winter months.

The *Semi-Tropical Zone* may be said to extend from 37° to 25° north latitude, or from Norfolk, Va., to the mouth of the Rio

Grande, in Texas. Here is another marked feature in the vegetable kingdom, the trees becoming dwarfed with a rank undergrowth of vegetation.

The *Tropical Zone* extends from 25° north latitude to the Equator, where the animal and vegetable kingdoms are of a tropical or equatorial character, contrasting strongly with the productions of the more northern zones, or sections of country.

Thus, North America, *east* of the Rocky Mountains, may be justly divided into *five* grand Climatic Divisions, being marked by different races of men, or traits of character, and different vegetable productions. *West* of the Rocky Mountain range is found a climate *peculiar* to the region of the Pacific coast of America. Within this wide extent of country, running through 90 degrees of latitude, from the Equator to the North Pole, may be found almost every variety of the vegetable kingdom—from the stunted growth of the Arctic Circle to the lofty white pine of the North or Temperate Zone, the yellow or pitch-pine of the South, or Sub-tropical Zone, and the rank evergreens of the Torrid or Equatorial Zone.

To the extreme north the Esquimaux alone reside, unused to all the comforts and luxuries of civilized life; while south of Hudson Bay and the coast of Labrador, and north of Quebec, the European settler and the more hardy tribe of Indians are found in limited numbers, being mostly engaged in hunting and fishing. South of 47° north, running through the temperate and semi-tropical climate to the 25th parallel, the pure white race holds dominion on this continent, although the Indian and African races are intermingled.

South of the Rio Grande, within the confines of Mexico, the climate becomes tropical. Here are to be found the mixed Spanish race, their blood being so far intermingled with the Indian and African races as to make them an inferior people, when compared with the pure Castilian race, whose blood flowed in the veins of the early conquerors of Mexico and Central America.

Another distinct feature of the nature that marks these great climatic divisions of North America, are, that the tropical region is divided into but two seasons, the *wet* and the *dry*, where by far the greatest amount of moisture falls during the wet season, often in deluging torrents of rain, accompanied by fierce winds. The temperate region has four regular seasons, Spring, Summer, Autumn, and Winter. These different seasons are usually attended with about the same amount of moisture, a portion, however, falling in the shape of snow—this latter phenomena showing an exemption from malignant contagious fevers, which are the scourge of the semi-tropical and tropical portion of the continent. The extreme northern or cold region, which may be termed *hyperborean*, although free from malignant diseases, yet unfits man to acquire the full development of his higher faculties—hence we see in this dreary section a half-savage race, which are alone fitted by nature to live in northern latitudes.* Again, the food and the clothing necessary for the comfort of these different races of men are as distinct as the climate, showing plainly that it is at the risk of life to change suddenly the residence of man, as well as the inferior animals, from one extreme temperature to one of a marked opposite influence.

A late English writer says, "In a continent embracing 75 degrees of latitude, and nearly twice as many of longitude, the varieties of climate must be very great. Speaking generally, however, we find that its various portions have a lower average temperature than the corresponding latitudes of the Old World.

"The western side of the continent, however, is greatly warmer than its eastern. For example, in Russian America, the Island of Sitka has a mean annual temperature of 42° Fahr., while Nain, in Labrador, near the same latitude, 57° north, has a mean temperature of only 29°. The difference between the mean summer and the mean winter temperatures is still greater; for while at Sitka and San Francisco it amounts to only 22°, at Nain and Quebec it amounts to 44° and 54°.

"The hottest portion of the New World lies mainly within this continent, being embraced within the isothermal line of 81° Fahr., which encloses the Gulf of Mexico and the Caribbean Sea—that great cauldron of heated waters which originates the Gulf Stream—the West Indies, the eastern side of Mexico and Central America, and the northern part of New Granada on the

* *Man* is found from the 75th degree of north latitude (in North America) to Terra del Fuego, 55° south (in South America), with this marked difference, that the dwarfed Esquimaux of the northern continent is strangely placed in contrast with the full-grown race of the Patagonians of the southern.

Isthmus of Panama. The coldest region of North America and of the New World is embraced within the isochimenal line of 26° below zero—a line which, commencing at Cape Bathurst, near the mouth of the Mackenzie, deflects southeastwardly to the head of Chesterfield Inlet (south of the Magnetic Pole), and thence northwards to Lancaster Sound and North Devon. North America is also more humid than the corresponding latitudes of the Old World. It is calculated that 110 inches of rain fall annually in tropical America, while in tropical Asia and Africa the amount does not exceed 80 inches. In the temperate regions of the eastern continent the annual average is estimated at 34 inches, while it amounts to 38 inches in the corresponding zone of the western. The rainiest region of this continent corresponds with the region of highest temperature above described, in which, as well as in Lower California, Southern Mexico, and all Central America, snow never falls."

BOTANY.—The same writer remarks: "The New World has long been famed for the prodigious luxuriance and variety of its vegetation, as well as its peculiar climate. When the northern continent was discovered, one vast continuous forest covered the whole surface, from the St. Lawrence and the Great Lakes to the Gulf of Mexico, along the Appalachian range, and from the Rocky Mountains on the north to the Atlantic, embracing an area of upwards of two millions of square miles. Much of this ocean of vegetation has since been cleared away, though, to this day, hundreds of miles of unbroken forest exist in several localities, while boundless prairies, destitute of trees, but covered with tall grasses, occupy vast tracts in the centre of the continent, and on the eastern side of the Rocky Mountains. The forest trees are extremely numerous in species, embracing many varieties of oak and pine, with the ash, birch, beech, cedar, chestnut, cotton-wood, cypress, juniper, hickory, locust, maple, mulberry, poplar, and walnut. As the traveller passes northwards into the British territories, the variety of species is smaller, embracing mainly pines, cedars, larches, aspens, poplars, alders, hazels, birch, and willows; while towards the shores of the Arctic Ocean the trees become fewer in number and more stunted in size, till at length the dwarf willow, six inches in height, is the sole representative of the gigantic trees of the temperate and tropical regions."

"WHEAT is the cereal which requires most heat of those usually cultivated in England. Its culture is said to ascend to 62° or 64° north latitude, on the west side of the Scandinavian peninsula, but not to be of importance beyond the 60th. On the route of the Expedition,* it is raised with profit at Fort Liard,

* Sir John Richardson in search of Sir John Franklin, 1848.

British America, in latitude 60° north, longitude 122° 31′ west, and having an altitude of between 400 and 500 feet above the sea. This locality, however, being in the vicinity of the Rocky Mountains, is subject to summer frosts, and the grain does not ripen perfectly every year, though in favorable seasons it gives a good return. It grows, however, freely on the banks of the Saskatchewan and Lake Winnipeg, except near Hudson Bay, where the summer temperature is too low.

"At Fort James, on the borders of Stuart's Lake, in latitude 54½° north, in a mountainous region, near the source of Frazer's River, wheat continues to grow, but often suffers from the summer frosts. In these quarters the grain comes to maturity in about four months. In the colony of Red River its growth is luxuriant, though the upper part of that country, which touches the 49th parallel of latitude, is elevated about 900 feet above the sea. Periodical ravages of grasshoppers, however, frequently destroy the hopes of the husbandman.

"At Fort Francis, situated on the banks of Rainy River, in latitude 48° 35′ north, 93° 28′ west, wheat is generally sown about the first of May, and is reaped in the latter end of August, after an interval of about 120 days."

"*Potatoes*, which have been cultivated from time immemorial on the banks of Lake Titicaca, South America, yield abundantly at Fort Laird, and grow, though inferior in quality, at Fort Simpson, 62° north latitude.

"On the Island of Sitka, lying in 57°–58° north latitude, though the forest, nourished by a comparatively high mean temperature and a very moist atmosphere, is equal to that of the richest woodlands of the Northern United States, yet corn does not grow."

The climate of South America is generally superior to that of the northern part of the continent in all the districts north of the 50th parallel of southern latitude; but to the *south* of that line, the cold increases more rapidly than it does as we approach the Pole in the Arctic regions. Among other causes which powerfully influence the temperature in South America and Central America, must be reckoned the extraordinary elevation of the surface in many places. Thus the city of Mexico (19° 25′ north), which according to its latitude, should be excessively hot, being elevated 7,500 feet above sea-level, enjoys a climate of perpetual spring; and Quito, elevated 9,543 feet, which lies under the Equator, has a similar climate, though within sight of that city are regions, at an elevation of 16,000 or 18,000 feet,* which are covered with never-changing masses of snow and ice; and at the distance of a few miles, the inhabitants of Guayaquil, on a low and level margin of the sea, experience an intense and sickly degree of heat.

"The three zones of temperature which originate in America," says Malte Brun, "from the enormous difference of level between the various regions, cannot by any means be compared with the zones which result from a difference of latitude. The agreeable, the salutary vicissitudes of the seasons are wanting in those regions that are here distinguished by the denominations of *frigid, temperate, hot* or *torrid*. In the Frigid Zone it is not the intensity but the continuance of the cold—the absence of all vivid heat—the constant humidity of a foggy atmosphere, that arrest the growth of the great vegetable productions, and, in man, perpetuate those diseases that arise from checked perspiration. The Hot Zone of America does not experience excessive heat; but it is a continuance of the heat, together with exhalations from a marshy soil, and the miasmata of an immense mass of vegetable putrefactions, added to the effects of an extreme humidity, which produces fevers of a more or less destructive nature, and spreads through the whole animal and vegetable world the agitation of an exuberant but deranged vital principle. The Temperate Zone, by possessing only a moderate and constant warmth like that of a hot-house, excludes from its limits the animals and vegetables which delight in the extremes of heat and cold, and produces its own peculiar plants, which can neither grow above its limits, nor descend below them. Its temperature, which does brace the constitution of its constant inhabitants, acts like spring on the diseases of the hot regions, and like summer on those of the frozen regions: accordingly, a mere journey from the Andes to the level of the sea, or *vice versa*, proves an important medical agent, which is sufficient to produce the most astonishing changes in the human body. But, living constantly in either one or the other of these zones, must enervate both the mind and the body by its monotonous tranquillity. *Summer, Spring* and *Winter* are here seated on three distinct thrones, which they seldom quit, and are con-constantly surrounded by the attributes of their power."

Mountains and Rivers.

The *Mountain* and *River Systems* of North America, when viewed in their proper light, in connection with a healthy climate, are of the most grand and beneficent character, as regards

* Height of the snow-line in the Andes of Quito, 15,800 feet.

their influence on the animal and vegetable kingdoms, adapting most of this immense region to the abode of civilized man—now being fast peopled by different races emigrating from Europe and Asia.

The two great mountain systems which give rise to almost innumerable streams, are the Alleghany or Appalachian range of mountains, traversing the Atlantic States, and the Rocky Mountains and coast range of the Pacific coast. The Rocky Mountain range, forming the great axis or vertebral column of the continent, is by far the most extensive, running through the entire length of North America, under different names, a distance of five or six thousand miles, presenting in their course fertile plains, and mountain peaks extending upward above the line of perpetual snow.

On the western slope of the Rocky Mountains, the Columbia and other rivers of considerable magnitude rise and flow into the Pacific Ocean and Gulf of California—all being fed and supplied with water by a system of rain and snow, operating alike, in a greater or less degree, across the continent, varying from twenty to sixty inches in depth during the year. This abundant and constant supply of moisture, falling in the shape of rain or snow, most plentiful in hilly or mountainous tracts of country, goes to swell the numerous streams flowing onward to the ocean.

The vast regions east of the summit of the Rocky Mountain range, extending from the Arctic Circle to the Tropic of Cancer, embracing the great valleys of Hudson Bay, the Mississippi, and St. Lawrence, are drained by noble rivers entering the Arctic Ocean, Hudson Bay, and the Gulf of St. Lawrence on the north, and the Gulf of Mexico on the south.

The Great Lakes or Inland Seas of America, embraced in this region, are of themselves the largest and purest body of water on the face of the globe, having an estimated area of 90,000 square miles of surface. The three great upper lakes, standing from 565 to 600 feet above the ocean level, with a depth varying from 100 to 800 feet, are surrounded by a healthy and fertile region of country. The St. Lawrence River, the outlet of those mighty waters, flows northeast into the Gulf of

St. Lawrence, by a succession of falls and rapids, the most remarkable of any stream on the face of the globe.

The Appalachian range commences in Alabama and Georgia, extending in a northeast direction some one or two hundred miles from the Atlantic coast, through the States of South and North Carolina, Eastern Tennessee, Virginia, Maryland, and Pennsylvania to the confines of the State of New York. Here the system is continued by the "Highlands," and Taghkanic Mountains of New York and Massachusets, when the Green Mountains of Vermont are reached, and the water-shed or system is continued through Lower Canada to the District of Gaspé, terminating on the Gulf of St. Lawrence. The highest peak of the whole extended range is the Black Mountain in North Carolina, which rises to an elevation of 6,476 feet above the ocean.

The streams and navigable rivers rising in this mountain range, and flowing into the Atlantic Ocean on the east, and the Gulf of Mexico on the south, are very numerous.

The *Adirondack* or Clinton range of mountains lying in the State of New York, extending from the Canada border to the Mohawk Valley, are an elevated range, varying from 1,000 to upwards of 6,000 feet, while the Catskill Mountain rises west of the Hudson River, forming an independent spur, being elevated about 3,000 feet.

The *White Mountains* of New Hampshire are an independent group of great interest, attaining an elevation of 6,428 feet above the ocean. The waters flowing from these several mountains find their way into the Gulf of St. Lawrence and the Atlantic Ocean by numerous streams.

For a healthy climate, fertile soil, rivulets and inland navigation, no stretch of country on the globe, for the same extent of territory, equals that lying between the Gulf of Mexico on the south, and the Gulf of St. Lawrence on the north, being drained by the Appalachian range of mountains and hills, abounding in noble forests and rich mineral deposits.

Great River Basins and Valleys.

1. *Inclined to the Arctic Ocean.*

	Length in Miles.	Area.
Mackenzie River, ...	1,200	440,000
Saskatchewan and Nelson, including Red River of the North, ...	1,500	450,000

2. *Inclined to the Pacific.*

Frazer River, ...	450	30,000
Columbia	1,000	200,000
Sacramento	350	25,000
Rio Colorado, ...	750	170,000

3. *Inclined to the Atlantic.*

St. Lawrence River and Great Lakes,	1,500	300,000
Susquehanna and Chesapeake Bay, .	450	20,000

4. *Inclined to the Gulf of Mexico.*

Mississippi River, ...	2,000	1,000,000
Rio Grande del Norte, ...	1,000	180,000
Tabasco, ...	250	20,000

The lesser valleys are those of the Connecticut, Hudson, Delaware, &c., along the Atlantic and Gulf coasts.

The great basins and valleys modify the climate in their different regions in connection with prevailing winds—thus we find the valley of the Red River of the North, in 50° north latitude, comparatively warmer than the Lower St. Lawrence on the same parallel.

Mountain Peaks.

	°	′		Altitude in Feet.
RUSSIAN AMERICA.				
Mount St. Elias, coast range . .	60		N. Lat.	16,860
Mount Fairweather, coast range, .	59		"	14,783
BRITISH AMERICA.				
Mount Brown, Rocky Mountains, .	52	30	"	16,000
Mount Hooker, Rocky Mountains, .	52		"	15,700
UNITED STATES.				
Mount Hood, " "	45	20	"	16,500
Mount St. Helens,* Cascade Mountains,	46	10	"	15,500

* Active volcano.

	°	′		Altitude in Feet.
Mount Rainer, Cascade Mountains, .	46	40	N. Lat.	14,500
Mount Baker, " "	48	40	"	14,000
Mount Adams, " "	46		"	12,000
Mount Jefferson, " "	44	30	"	12,000
Fremont's Peak,* Rocky Mountains,	43		"	13,570
Sierra Nevada, California, (*est.*) . .	38		"	12,500
Mount Shasta, "	41		"	14,500
Pike's Peak, Colorado Ter., . .	38	30	"	12,000
Long's Peak, " . . . ,	40	30	"	11,500
WHITE MOUNTAINS, N. H.				
Mount Washington, ...	44	15	"	6,285
Mount Adams, ...				5,960
Mount Jefferson, ...				5,860
Mount Madison, ...				5,415
Mount Monroe, ...				5,350
Mount Franklin, ...				4,850
ADIRONDACK GROUP, NEW YORK.				
Mount Marcy, ...	44	10	"	5,467
Mount McIntyre, ...				5,183
Mount Seward, ...				4,000
Mount Katahdin, Maine, ...	45	45	"	5,335
Mount Mansfield, Green Mountains, Vt.,	44	20	"	4,280
Alleghany Mountains, Pennsylvania and Virginia, ...				3 to 4,200
Black Mountain, North Carolina, .	35	40	"	6,420
MEXICO AND CENTRAL AMERICA.				
Popocatepetl, Volcano, Mexico, . .	19		"	17,720
Orizaba, " " . .	19	20	"	17,500
Line of perpetual snow, . . .	19		"	14,500
Iztacihuatl, " " . .				15,700
Toluca, " " . .				15,168
Cofre de Perote, Mexico, . . .	19	45	"	13,415
Agua, Volcano, Guatemala, . . .	14	15	"	13,758
Amilpas, " . .	15	10	"	13,160
Atitlan, Volcano, " . .				12,506
Irasu, " " . .				11,478
Votos, " " . .				9,848
Mount Omoa, " Honduras, .	15	40	"	7,000

* Elevated 1,000 feet above the line of perpetual snow.

PART II.

ARCTIC LANDS AND OCEAN.

THIS is a vast dreary region, lying within the Arctic Circle, north of the parallel of 66° 32′, embracing the northern portion of the Continent of America, including all the numerous islands discovered near the North Pole by English and American navigators. The *Arctic Circle* is one of the smaller circles of the sphere, running parallel with the Equator, and 23° 28′ distant from the Pole. In crossing North America, it passes Behring Strait, Great Bear Lake in British America, Back River, south of the Magnetic Pole, Fox Channel, and Davis Strait, striking Greenland at Holsteinberg, and passing immediately north of Iceland. Within these bounds lie Grinnell Land, Washington Land, Parry Islands, Melville Island, Prince of Wales Island, Prince Albert's Land, King William's Land, Boothia Felix, Cockburn Land, and a portion of the continent proper, together with the greater part of Greenland.

The Arctic Ocean lies within this but partially explored region, also, the open *Polar Sea*, which is supposed to exist, and is now engaging the attention of explorers and scientific men, both in Europe and America. A portion of this *terra incognita*, or "Arctic Highlands," is thus described by Capt. Ross, R. N., in 1818, lying between 76° and 78° north latitude, "bounded on the south by an immense mountain barrier, covered with ice, with cliffs 1,000 feet or upwards in height, and spurs of solid ice projecting for miles into the sea. The vegetable productions are heath, moss, and coarse grass, which afford shelter to hares and other game." The thermometer in these regions, during the month of July, sometimes rises to 40° and 50° Fahr., and in winter falls to 50° and 60° below zero, often freezing the mercury.

Dr. Kane, of the American Navy, who penetrated as far north as 81°, in 1854, describes the formidable barriers of ice and snow in a still more vivid and dreary form, having seen mountains of glaciers extending for many miles along the north coast of Greenland, in the vicinity of 79° and 80° north latitude. Here the seasons are divided into six months of daylight and six months of darkness. It was left for two of his adventurous companions to penetrate to about 81° 20′ north latitude, where an open *Polar Sea* was discovered of vast extent.

Within the Arctic Circle the only permanent human inhabitants are the *Esquimaux*, a race scattered along the coasts of Labrador and Hudson Bay, and the Arctic shores, extending from Behring's Strait to Lancaster Sound, and Greenland; they are entirely dissimilar in manners and character to the other Indian tribes of North America. They are generally located to the north of the parallel of 68°, but are met with as far south as 52° north, in Labrador. The name which has been given them by the Indians of the north, signifies "eaters of raw flesh." Their color is not that of copper, but approaches the tawny brown which distinguishes the inhabitants of the more northern parts of Europe. They have beards, and some of them have been observed with hair of different colors, in some fair, in others red, though the prevailing color is black. These marks, by which they are evidently distinguished from the American Indian, have inclined several savans to believe that they are of European descent. "It is a singular fact that tribes of this description, agreeing in form, features, and manners, and apparently of kindred race, occupy the whole shores of the Polar Sea in Europe, Asia, and America. One would almost suppose that this variety of the human species had been created expressly to tenant those frozen regions to which their mode of life appears to attach them, as the Negro seems adapted by an opposite organization to the scorching heats of the Torrid Zone." The Esquimaux, when first met by Captain Ross's exploring party in high latitudes, supposed themselves to be the only people in the world, showing that they had lost all knowledge of even kindred races in other parts of the globe. They are lower in stature than the other Indian races, as well as Europeans, if we except the Laplanders. Their visage presents the peculiar form which the face assumes in intensely cold weather; their dress is more ample and prepared with more care than is usual among the savages. They subsist mostly on fish and animal food, which abounds in northern latitudes, beyond the limits of vegetation.

"The meteorological phenomena of the Arctic regions are of a highly interesting character. Besides the recurrence of day and night produced to a length very different from that which is usually indicated by those terms, and varying in proportion to the latitude, the Aurora Borealis cannot fail to strike the observer with wonder and admiration during the winter nights. These beautiful streams of light, forming or tending to form an irregular arch, with 'showers of rays' shooting in every direction, brilliant and rapid as lightning, supply, in some degree, the absence of the sun, and impart an air of fairyland to the scene. They are variously described—by some they have been supposed to be attended with a 'hissing and cracking noise;' but this has been contradicted by others. To Capt. Parry the light appeared to be tinged with yellow and blue; while to Captain Lyon it resembled the milky way or vivid sheet-lightning. From the observation of Captain Back, the Aurora Borealis appears to exert a sensible influence on the magnetic needle. The sun and moon are often surrounded here with halos tinted with the brightest hues. Parhelia, or mock suns, shine in different parts of the firmament, several in number, and most brilliant about the period of sunrise and sunset; and various luminous meteors, due probably to the refraction of the light by the crystals of ice, adorn at uncertain intervals the northern sky.

"The bleak and inhospitable regions of the Frozen Zone teem with animal life: the whale, the walrus, and the seal not only afford light and clothing to the inhabitants, but invite the adventurous mariner of more favored climes to visit these ice-bound coasts. The polar bear, the wolf, and the fox, are the antagonists, the dog and the reindeer are the friends and coadjutors, the musk ox and the elk, or moose deer, and hares of various kinds, supply the principal nourishment of the inhabitants. Innumerable flocks of birds, the auk, the petrel, and the gull, the swan, the goose, and the duck, the tern, the plover, and the ptarmigan, animate the air or the waters; and the spring and summer are cheered with the notes of many song birds. Insect species are few; but in the short and sudden summer even mosquitoes are found. The vegetable world is likewise very limited. The American spruce or white-pine is said by Dr. Richardson to be 'the only tree that the Esquimaux of the Arctic Sea have access to while growing.' They are chiefly supplied by drift-wood. The mosses and the lichens are the principal products of the soil. There are mushrooms, ferns, algæ, and confervæ. The summer is not wholly destitute of the ranunculus, the anemone, and the poppy, the strawberry, the raspberry, and some other edible fruits. The *Protococcus nivalis*, a minute cryptogamic plant, imparts a general rosy tinge, which has occasioned the inconsistent appellation of red snow." —*English Gazetteer*."

The general character of the climate in these regions is that of intense cold, ranging so low as 50° below zero, or even much lower. In a journal kept by Captain Back, at Fort Reliance, north latitude 62° 50′, the thermometer is recorded to have stood at 70° below zero. It has been observed that the extreme point of cold, as indicated by the thermometer, is often less severely felt in these northern latitudes than in a more moderate temperature; the air in such circumstances being calm and clear.

The Magnetic Pole.

It may gratify our readers to give the result, in a simple and plain manner, of Captain Ross's attempt to reach and discover the *Magnetic Pole*, where is supposed to be found the greatest degree of cold on the earth's surface: "The place of observation," he says, "was as near the Magnetic Pole as the limited means which I had enabled me to determine. The amount of the dip was 89° 59′, being within one minute of the *vertical*, while the proximity of the Pole, if not its actual existence where we stood, was further confirmed by the total inaction of the several horizontal needles in my possession. These were suspended in the most delicate manner possible, but not one showed the least effort to move from the position it was in—a fact, which the most uninformed on the subject must know, proves that the centre of attraction lies at a very small horizontal distance, if at any." Captain Ross proceeds to state, "that a learned professor in England had, in the absence of the expedition, laid down all the curves of equal variation to within a few degrees of the point of concurrence, leaving that point of course to be determined by observation, should the observation ever fall within the power of navigators. It was most gratifying on our return to find that the place I had fixed upon was precisely the one where these curves should have coincided in a centre, had they been protracted on his magnetic chart."

"A few days after, and on returning to their more permanent station, Captain Ross examined his instruments, and his experiments served to convince him that his observations on the celebrated spot were correct. The theory previously adopted was, that the place of the Magnetic Pole was at 70° north latitude, and 98° 30′ 45″ west longitude;—and the spot where Captain Ross supposes it, is 70° 5′ 17″ north latitude, and 96° 46′ 45″ west longitude. A Committee of the Admiralty, to whom Capt. Ross' narrative and statement was submitted, reported (among other things) that "they have no reason to doubt that Captain Ross actually reached the Magnetic Pole."

As to the Magnetic Pole, or the substance which produces the phenomena of the magnetic needle, some suppose "that there are great magnets in the earth, which move periodically." Professor Steinhauser was of opinion "that an interior planet revolved around the centre of the earth once in 440 years, and produced the magnetic effects at the surface." Prof. Sander contended these phenomena are to be ascribed to a *magnetic planet* beyond the newly discovered planet Herschel, or Uranus, performing its revolution in 1,720 years. Truly, the subject requires further explanation!

Appearance of the Sun from the North Pole.

To a person standing at the North Pole, the sun appears to sweep horizontally around the sky every twenty-four hours, without any perceptible variation during its circuit in its distance from the horizon. On the 21st of June it is 23° 28′ above the horizon—a little more than one-fourth of the distance to the zenith, the highest point that it ever reaches. From this altitude it slowly descends, its track being represented by a spiral or screw with a very fine thread; and in the course of three months it worms its way down to the horizon, which it reaches on the 23d of September. On this day it slowly sweeps around the sky, with its face half hidden below the icy sea. It still continues to descend, and after it has entirely disappeared, it is then so near the horizon that it carries a bright twilight around the heavens in its daily circuit. As the sun sinks lower and lower, this twilight gradually grows fainter, till it fades away. On the 20th of December, the sun is 23° 28′ below the horizon, and this is the midnight of the dark winter of the Pole. From this date the sun begins to ascend; and after a time his return is heralded by a faint dawn, which circles slowly around the horizon, completing its circuit every twenty-four hours. This dawn grows gradually brighter; and on the 20th of March the peaks of ice are gilded with the first level rays of the six months' day. The bringer of this long day continues to wind his spiral way upward, till he reaches his highest place on the 21st of June, and his annual course is completed. Such is one of the most wonderful works of God.

Climate of Greenland.

GREENLAND, the most northern country of the Western Hemisphere of the globe, lying between Iceland and the American Continent, reaches, as far as the land is discovered, from Cape Farewell, in latitude 59° 49′ to the 80th parallel of north latitude; further than that, it has been found unapproachable toward the northeast by reason of an immense barrier of ice stretching along the coast. The intrepid Dr. Kane has explored the farthest in this direction, when on his last expedition in search of an open Polar Sea.

"The whole coast of Greenland, receiving the beams of the sun in a very oblique direction, is deprived of that genial warmth which most other parts of the earth enjoy. The soil being shallow, is frozen during the greater part of the year, and the ice having taken possession of all the valleys of this barren and rocky land, the winds which blow over these are, even in summer, extremely cold. The climate on the east coast of Greenland, however, is undoubtedly more severe than on the west, where are to be found its only inhabitants.* Snow falls in an extraordinary quantity everywhere along the east coast, causing the glaciers with which the land is covered to increase perpetually, the loose snow upon the surface melting when it chances to thaw, and pressing down upon the strata below, or sinking through it, till the whole becomes one solid mass of ice, which never melts or undergoes change, until, in lapse of time, possibly not until some centuries pass by, it yields to the vast superincumbent pressure, and is precipitated into the sea in the form of icebergs. The winds which blow directly along the coast of Greenland from the sea, or Davis' Strait, are moist, and generally attended with rain, and in winter with snow and sleet, and are more boisterous in spring and autumn than in other seasons. Strong stormy winds from the west or southwest always break the sea-ice, even in the middle of winter. The severe cold sets in with the month of January, but it is accompanied with little snow, which generally falls either before or

* In regard to diseases prevalent in northern latitudes, a late writer remarks: "There are some regions of the globe which enjoy a complete immunity from consumption; such regions are generally situated in a high latitude, and, indeed, we find, in proportion to the intensity and long continuance of cold, a proportionate decrease in the amount of consumption. In Iceland, from 1827 to 1837, there was not a single case, and, according to a writer who has made the diseases of that country his study, it is entirely unknown. Those who have passed a long time also in Greenland, and along the coast of Labrador, state that they had never known among the Esquimaux a single case of consumption, and that catarrhal and bronchial affections are almost unknown."

after that time. More snow falls in the south than in the north. Of all the atmospheric phenomena peculiar to this country, the *Aurora Borealis* is the most beautiful. It streams here with peculiar lustre, and with a variety of colors, which, having great brilliancy, sometimes fill the whole horizon with the most beautiful tints of the rainbow. It is more frequent and more powerful from the 60th to the 67th parallel than in higher latitudes.

"Although Greenland affords a great variety of objects to the mineralogist, yet it offers but few to the botanist, as compared with other countries; vegetation being here repressed by the barrenness of the soil, and the want of the sun's genial influence. Those trees and shrubs, therefore, which in milder climates afford a comfortable shade to the wanderer, creep, in this forlorn land, under scattered rocks, to find shelter from the storm, snow and ice. There are a series of plants, however, which probably could not subsist in a milder climate; and in the interior of the inlets and firths may be found many species hitherto unknown in other countries. There are spots which even boast a luxuriant verdure, but they are only such places as, being in the neighborhood of dwellings, have been improved for many years by the blood and fat of seals and other animals. Vegetation commences very late in Greenland, not till the end of May or June, in proportion to the different latitudes, and is over by the end of August or September. The bottom of the sea in these climates appears to be better suited to vegetation than the surface of the land, presenting a great variety of fuci, ulvæ, and confervæ.

"Among the marine animals the whale tribe is here very conspicuous. The porpoise, the sword-fish, and the narwal, or sea-unicorn, frequently appear on the Greenland coast. Various species of seal inhabit the surrounding seas, and are of immense importance to the inhabitants in supplying them with food and clothing, as well as with various articles useful in their simple arts and domestic economy."—*English Gazetteer*.

The mean annual temperature of the southern part of Greenland is probably from 26° to 30° Fahr.; but the difference between the highest and the lowest temperatures (124°) is perhaps without a parallel. In July, the thermometer sometimes stands as high as 84°, even in the shade, while in January it often sinks as low as 40° below zero. July is the only month in the year in which no snow falls; but the seas do not usually begin to freeze till December.

East Coast of Greenland.

Captain W. A. Graah, of the Danish Royal Navy, who wintered at Nukarbik, north lat., 63° 22′ west long., 40° 50′ from Greenwich, remarks: "The climate on the east coast of Greenland is undoubtedly somewhat more severe than on the west. The summer of 1829 began late, and passed away without a single day that could properly be called warm. As early as the close of August, the sea was every night covered with a crust of new ice, which, by sunrise, attained such a thickness that it was no easy matter, nay, sometimes impossible, to break through it with the oars; and by the middle of September all the bay and firths were covered with sheet-ice from an inch to two inches thick. The winter of 1828–9 had been, it was said, unusually mild, and yet the winter ice lay still undissolved, when the new ice began to form. Towards the end of October sledging and hunting on the ice was in full train, and in November and December there were several days from eight to ten degrees of cold. Subsequently, indeed, and until the close of February, the weather was particularly mild; but at that date, it again became severe, and the cold increased to as much as 16° or 17° Réaumer; —4° or —6° of Fahr."

Temperature within the Arctic Circle.

Yearly Temperature of Rensselaer Harbor in north latitude 78° 37′, west longitude 70° 40′ from Greenwich, being the winter quarters of the brig *Rescue*, under the command of Dr. Kane, in 1853–54–55.

Months.	Mean Temp. ° Fahr.	Months.	Mean Temp. ° Fahr.
January,	—29.42	July,	+38.40
February,	—27.40	August,	+31.35
March,	—26.03	September,	+13.48
April,	—11.30	October,	—5.00
May,	+12.89	November,	—23.02
June,	+29.33	December,	—29.56

Yearly Mean Temperature, 3.22° below zero.

FOUR SEASONS.

Spring,	—11.48°	Autumn,	—4.85°
Summer,	+33°	Winter,	—29.56°

Warmest day, July 4, 1854, . . 53.9° Fahr.
Coldest day, February 5, 1854, . . —68° "

Summer and Winter Temperature.

Stations on the West Coast of Greenland.

Stations.	N. Lat.	Sum. Temp.	Win. Temp.	Difference.
	° ′	° Fahr.	° Fahr.	° Fahr.
Rensselaer Harbor,	78 37	33	—29.6	62.6
Wostenholm, . .	76 33	38	—28.7	66.7
Upernavik, . . .	72 48	35.2	—12.5	47.7
Omenak, . . .	70 41	40.7	— 5.1	45.8
Jacobshaven, . .	69 12	42.4	+ 0.8	41.6
Stations West of Baffin's Bay.				
Melville Island, .	74 47	37	—28.2	65.3
Assistance Bay, .	74 40	35.9	—26.7	62.6
Port Bowen, . .	73 14	37	—25.1	62.1
Boothia Felix, . .	69 59	38	—27.7	65.7
Igloolik, . . .	69 21	35.2	—21.3	56.5
Fort Hope, . . .	67 25	39.7	—25.1	64.8
Winterinsel, . .	66 11	35.1	—20.5	55.6
Fort Franklin, . .	65 12	50.2	—17.	67.2

1.—METEOROLOGICAL ABSTRACT kept on board Her Majesty's Ship "INVESTIGATOR," Captain McClure, wintering in Bay of Mercy—north latitude 74° 6′, west longitude 117° 12′.

Months—1851.	Mean Temp. ° Fahr.	Months—1852.	Mean Temp. ° Fahr.
September, . .	+24.4	March, . . .	—28.4
October, . . .	+ 3.4	April, . . .	— 1.3
November, . . .	—14.6	May, ...	+10.2
December, . . .	—20.0	June, ...	+31.5
January, 1852, . .	—27.2	July	+36.7
February, . . .	—25.9	August, . . .	+33.2

Mean Annual Temperature, 1.8° Fahr.

2.—METEOROLOGICAL ABSTRACT kept on board Her Majesty's Ship "RESOLUTE," wintering in Bridport Inlet, Melville Island —north latitude 74° 56′, west longitude 108° 48′.

Months—1852.	Mean Temp. ° Fahr.	Months—1853.	Mean Temp. ° Fahr.
September, . .	+19.2	March, . . .	—19.0
October, . . .	+ 0.6	April, . . .	— 2.5
November, . . .	—10.2	May, ...	+ 2.0
December, . . .	—26.0	June, ...	+21.0
January, 1853, . .	—40.7	July, ...	+36.4
February, . . .	—33.2	August, . . .	+33.1

Mean Annual Temperature, *Zero.*

Temperature of the Arctic Sea.

TABLE I.—Showing the Temperatures of the Sea at the Surface, and at various Depths, from Mr. SCORESBY'S Observations.

Temperature at Surface.	Below the Surface.		Remarks.
	Temperature.	No. Fathoms.	
	° Fahr.		
31 ° Fahr.	31	13	In lat. 79° N.
"	33.8	37	
"	34.5	57	
"	36	100	
"	36	400	
"	37	730	
29.7	36.3	120	In lat. 80° N.
32	38	760	In lat. 78° N.

TABLE II.—Comparisons of Temperature of the Sea at Depths, and at the Surface, from Experiments made by the Expedition under Captain DAVID BUCHAN, R. N., in 1818, between latitudes 79° 45′ N., and 80° 27′ N.

Date. 1818.	Temperature at Surface.	Below the Surface.		Variation.
		Temperature.	No. of Fathoms.	
	° Fahr.	° Fahr.		° Fahr.
July,	34	34.5	35	.5
"	33	34	60	1
"	32	36.7	73	4.7
"	31	35.6	83	4.6
"	32	36	94	4
"	31.5	36.5	103	5
"	30.5	36	120	5
"	30.5	36.5	142	6
"	32.5	36.5	173	4
"	32.5	36.5	185	4
"	31.5	37	237	5.5
"	32	36	330	4
May,	33	43	700	10

These results show an increase of temperature with an increase of depth, in the latter observation, making a difference of 10° Fahrenheit.

PART III.

THE COLD ZONE OF AMERICA.

The second climatic division of America, or the *Cold Zone*, lies mostly between the 50th and 60th degrees north latitude, embracing Labrador, the central portion of British America, and most of Russian America, where is to be found the widest part of the American Continent, running through upwards of one hundred degrees of longitude. This immense region, which is sparsely populated by Indians and Europeans in the employ of the Hudson Bay Company and Russian Fur Company, has a mean annual temperature ranging from 20 to 42 degrees Fahrenheit—being subject, for the most part, to killing frosts during every month of the year; yet still wheat and other cereals, and hardy vegetables come to perfection in the vicinity of Lake Winnipeg, the Saskatchewan country, and of the Mackenzie River, as far north as the 60th parallel of latitude; the summer months in this section have a mean temperature of 60° Fahrenheit, and upwards, with an abundance of rain, while the winter months are excessively cold and dry. On the northern limit of this cold belt, the subsoil is permanently frozen, and vegetation becomes dwarfed, or entirely disappears.

Land in the greater part of this region may be said to be almost valueless for agricultural purposes; the forests, the mines, the fur-bearing animals, and the fisheries, affording the only reliable means of support to the inhabitants, made up of Indians, half-breeds, and Europeans. These people do not aspire to maintain a separate national organization, but content themselves by being ruled by the governors and agents of British and Russian chartered companies, who have here held sway for some two hundred years.

The climate, throughout the western portion of this country, is remarkable for its healthy and invigorating influence, being in a great degree free from pulmonary complaints and fevers of every type. Like Northern Russia, in time, no doubt, much of this country will be reclaimed, and found capable of sustaining a hardy and dense population: the lumber trade, the mines and fisheries being of themselves an inexhaustible source of wealth to the white race, while the Indian will follow his favorite pursuit of the chase, and furnish rich furs, which are found in all cold regions.

Russian America.

This is an extensive and little known region, forming the northwest portion of the Continent of America, running north of latitude 54° 40′, to Behring Strait and the Arctic Ocean. Being mostly valuable for its fur-bearing animals, and having early attracted the notice of the Russian navigators, it was taken possession of by the Russian Government, as well as the Aleutian Islands, lying in the sea which divides Asia from North America.

"That part of the continent comprehended under the name of Russian America is generally of a very alpine and sterile appearance. The celebrated mountain of St. Elias, which is probably a volcanic peak, is calculated to have an elevation of 16,860 feet, while other peaks and ranges rise to 10,000 feet and upwards. Between the foot of the mountains and the sea extends a strip of low land, the soil of which is almost everywhere a black and marshy earth, only calculated for producing coarse though numerous mosses, short grasses, a few vaccinias, and other small plants. Some of these marshes on the side of the hills retain the water like a sponge, while their verdure makes them appear to be solid ground; but in attempting to pass them, the traveller sinks up to the mid-leg. Nevertheless, the pine and other trees acquire a large size on this gloomy soil. Next to the fir, the most common species of tree, is that of the alder; but in many places on the rocks the vegetation is confined to dwarf trees and shrubs. The proximity of mountains covered with eternal snow, and the extent of the American Continent in the latitude of 58°, render the climate northward excessively cold and inimical to vegetation."

"Aliaska, or Alashka, is a long peninsula on the northwest coast of America, running out in a southwest direction from the continent into Behring Sea. It chiefly consists of a ridge of steep rocks, which, in some parts, attain a great elevation, and has an average breadth of fifteen or twenty miles, while its length, if measured from the head of Cook's Inlet, exceeds 400 miles. That division of the Aleutian chain, called the Fox Islands, commences at its western extremity, a very narrow channel here dividing the volcanic Island of Unimak, which rises to an altitude of 8,083 feet, from the extremity of the peninsula. This great uninterrupted sea-wall, as it may be termed, with its insular continuation in the Aleutian range, prevents ice brought from the Arctic Ocean, and down the large rivers which discharge themselves into the Sea of Behring, from flowing into the North Pacific Ocean; while the accumulation of ice on the former sea absorbs a large proportion of the heat of the atmosphere. Hence navigators, in passing from the Pacific into the Sea of Behring, seldom fail to remark a great change of temperature; and while all the coast to the southeast of the peninsula is clad with dense forests of noble timber, on that of the northwest only a few stunted shrubs are seen. The Island of Kadiak, sheltered from the Arctic influences by the peninsula, has an abundance of timber; while the Aleutian Islands, though several degrees further south, have not a single tree; and the difference in the animal is as striking as in the vegetable world on the two sides of the peninsula. On the one we find the walrus, on the other the humming-bird; and while the ice-fox is often met with on the Aleutian Islands, it has never been encountered at Kadiak."

SITKA, Russian America, is a large island, eighty miles in length, and from ten to twenty miles in breadth, separated from the mainland by the narrow inlet of Norfolk Sound. Here is situated the principal settlement of the Russian American Company, called *New Archangel*, in north latitude 57° 3′, west longitude 135° 18′. It was formerly called Sitka. The mean temperature for the year at New Archangel is about 42° Fahrenheit, spring 40°, summer 54°, autumn 44°, and winter 32° Fahrenheit. This is about the same mean annual temperature as Quebec, about 10° south, and Bergen in Norway, 5° north.

The climate of Russian America is described as very humid, especially on the southwest coast and Sitka Island. At the latter place, Wrangell found, in 1828, that only sixty-six days were dry in the whole year, while for 128 days rain fell without intermission. Snow was frequent, but did not last long. The temperature, though warmer than on the eastern coast of Asia, is considerably more severe than in corresponding latitudes on the western coast of the Old World: for example, while the mean annual temperature at Sitka, is 42° Fahr., at Bergen, in Norway, it is 47.°

The soil is generally sterile; grain crops refuse to grow, except in the Sitka archipelago, and the adjacent coast of the mainland, where a little barley, rye, and oats are raised.*

British America.

This vast region, exclusive of Canada and the other provinces bordering on the Atlantic Ocean, extends from the Pacific Ocean along the parallel of the 49th degree north latitude to the Lake of the Woods, near the centre of the continent, thence to the mouth of Pigeon River, near the head of Lake Superior, in about 48° north latitude. From thence, in a north and northeasterly direction, along the water-shed that divides the waters flowing into Hudson Bay from the waters of the Great Lakes and the St. Lawrence River, to the coast of Labrador, striking the Atlantic on about the 50th parallel of latitude, forming an extended line of about 4,000 miles in length from ocean to ocean. The Arctic Ocean forms its northern boundary, and thus it is bounded on three sides by the greatest oceans of the globe, while Hudson Bay and Strait form an inland sea of great magnitude.

This region includes all the country lying between the meridians of 55° and 141° west longitude, excepting a strip of Russian America on the Pacific Ocean, between 54° 40′ and 60° north latitude, where stands Mount St. Elias. The dividing line between the Russian and British Possessions runs due north from that point to the Arctic Ocean, terminating on about the 70th degree of north latitude. From south to north it extends through upwards of twenty-one degrees of latitude, or

* The Russo-American Telegraph Expedition.—*The Alta California*, of October 26, 1865, publishes the following concerning Russian America: "Col. Bulkley reports that the river laid down on the map as the Kwichpak, in Russian America, is identical with the Youcan, and is navigable for small vessels as far as English Fork. A party was sent to explore the Youcan and Kwichpak in a small steamer. They will proceed to the head of navigation, and then cross with reindeer or on foot over the ice and snow, until they strike the settlements in British Columbia. Colonel Bulkley's party found the earth, on the American side, thawed to an average depth of ten inches, but frozen solid below to an unknown depth. On their arrival in September, the country on the American shore was rolling and breaking, but it is not high, and was destitute of timber.

about 1,500 miles, and contains an estimated area of 3,000,000 square miles.

Canada proper claims all the territory drained by the waters flowing into the Great Lakes and the St. Lawrence River, north of the United States boundary as defined by treaty.

The *climate* of this immense country is, for the most part, cold and forbidding, with the exception of Vancouver's Island, Queen Charlotte's Island, and other islands which lie off the main coast of British America, between the 48th and 55th degrees of north latitude, having a front of about 450 miles on the North Pacific Ocean. These islands possess a genial and healthy climate, good soil, and are said to be rich in mineral productions.

The shores along the mainland, and inland to the Rocky Mountain range, embracing *New Caledonia*, are alike favored as regards climate and the productions of the earth; here being recently discovered large and valuable *gold fields*, which are attracting great attention, and filling the country with a hardy and industrious class of inhabitants. This portion of British America, as well as Washington Territory, north of the Columbia River, affords pure water, pure air, and a fruitful soil, with rich mineral productions, possessing all the requirements in order to sustain civilized life in ease and comfort.

The Saskatchewan region, embracing the Assinniboine and Red River valleys, and the shores of Lake Winnipeg, together with the Lake of the Woods and Rainy Lake, is all alike a fine region of country, producing grasses, cereals, and vegetable productions in great abundance and of excellent quality.

All the above enumerated regions, being about one-fourth of British America, is capable of sustaining a dense and hardy population. The climate and soil being in every respect equal to a great part of Russia in Europe, where large cities are found.

The country drained by the head branches of Mackenzie River, running north into the Arctic Ocean, including Athabasca Lake, as well as the extensive country around the south shore of Hudson Bay and James' Bay, together with Labrador, extending from the Rocky Mountains to the Atlantic Ocean, and lying between the 52d and 60th degrees of north latitude, is mostly unfit for settlement or habitation, except by a hardy race of trappers and Indians, who are now found in the employ of the Hudson Bay Company.

This extensive region, south of the Esquimaux country, seems peculiarly adapted to a race of men essentially different from those who inhabit the temperate climate of America—thus marking the second grand division of people, whose habits seem distinct from those of more southern climes. The chase of winter for game, for food, and furs for clothing and sale, is succeeded by fishing, and a short season for raising a few hardy vegetables during the months of June, July and August. The winters here averaging eight or nine months, while frosts occur every month during the year.

HUDSON BAY, an extensive inlet of the Atlantic Ocean, forming a Mediterranean Sea on the east side of the North American Continent, lying between the parallels of 51° and 66° north latitude, and measuring upwards of 1,000 miles from north to south; while its breadth varies from 150 to 500 miles. It is navigable during June, July, August and September, but is filled all the rest of the year with shoals of ice. The main entrance of the bay from the Atlantic, known as Hudson's Strait, is 350 miles in length, with a breadth varying from 75 to 150 miles. This is a cold, inhospitable region. "At York Factory, situated on the southwest shore, the land seems to have been thrown up by the sea, and is never thawed during the hottest summer, with the thermometer at 90° in the shade, more than ten or twelve inches, and then the soil is of the consistency of clammy mud." At Fort Franklin, on north latitude 65° 12′, west longitude 132° 13′, the mean annual temperature is 17.50° Fahr.; the maximum of heat 80°, the minimum —58°. "To account in some degree for this extreme cold, it may be observed, that very little of the coast of this country is bounded by the ocean; the chief parts of the sea upon which it borders are Davis' Strait, Hudson's Strait and Bay, and James' Bay. Now, these bodies of water, though of considerable magnitude, are not sufficiently large to check the influence of the wind proceeding from the frozen region in the northwest; the consequence is, that they are almost entirely covered with ice during six months of the year, and thus, instead of mitigating the cold, they add considerably to its force. It has further been observed, that much of the inland is elevated and dry, unacquainted with fogs, and accordingly healthy; while the coasts are low, marshy, exposed to frequent and dense fogs and moist weather, and consequently highly noxious to the human frame. The former is abundantly fertile in spontaneous productions, and, by being cultivated in favorable places, becomes a very agreeable country; the latter is dreary and unproductive, and scarce affords either food or shelter to those wild beasts by which it is frequented. In the northern parts, the land is barren and comfortless; in the southern parts, it is more fertile, and offers sufficient encouragement to him who would bestow the proper cultivation. On the coasts, the country chiefly produces pines, birch, larches and willows, but the trees are stunted and knotty. In the south interior the same kinds of trees are more abundant, and of great size."

Monthly and Yearly Mean Temperature of Posts on the Shores of Hudson Bay.

Stations.	York Factory. 57° N.	Fort Churchill. 59° N.	Fort Hope, 62.50° N.
	° Fahr.	° Fahr.	° Fahr.
January, . .	—5.12	—21.21	—29.32
February, . .	—6.60	— 7.31	—26.68
March, . . .	—4.77	— 4.63	—28.10
April, . . .	19.21	—16.29	— 3.95
May, . . .	33.53	28.42	17.88
June, . . .	47.66	44.69	31.38
July, . . .	60.00	56.80	41.46
August, . . .	54.85	53.39	46.32
September, . .	41.90	36.03	28.57
October, . . .	33.43	26.50	12.56
November, . .	25.17	3.31	0.68
December, . .	3.73	—14.00	—19.27
Yearly mean, . .	25.63	18.20	6.14

York Factory, situated on Hudson Bay, in north latitude 57°, is the principal depot of the northern department of the Hudson Bay Company, from whence all the supplies for the trade are issued, and where all the returns of the department are collected and shipped for England.

"As the winter is very long, nearly eight months, and the summer consequently very short, all the transport of goods to, and returns from, the interior must necessarily be effected as quickly as possible. The consequence is, that great numbers of men and boats are constantly arriving from inland, and departing again during the summer; and, as each brigade is commanded by a chief factor trader, or clerk, there is a constant succession of new faces, which, after a long and dreary winter, during which the inhabitants never see any stranger, renders the summer months at York Factory the most agreeable part of the year. The arrival of the ship from England, too, delights them with letters from *home*, which are only received twice a year.

"The Fort (as all establishments in the Indian country, whether small or great, are called) is a large square, I should think about six or seven acres, enclosed with high stockades, and built on the banks of Hayes River, nearly five miles from its mouth. The country is flat and swampy, and the only objects that rise very prominently above the rest, and catch the wandering eye, are the lofty "outlook" of wood, painted black, from which to look out for the arrival of the ship, and a flagstaff, from which on Sundays the snowy folds of St. George's flag flutter in the breeze.

"The trade carried on by the company is in peltries of all sorts, oil, dried and salted fish, feathers, quills, &c.; and a list of some of their principal articles of commerce is subjoined:—Bear-skins, black, brown, and white or polar; beaver-skins; badger-skins; moose or elk-skins; buffalo or bison robes; deer-skins; parchment; fox-skins, black, red, silver, cross, white, and blue; marten-skins, lynx-skins, musquash-skins, otter-skins, wolf-skins, seal-skins, &c., &c.

"The most valuable of the furs mentioned in the above list is that of the *black fox*. This beautiful animal resembles in shape the common fox of England, but it is much larger, and jet-black, with the exception of one or two white hairs along the back-bone, and a pure white tuft on the end of the tail. A single skin sometimes brings from twenty-five to thirty guineas in the British market; but, unfortunately, they are very scarce. At present the most profitable fur in the country is that of the marten; it somewhat resembles the Russian sable, and generally maintains a steady price. These animals, moreover, are very numerous throughout most part of the Company's territories, particularly on Mackenzie River, from whence great numbers are annually sent to England. Most of the above animals are caught in steel and wooden traps by the natives, while deer, buffaloes, &c., are run down, shot, and snared in various ways."

York Factory to Norway House.

[Extract from "Ballantyne's Hudson Bay Territory," London, 1848.]

June 23, 1845, leave York Factory.—"The banks of Hayes' River were covered with huge blocks of ice, and scarcely a leaf had as yet made its appearance; not a bird was to be seen, except a few crows and whiskey-jacks, which chattered among the trees; and nature appeared as if undecided whether or not she should take another nap, ere she bedecked herself in the garments of spring. My Indians paddled slowly against the stream, while I was watching the sombre pines as they dropped slowly astern. During the day we passed a brigade of boats bound for the Factory; but, being too far off, and in a rapid part of the river, we did not hail them. About nine o'clock, we put ashore for the night, having travelled nearly twenty miles.

June 24.—"Everything was obscure and indistinct till about six o'clock, when the powerful rays of the rising sun dispelled the mist, and nature was 'herself again.' A good deal of ice still lined the shores; but what astonished me most was the advanced state of vegetation apparent as we proceeded inland. When we left York Factory, not a leaf had been visible; but here, though only thirty miles inland, the trees, and more particularly the bushes, were pretty well covered with beautiful light green foilage, which appeared to me quite delightful, after the patches of snow and leafless willows on the shores of Hudson Bay. At noon, the day, which had hitherto been agreeable, now became oppressively sultry; not a breath of wind ruffled the water; and as the sun shone down with intense heat from a perfectly cloudless sky, it became almost insufferable.

June 25.—"The weather was still warm, but a little more bearable, owing to a light, grateful breeze that came down the river. After breakfast, which we took at the usual hour, and in the usual way, while proceeding slowly up the current, we descried, on rounding a point, a brigade of boats close to the bank, on the opposite side of the river; so we embarked our man, who was tracking us up with a line (the current being too rapid for the continued use of the paddle), and crossed over to see who they were. On landing, we found it was the Norway House brigade, in charge of a Red River settler. Shortly after we arrived at the mouth of Hill River, which we began to ascend. The face of the country was now greatly changed, and it was evident that here spring had long ago dethroned winter. The banks of the river were covered from top to bottom with the most luxuriant foilage, while the dark clumps of spruce-fir varied and improved the landscape.

June 26.—"On the following morning, we started at an early hour. The day was delightfully cool, and mosquitoes were scarce, so that we felt considerably comfortable as we glided quietly up the current.

July 5.—"Arrived at Norway House, situated in north latitude 54°, west longitude 98°, being about 300 miles southwest of York Factory."

NORWAY HOUSE.—This fort, situated about twenty miles below Lake Winnipeg, in north latitude 54°, is built at the mouth of a small and sluggish stream, known by the name of Jack River, emptying into Playgreen Lake. The houses are ranged in the form of a square, none of them exceed one story in height, and most of them are whitewashed. This is one of the oldest and most important establishments belonging to the Hudson Bay Company. The climate is dry and salubrious, and although (like nearly all other parts of the country) extremely cold in winter, it is very different from the damp, chilling cold of that season in Great Britain. The country around is swampy and rocky, and covered with dense forests. Many of the Company's posts are but ill provided with the necessaries of life, and most of them entirely destitute of luxuries. Norway House, however, is favored in this respect. We always had fresh meat of some kind or other; sometimes, beef, mutton, or venison, partridge, wild ducks in their season, and occasionally buffalo meat, was sent us from Swan River District; and, besides the produce of our garden in the way of vegetables, the river and lake contributed white-fish, sturgeon, and pike, or jack-fish in abundance. This is also an aggreeable and interesting place, from its being in a manner the gate to the only route to Hudson Bay. Often might be seen a strange and noisy collection of human beings, half-breeds, and Indians, who rested here awhile ere they started for the shores of Hudson Bay, for the distant region of Mackenzie's River, or the still more distant lands bordering on the Pacific Ocean. The mean annual temperature of Norway House is 30° Fahr.; spring 28°, summer 60°, autumn 30°, winter 2° below zero.

FORT ALEXANDER, situated near Lake Winnipeg, 50° north latitude, says the Rev. J. Ryerson, "belongs to the Lac La Pluie District, and, in my judgment, is not surpassed, if equalled, in beauty and pleasantness, by any station belonging to the Hudson Bay Company, between St. Marie and Red River. It is situated on an elbow of land made by a bend in the River Winnipeg, three miles from the river's mouth. The scenery for many miles around is strikingly beautiful. The climate for Hudson Bay Territory is remarkably fine and salubrious, the land amazingly rich and productive. The water in the lakes Lac La Pluie, Lac Du Bois, Winnipeg, &c., is not deep, and because of their wide surface and great shallowness, during the summer season, they become exceedingly warm; this has a wonderful effect on the temperature of the atmosphere in the adjacent neighborhood, and no doubt makes the great difference in the climate (or at least is one of the principal causes of it) in these parts, to the climate in the neighborhood of Lake Superior."

He further says: "In many parts of this immense country there is a great deal of excellent land, very suitable for agricultural purposes; this is especially the case in many localities south of lat. 55°, where almost every kind of summer grain and useful vegetables can be grown with the greatest facility, and in great abundance. There is not to be found in British America finer, richer, and a more productive soil, than there is in the Selkirk Settlement, on the Assinniboine and Red Rivers; and in the bounds of Rupert's Land there are millions of acres equally rich and fertile, and especially suited, from climate and locality, for farming and agricultural purposes."

Climate of a Portion of British America, or Rupert's Land.

[Extract from Professor Hind's "Report on the Assiniboine and Saskatchewan Country." Official Document, 1859.]

"The climates of Canada and Rupert's Land (or Hudson Bay Company's Territory), under the same parallels of latitude, vary to a considerable extent with the rock formation of the country. Throughout the undulating region of the Laurentides, the proportion of water to dry land is about one to two, not collected into one large water area, but distributed over the surface of the country in the form of countless thousands of lakes, ponds, and marshes. The intense cold of winter is sufficient to solidify the deepest lakes for a depth of several feet, and the thawing of so much ice in spring has the effect of absorbing and rendering latent the heat which would be otherwise expended in warming the soil and advancing vegetation.

"Lakes Winnipeg,* Manitobah, and Winnipego-sis, together with the smaller lakes belonging to the Winnipeg basin, are deeply frozen every winter, and ice often remains in their northern extremities until the beginning of June, greatly retarding the progress of navigation on their immediate shores. Hence one reason that north of the 48th parallel the mildness of the seasons increases rapidly as we advance toward the west, after leaving Red River. The improvement arises not only from greater longitude, but also from the character of the rock for-

* Lake Winnipeg, lying between the 50th and 54th degrees of north latitude, is about as long as Lake Erie, and receives the waters of Red River, Assiniboine, Saskatchewan, and the Winnipeg; the latter stream being the outlet of the Lake of the Woods, lying on the 49th parallel.

mations, by which the country is underlaid and surrounded. The soil of the prairies is in general dry, and is rapidly warmed by the rays of the sun in spring. The prairies enjoy, too, north of the 48th parallel, the genial, warm and comparatively humid winds from the Pacific, which are felt as far north as the latitude of Fort Simpson.*

"The country embraced within the limits of this exploration may be divided into two regions in relation to climate; the *arid* and the *humid* region. The vast treeless prairie west of the Little Souris (101° west longitude) lies within that part of the area which receives comparatively a small annual rain-fall. Its northern limit is roughly shown by the Qu'Appelle Valley, or more accurately by an imaginary line drawn from Fishing Lakes to the Moose Woods, (52° north latitude). North and east of this area, the precipitation is considerably greater, and supplies the valley of the main Saskatchewan, the Touchwood Hill range, and the valley of the Assiniboine with an abundance of moisture, which is protected and treasured by forests."

The valley of the Red River, east of the Little Souris, receives much humidity from the moist winds coming from the Gulf of Mexico up the valley of the Mississippi, and over the *height of land* which separates the waters of Red River from those of the St. Peter (flowing southward into the Mississippi). The Touchwood Hill range, and the country generally north of the Qu'Appelle Valley, and in an easterly direction towards and beyond Lake Winnipeg, are made humid by the southwest Pacific wind, in concurrence with the prevailing east wind of this region. These phenomena are referred to in detail in succeeding paragraphs.

Lakes in the Great Basin of Lake Winnipeg.

Name.	Area. Square Miles.	Above the Sea. Feet.
Lake Winnipeg,	8,500	650
Lake Manitobah,	1,900	670
Lake Winnipego-sis,	1,936	692
St. Martin Lake,	316	660
Cedar Lake,	312	688
Dauphin Lake,	170	700
Total,	13,134	

* Colonel Lefroy—"Meteorological Observations at Lake Athabasca and Fort Simpson."

Observations on the Temperature of the Air in Different Parts of British North America.

FORT FRANKLIN, in lat. 65° 12′ N. Months.	Deg. Fahr.	FORT CHEPEWYAN, in lat. 58° 43′ N. Months.	Deg. Fahr.
January, . . .	—23.78	January, . . .	—9.56
February, . .	—12.70	February, . .	—4.26
March, . . .	— 8.26	March, . . .	—0.55
April, . . .	15.21	April, . . .	25.86
May, . . .	36.35	May, . . .	46.50
June, . . .	48.00	June, . . .	55.70
July, . . .	52.10	July, . . .	63.42
August, . . .	51.09	August, . . .	58.10
September, . .	42.92	September, . .	43.53
October, . . .	20.28	October, . . .	32.00
November, . .	2.97	November, . .	26.70
December, . .	13.96	December, . .	—2.82
Annual Mean, .	17.50	Annual Mean, .	31.50

LAKE WINNIPEG, lying between 50½° and 54° degrees north latitude, is larger than either Lake Erie or Ontario, though the water is very shallow; it is 300 miles long and about 50 miles wide, being 650 feet above the level of the sea (or 50 feet higher than Lake Superior). There appears to be but little land that would admit of cultivation, though I was told there were localities in the neighborhood of the lake where are found large tracts of land of the most excellent quality. Along the eastern shore the granite and trap rocks are everywhere exposed, the latter being the most extensive, and nowhere do these masses rise to the altitude of hills. Lake Winnipeg is very much subject to winds and storms, which many times rise so suddenly as to give the mariner no warning of their approach, until, like a giant in his strength, they are upon him. Imagination cannot paint, much less language describe, the sublimity and grandeur of a thunder-storm, as seen in the forest on the shore of the lake, when the wild waves are raging; the lurid glare of the vivid lightning seems brighter, and the claps and roarings of the thunder seem louder and deeper than anywhere else.

NORWAY HOUSE, an important Hudson Bay Company's post, situated in north latitude 54°, is represented as possessing a fine healthy climate, where grains, vegetables, and flowers are raised in great abundance.

Seasons of the Valley of Lake Winnipeg.

The natural division of the seasons in the Lake Winnipeg Valley are as follows:

Spring.—April and May.

Summer.—June, July, and August.

Autumn.—September and October.

Winter.—November, December, January, February, and March.

The natural division of the seasons is strikingly represented by the early and rapid advance of temperature in May in the valley and prairies of the Saskatchewan; and it is also indicated in a very marked degree by the extension northwards to the same valley, between the 95th and 105th degrees of longitude, of numerous plants, whose geographical distribution, east and west of those limits, has a much more southern climatic boundary. The limits of trees rise with the isothermal lines, and these attain a much higher elevation in the interior of British America than on the Atlantic coast.

In relation to agriculture, the intensity of winter cold is of comparatively little moment. The elevated spring and summer temperature, combined with the humidity of the humid region in the valley of Lake Winnipeg, enable Indian corn and the melon to ripen with certainty, if ordinary care is taken in selecting soil and in planting seed.

SUMMER TEMPERATURES.

Fort Gary, Red River,	67° Fahr.	Kingston, C. W.,	67° Fahr.
Norway House, Lake Winnipeg, . .	60° "	Toronto, "	65° "

"In the absence of instrumental observations, the progress of vegetation affords the best indications of climate, apart from latitude and elevation above the sea. The period of flowering and fruiting is about three weeks earlier in latitude 51°, than between the 53d and 54th parallels west of the 100th degree of longitude. The prairies of the Assiniboine, of the Qu'Appelle, and of the South Branch of the Elbow, are decorated with brilliant spring flowers, and covered with luxuriant herbage, at a time when the ice still lingers at the head of Lake Winnipeg, or chills the air and arrests vegetation in Cedar and Cross Lakes on the Main Saskatchewan. At Touchwood Hills, in north latitude 51° 30', horses and cattle are allowed to remain in the open air all winter, finding sufficient pasture under the snow to keep them in good condition.

"The growth of forests is very intimately connected with the climate of a large extent of country. That forests once covered a vast area in Rupert's Land there is no reason to doubt. Not only do the traditions of the natives refer to former forests, but the remains of many still exist as detached groves in secluded valleys, or on the crests of hills, or in the form of blackened prostrated trunks covered with rich grass, and sometimes with vegetable mould or drifted sand. The agent which has caused the destruction of the forests, which once covered many parts of the prairies, is undoubtedly fire; and the same swift and effectual destroyer prevents the new growth from acquiring dimensions which would enable it to check their annual progress."

Prevailing Winds in the Northwest Territory.

[From Hind's "Report on the Assiniboine and Saskatchewan."]

"All the thunder-storms we encountered in 1858, in the valley of Lake Winnipeg, came from the west, southwest, or northwest, with one exception. I do not find a single record of thunder-storms with heavy rain coming from the south. This may have been an exceptional year, but the warmth and dryness, often oppressive, of the south wind, west of the 100th degree of longitude, contrasted strongly with the humidity and coldness of winds from the west. This phenomena is directly opposed to those which prevail in lower latitudes, and may probably be explained as follows:

"Warm air from the Pacific, loaded with moisture, passes at certain periods of the year over the whole range of the Rocky Mountains, in British America and in the United States. These Pacific winds occasion but a very small precipitation of rain or snow on the eastern flank of the Rocky Mountains, south of the great Missouri Bend. Similar winds from the Pacific do occasion a considerable precipitation in the northern part of the Saskatchewan valley. Whence, then, this apparent anomaly? It probably arises from the difference in the temperature of the two regions, the direction of the prevailing winds, and lowness and comparatively small breadth of the Rocky Mountain ranges in that latitude. In spring and summer, warm westerly winds, ladened with moisture, in passing over the mountain range south of, say the 46th parallel, are cooled to a certain temperature, and precipitate the greater portion of their moisture, in the form of rain or snow, upon the mountain ridges. On arriving at the eastern flank of the Rocky Mountains, their temperature rises to that of the region over which they pass, being elevated by the deposition of their moisture in the form of rain or snow, and continually increasing density as they descend; but the capacity of air for moisture is well known to be dependent upon its temperature, within certain limits, hence the westerly Pacific winds become more warm and more dry as they descend the eastern Rocky Mountain slope, until they meet the moist winds from the Gulf of Mexico passing up the valley of the Mississippi, towards and through the region of the Great Canadian Lakes and over the low height of land separating the waters flowing into Lake Winnipeg, from the Mississippi Valley.*

"In the latitude of the Valley of the Saskatchewan, however, the moist southwest winds from the Pacific find a broad depression in the Rocky Mountain range, and losing less humidity than those passing over the higher ranges to the south, meet with a prevailing northeasterly wind as they begin to descend their eastern flank, their temperature is consequently lessened, instead of being elevated, and their capacity for moisture diminished, hence precipitation in the form of rain and hail takes place as they descend the slope towards Lake Winnipeg. Hailstorms are not unfrequent during the summer months, and the prairies sometimes retain the record of their occurrence for many weeks.

"There is no doubt that the southern Pacific winds, passing through the broad depression in the Rocky Mountains near the 49th and 51st parallels, without losing the whole of their moisture, give humidity to the large portion of Rupert's Land over which they traverse."

Passes over the Rocky Mountains.

Summit of the Lowest Passes above the Ocean from the 32d to the 51st parallel North Latitude.

	Feet.
32d parallel, near El Paso, Mexico, ...	5,717
35th parallel, near Albuquerque, New Mexico, . .	7,472
38th and 39th parallels (Coochecopa Pass), . . .	10,000
41st and 42d " (South Pass), ...	7,085
47th and 49th " (Cadotte's Pass), ...	6,044
Kutanie Pass, latitude 49° 30′, British America, . .	6,000
Kananaskis Pass, near 50th parallel, ...	5,985
Vermillion Pass, latitude 51° 10′, ...	4,944

* See "Meteorology in its Connection with Agriculture," by Prof. Henry.

Mean Temperatures.

Monthly and Yearly Mean Temperature of *Cumberland House,* situated on the North Shore of the Saskatchewan River, in north latitude 53° 57′; altitude 900 feet above the ocean.

Months.	° Fahr.	Months.	° Fahr.
January, ...	—1.	July, ...	65.00
February, . . .	—8.	August, ...	62.84
March, ...	18.30	September, . . .	44.50
April, ...	27.00	October, ...	33.15
May,*	52.60	November, . . .	21.48
June, ...	60.00	December . . .	7.94

Yearly Mean, 33.20° Fahr.

FOUR SEASONS.

Spring, ...	32.70°	Autumn, . . .	33.04°
Summer, . . .	62.62°	Winter, ...	—0.17°

Climate of the Valley of the Red River of the North.

"The climate of the valley of Red River exhibits the extremes of many characteristics which belong to the interior of continents in corresponding latitudes. High summer temperatures, with winter cold of extraordinary severity, appear to prevail in the district called the Assiniboine, as in the interior of north-eastern Europe and Asia. It cannot fail to be noticed, however, that the general absence of late spring and early autumn frosts, with an abundant fall of rain during the agricultural months, are its distinguishing features in relation to husbandry, The melon growing in the open air and arriving at perfect maturity in August and September; Indian corn succeeding invariably when due precautions are used to ensure the ripening before the middle of September, are strong proofs of the almost uniform absence of summer frosts.

"A comparison with the climate of Toronto, Canada West, for corresponding months of the years 1855 and 1856, reveals some very curious and interesting facts, which may possess importance. Limiting our attention at present to the summer months, we find from inspection of the following table of comparison, that the summer on Red River during the above years was more than three degrees warmer than the summer at Toronto, and with this excess of temperature there occurred the unexpected difference of 21.74 inches of rain in favor of Red River during that year.

* The mean temperatures of May, June, July, and August are about the same as Toronto.

"It must be borne in mind, however, that the results of one year's comparison *are not of much value* in estimating the relative climatic adaptation of regions far apart; nor do they afford sufficient data for a fair estimate of the climate of the locality where the observations were made."—*Hind's Report.*

The following comparisons refer to corresponding months of the same years, and are of course liable to those annual fluctuations to which the climatic elements of all countries are subjected. It is very probable that more extended observations will reduce the extremes.

Temperature, Rain, &c.

Comparison of the Meteorology of Red River Settlement with Toronto, Canada West, with reference to Temperature, Depth of Rain and Snow, from corresponding Observations in the Years 1855 and 1856.

Months.	Mean Temp.		Rain in Inches.		Snow in Inches.	
	Red River.	Toronto.	Red River.	Toronto.	Red River.	Toronto.
	° Fahr.	° Fahr.				
March, . .	9.09	23.06	0.0	0.0	6.5	16.2
April, . .	39.83	42.27	6.5	2.8	3.0	0.1
May, . .	58.46	50.52	4.0	4.5	2.0	0.0
Spring, .	35.79	38.62	10.5	7.3	11.5	16.3
June, . .	69.10	59.93	6.0	4.0	0.0	0.0
July, . .	71.16	67.95	12.0	3.2	0.0	0.0
August, .	63.03	64.06	12.5	1.5	0.0	0.0
Summer, .	67.76	64.00	30.5	8.7	0.0	0.0
September, .	59.26	59.49	5.0	5.6	0.0	0.0
October, .	42.20	45.39	0.0	2.4	2.0	0.8
November, .	21.19	38.58	2.5	4.6	7.0	3.0
Autumn, .	40.88	47.82	7.5	12.6	9.0	3.8
December, .	— 8.31	27.00	0.0	1.8	8.0	29.5
January, .	—10.00	16.02	0.0	0.0	5.0	13.6
February, .	— 1.71	15.69	0.0	0.0	6.0	9.7
Winter, .	—6.85	19.57	0.0	1.8	19.0	52.8
Annual Temp.,	35.00	44.00	48.5	30.6	39.5	72.9

The summer temperature of Red River, and the absence of frosts during that season, determine the fitness for agricultural purposes. The following table exhibits a comparison, based upon *one year's observations only*, between the summer temperature of the settlement and other well known places in Canada:—

	Sum. Temp.
Red River Settlement, ...	67.76°
Montreal, Canada East, ...	66.62°
Quebec, " "	62.91°
Toronto, Canada West, ...	64.00°

Mr. Ballantyne, in his very interesting work on the "Hudson Bay Territory," after a residence there of six years, remarks:— "The climate of Red River (between north latitude 49° and 50°) is salubrious and agreeable. Winter commences about the month of November, and spring generally begins in April. Although the winter is very long and extremely cold, yet, from its being always *dry* frost, it is much more agreeable than people accustomed to the damp thawy weather of Great Britain might suppose. Winter is here the liveliest season of the year, affording the most enjoyment.

"During the summer months, there are often very severe thunder-storms, frequently accompanied with tremendous showers of hail, which do great mischief to the crops and houses. Generally speaking, however, the weather is serene and calm, particularly in autumn, and during the delicious season peculiar to America, called the Indian Summer, which precedes the commencement of winter."

From the above authentic account, it seems that the year is about equally divided, giving six months for the production of all kinds of vegetation, and six months of frost and winter, not differing materially from the climate found in different parts of the State of New York, above the "Highlands," or in the New England States.

The Lake of the Woods, which empties its surplus water into Lake Winnipeg, sixty-eight miles in length, and from fifteen to twenty-five miles wide, is a splendid sheet of water, dotted all over with hundreds of beautiful islands, many of which are covered with a heavy and luxuriant foliage. Warm and frequent showers occur here in May and June, bringing forth vegetation at a rapid rate, although situated on the 49th parallel, from whence extends westward to the Pacific Ocean the boundary line between the United States and Canada.

"Not a trace of civilization is anywhere observable, but the Indians are numerous; and, indeed, this lake seems to be their favorite resort in summer; the wild rice on its borders, and the fish which abound in its waters, afford them an easy means of subsistence, not to mention the maize which they grow on the islands. They are a fine looking race, and if removed from the humanizing influences of civilization, they are also strangers to the vices which it brings. The men are generally tall and well formed, and some of the women remarkably comely, but they are not very cleanly in their habits, and there can be nothing more suggestive of indolence than their mode of life, which, however, has one feature to recommend it, in the entire exemption from care with which it seems to be attended. Gliding in their light canoes from island to island, basking in the sunshine on some pebbly strand, and merely exerting themselves to an extent sufficient to supply their immediate wants, the future affects them not, and they appear to be supremely happy; but the winter brings its troubles, and they have to betake themselves to the forests in single families, where, having only game to depend upon, they are sometimes sadly straitened."—*Dawson's Report.*

RAINY LAKE, or Lac la Pluie, forming also, in part, the boundary between the above countries, lying to the eastward, is another most beautiful sheet of water; it is forty-eight miles long, and averages about ten miles in breadth. It receives the waters flowing westward from the dividing ridge separating the waters flowing into Lake Superior from those flowing northwest into Lake Winnipeg and Hudson Bay.

"There is nothing, I think," says Ballantyne, "better calculated to awaken the more solemn feelings of our nature than the noble lakes, studded with innumerable islets, suddenly bursting on the traveller's view as he emerges from the sombre forest rivers of the American wilderness. The pure and clear unruffled water, stretching out on the horizon—here intersecting the heavy and luxuriant foliage of an hundred woody isles, or reflecting the wood-clad mountains on its margin, clothed in all the variegated hues of autumn; and there glittering with all the dazzling brilliancy in the bright rays of the evening sun, or rippling among the reeds and rushes of some shallow bay, where thousands of wild-fowl chatter as they feed with varied cry, rendering more apparent, rather than disturbing, the solemn stillness of the scene; all tend to raise the soul from nature up to nature's God, and remind one of the beautiful passage of Scripture, 'O Lord, how marvellous are thy works: in wisdom hast thou made them all: the earth is full of thy riches.'"

Climate of Labrador.

LABRADOR is a triangular peninsula, bounded on the east by Davis' Strait, on the south by Canada East and the Gulf of St. Lawrence, and on the west by Hudson Bay; but it is as excessively cold and barren as the countries to the west of the bay, and is, besides, almost constantly enveloped in fogs. The climate is very rigorous, and the winter lasts nine months of the year. It is too severe to ripen any of the cereals; but potatoes and several species of culinary vegetables are said to thrive and come to maturity. The climate, however, of the interior is somewhat milder than that of the coasts. The surface is mostly a mass of rocks and mountains, interspersed with innumerable lakes and rivers, which abound in fish. It swarms with beavers and other fur-bearing animals; reindeer, foxes, and bears also abound. The eider-duck and other birds in countless swarms frequent the eastern coast; also seals of different species. The northern and northeastern portions are inhabited by Esquimaux, among whom the Moravian brethren have established four settlements—at Nain, at Okak, Hoffenthal or Hopedale, and Hebron; and besides preaching the gospel, have taught the natives many of the useful arts of life.

The mean temperature of Okak, situated in north latitude 57° 30′, being about the same as Sitka, N. A., is as follows:—Spring, 25°; Summer, 50°; Autumn, 33°; Winter, 4°: Yearly mean, 28° Fahr.

Perhaps there is no region on this continent of which the general idea is more cloudy and indistinct than that of Labrador. American fishermen now more frequently extend their piscatorial visits to its coasts than formerly, and, so far as we understand, with good average success. Some interesting facts going to shed a little light on the general darkness, with respect to the natural history and meteorology of Labrador, are presented in the following extract from a letter written by one of a party engaged in cod fishing along the shore of that bleak and chilling land. He says:—

"From a thermometrical register kept at Rigolette, we find the lowest temperature of last winter to be but —37°, which is no colder weather than is found in New England; but that season was unusually warm. The average summer day temperature according to our own observations, is about 54° Fahr. In one instance the thermometer indicated as high as 80° in the shade, and in another 74°. The lowest winter day temperature was —33°, and the mercury several times sank to —40°. It is not intensity of cold, but the raw chilly atmosphere, impregnated with fogs and the moisture of melting snow, that renders necessary much thicker clothing than we wear at home in mid-winter. Although the climate is considered very healthy by the settlers, yet I could hardly recommend it to invalids. To those afflicted with weak or diseased lungs it is extremely injurious. Of this we had abundant evidence in three cases among our passengers, all of which have been aggravated by the exposure. The winters here are very long, and may be said to extend from the middle of September to the 1st of June. The cold is quite uniform, and the snow lies about four feet deep on a level. All that portion of the ocean embayed by the chain of islands that extends along the coast is frozen solid until May, and the ice-field sometimes extends beyond several miles to seaward. This is then the resort of the Arctic foxes and the white or polar bear. The 'water bear,' he is called here. They are often found upon the islands after the breaking up of the ice, where they have been left unexpectedly. They are quite numerous. Of foxes, large numbers are caught in traps. These live altogether upon the coast, and do not fraternize with the other species of fox in the interior. Snow may be said to disappear by the end of May, but it is found in gulleys and hollows all through the summer.

"Labrador water we cannot recommend. It is but the draining of melting snow and frozen earth, which, percolating through the moss, becomes a rank decoction, filled with vegetable matter, and of the color of whiskey. Natural springs are rare, but we have frequently found ponds and lakes upon the summits of rocky knolls, which appear to be fathomless, and whose water was pure and deliciously cold. As I have remarked, there are no roads in this country, and the only thoroughfares of travel in summer are the water courses which flow from the interior. I have referred at length to the Nor'west River and its immense outlet. The St. Francis or Alexis River is the only one of considerable size between that and the Straits of Belle Isle. It is a noble stream, but neither that nor the other is indicated on any maps yet made.

"Labrador furnishes little sport to the angler. Salmon are taken in immense quantities in nets, but will not meddle with the hook; neither will sea trout nor salmon trout. Brook trout afford the only sport. The winter hunting is good. The only deer found here is the caribou and reindeer. These are by some considered as one and the same, but the settlers make them quite distinct. The one is mottled with reddish spots; the other is of a mouse color in summer, and nearly white in winter. From all I can learn, the two animals are identical, and the difference in appearance is occasioned by the change of coat. The reindeer is not domesticated here, nor made to draw sledges, as in Lapland. Their time of fawning is two months later than in the States, and they are in the velvet until near the end of August. Trapping the fur-bearing animals affords a considerable profit, and quite frequently the hunter is rewarded by a black or silver fox, whose market prices are from forty to sixty dollars.

"We cannot but remark how carefully the animals of this icy country are protected by nature from their enemies. When man goes forth upon the snow to hunt, where upon the spotless mantle the smallest dark object would be readily revealed, then they are robed in white. The white partridge flies up from his very feet, where he perceived but lumps of feathery snow. The deer, bear, fox, ermine, all clad in white, pass him with impunity. Did not hunger lead them to the traps, or their deeply embedded tracks 'prate of their whereabout,' seldom would they fall victims to man. In the summer they are slaty and mouse-colored, like the rocks, or wood-colored, like the trees, and in many an imaginary rock, or stick, or stub, there is animal life, which will take to itself legs or wings when opportunity of easy escape offers."

ICEBERGS, which have a great influence on the climate of Labrador, are thus described by Ballantyne, when passing through Hudson Strait: "It is impossible to convey a correct idea of the beauty, the magnificence, of some of the scenes through which we passed. Thousands of the most grotesque, fanciful, and beautiful icebergs and ice-fields, surrounded us on all sides, intersected by numerous serpentine canals, which glittered in the sun like threads of silver, twining round ruined palaces of crystal. The masses assumed every variety of form and size, and many of them bore such a striking resemblance to cathedrals, churches, columns, arches, and spires, that I could almost fancy we had been transported to one of the floating cities of Fairy-land. The weather being pleasant, with a light breeze, not a sound disturbed the stillness of nature, save the gentle rippling of the vessel's bow as she sped on her way, or the occasional puffing of a lazy whale, awakened from a nap by our unceremonious intrusion on his domains."

Straits of Belle-Isle.

METEOROLOGICAL JOURNAL kept at Belle-Isle Lighthouse, by Capt. D. VAUGHAN, from May, 1859, to the end of April, 1860.

POSITION.—North latitude 51° 30′; West longitude 55° 30′,—off the coast of Labrador.

Date.		Temperature. Lowest.	Highest.	Remarks.
May,	1859,	24°	47°	May 27—The first steam-ship passed inward bound. May 29—120 icebergs were visible.
June	"	33°	65°	June 30—22 icebergs still visible. During the month there was constant hazy weather.
July,	"	39°	68°	Some fog almost every day this month, and several vessels wrecked.
August,	"	37°	64°	Icebergs in sight, except on eleven days this month.
September,	"	31°	61°	A few icebergs visible, except nine days this month.
October,	"	25°	46°	This was a very stormy month; on the 7th there was a hurricane.
November,	"	5°	39°	Nov. 1—The last steamer passed out. This was considered a pleasant month.
December,	"	—12°	30°	Dec. 5.—The Strait was one sheet of field-ice as far as can be seen.
January,	1860,	—20°	30°	Some icebergs grounded, and remained in sight until spring.
February,	"	—13°	32°	There was no thawing weather at Belle-Isle during the winter.
March,	"	—2°	38°	March 22.—The warmest day during the month.
April,	"	6°	38°	April 14.—A brig passed inward. The month was stormy and foggy.

Coldest day, 20° below zero; hottest, 68° above; variation, 88° Fahrenheit. Mean annual temperature, 35°.

Climate of Newfoundland.

"There is nothing in which the climate of Newfoundland differs more from that of Canada, and the adjacent provinces, than in its extreme vicissitudes. Spring comes more slowly than in Canada, the summer is shorter, the autumn less certain, the winter a series of storms of wind, rain, and snow ; the last rarely remains on the ground for any considerable length of time, and the frost is never, or very rarely, so intense as it is in Upper Canada, several degrees more to the south.

"All this may, perhaps, be accounted for by its insularity, and its lying at the embouchure of the great valley of St. Lawrence, whilst the frozen and desolate regions to the northwest of Labrador and Hudson Bay, cause the prevailing winds to sweep over it, loaded with a varying and reduced temperature of the air ; and then in the early spring vast masses of ice from Hudson Straits and East Greenland are forced along its Atlantic coast by a southerly current, where they consolidate or grind, until they are eventually forced off by milder air, and by the increasing warmth of the ocean, where they are sunk in the tepid waters of the Gulf Stream.

"If the laws of climate were regulated by the thermal zones which philosophers have drawn round the globe, Newfoundland would be an abode for man, equally free from great heats and from intense cold, as it lies in nearly the same parallels as France ; whereas, it has the general temperature of the European countries, situated fifteen or twenty degrees higher than the northern shores of that fertile country.*

"Various attempts have been made to account satisfactorily for this seeming anomaly between the climates of the Old and New World, as is most wonderfully exhibited on this island and on the coast of Labrador. The theory of winds is still, however, in its infancy, but that they are affected in their passage over bleak howling wildernesses, cannot be doubted. In Canada, and everywhere in North America, east of the Rocky Mountains, a wind from the northwest invariably lowers the thermometer, and in winter causes excessive cold. Canada, Labrador, and Newfoundland are the region of lakes ; and these, when frozen, of course increase the fury and bitterness of a storm from that quarter ; but although Newfoundland is but little removed from Labrador, the coldest country in the world, and from Cape Breton and Nova Scotia, where frost reigns in all its vigor in winter, it is not so cold as other parts of the American Continent lying several degrees further to the south. The ther-

* The mean annual temperature of St. John, N. F., 47½° north, is the same as St. Petersburg, Russia, 60° North latitude, 39° Fahrenheit.

mometer rarely falls to zero in winter, which lasts from the beginning of December until the middle of April; January and February being the coldest months, and the latter the most stormy.

"It is generally supposed that Newfoundland is constantly enveloped in fog and wet mist; nothing, however, can be further from the truth. The summers are frequently so hot and so dry, that for want of rain the grass perishes—and the nights are usually splendid; whilst, in winter, fog is very rarely seen. The fog being generated at sea, by causes which do not operate upon the land, is a true sea-mist, which may be observed in a voyage across the Atlantic, all the way from the west coast of Ireland, by keeping in a high latitude, until the vessel reaches Newfoundland. Fog on the shores, in summer, prevails with an easterly wind; west and southwesterly winds bring rain. The most remarkable features in the climate of Newfoundland are *the fogs on its banks*—which do not usually extend to the shores or inland, there frequently being a clear space or belt next the coast—and the prevalence of heavy winds."—*Bonnycastle's Newfoundland.*

The following is the mean temperature of the four seasons, and the yearly temperature of ST. JOHN, the capital of Newfoundland, situated on the southwest part of the island, in North latitude 47° 33°, West longitude 52° 43′.

MEAN TEMPERATURES.

Spring, . . .	32.30°	Autumn, . . .	43.80°
Summer, . . .	54°	Winter, . .	23.20°

Yearly Mean, 39° Fahrenheit.

Meteorological Table.

SHOWING THE SITUATION, ALTITUDE, MEAN ANNUAL TEMPERATURE, ETC., OF THE PRINCIPAL CITIES AND POSTS IN BRITISH AND RUSSIAN AMERICA.

Cities, etc	Latitude.	Longitude.	Altitude.	Yearly Mean.	Four Seasons.			
					Spring.	Summer.	Autumn.	Winter.
			Feet.	Fahr.				
Fort Good Hope, H. B. Ter..	67°00′	131°00′	....	12°	..	..	..	..
Yukon, Russian America....	66°00′	147°00′	....	16°	14°	58°	18°	—23°
Fort Franklin, H. B. Ter....	65°12′	124°00′	500	17°	15°	50°	28°	—16°
Fort Reliance, "	62°46′	109°00′	600	...	13°	..	..	—20°
Fort Hope "	66°32′	......	....	6°	-4°	58°	13°	—25°
Fort Enterprise, "	64°28′	113°06′	800	14°	8°	55°	17°	—24°
Fort Simpson, "	61°50′	121°57′	400	26°	27°	59°	28°	—14°
Fort Liard, "	59°30′	121°00′	500	30°	..	..	..	..
Fort Chepewyan, "	58°43′	111°48′	700	32°	24°	59°	34°	— 5°
Fort Churchill "	59°00′	93°10′	20	19°	13°	52°	23°	—14°
York Factory, "	57°00′	92°26′	20	26°	19°	54°	33°	— 3°
Sitka, Russian America.....	57°00′	135°18′	50	42°	40°	54°	44°	32°
Iluluk, Alaska; Russian Am.	53°52′	166°25′	....	40°	36°	52°	39°	32°
Hebron, Labrador..........	58°00′	64°00′	50	22°	18°	43°	31°	— 1°
Okak, "	57°30′	63°00′	....	28°	25°	49°	33°	4°
Nain, "	57°10′	62°00′	....	27°	2 °	50°	33°	0°
Oxford House, H. B. Ter....	54°55′	96°28′	350	27°	25°	..	..	—15°
Norway House, "	54°00′	98°00′	650	30°	28°	60°	30°	— 2°
Cumberland House, "	53°57′	102°20′	900	33°	33°	62°	34°	0°
Chesterfield House, "	51°00′	110°00′	....	39°	..	..	..	..
Moose Factory, "	51°15′	80°45′	....	27°	..	..	..	..
Fort Gary, "	50°15′	97°00′	760	36°	36°	68°	40°	8°
Pembina (Boundary Line)..	49°00′	99°00′	680	38°	..	..	..	..
Fort William, L. S.........	48°23′	89°27′	620	36°	35°	59°	38°	10°
Michipicoten "	47°56′	85°06′	620	38°	37°	59°	41°	16°
Saut Ste. Marie, C. W......	46°31′	84°43′	600	40°	38°	62°	44°	18°
Bruce Mines, "	46°15′	84°00′	576	40°	37°	62°	43°	19°
Quebec, C. E..............	46°49′	71°16′	100	40½°	38°	66°	44°	14°
St. John, N. F.............	47°33′	52°43′	140	39°	33°	54°	44°	23°
Charlottestown, Pr. Ed. Is...	46°15′	63°00′	....	40°	39°	63°	42°	18°
Frederickton, N. B.........	46°03′	66°08′	....	42°	40°	65°	47°	19°
St. John's, "	45°16′	......	....	43°	40°	63°	48°	21°
Pictou, N. S..............	45°34′	62°42′	....	42°	38°	63°	47°	20°
Halifax, "	44°39′	63°37′	....	43½°	39°	62°	49°	24°
Wolfville, N. S............	45°06′	64°25′	95	45°	..	..	..	..
Montreal, C. E............	45°30′	73°36′	60	45°	43°	70°	45°	18°
Stanbridge, "	45°08′	73°00′	....	43°	..	..	..	..
Ottawa City, C. W.........	45°23′	75°42′	300	42½°	..	..	..	..
Prescott, "	44°42′	75°36′	230	44°	..	..	..	..
Kingston, "	44°14′	76°34′	275	45°	44°	67°	45°	23°
Penetanguishene, C. W.....	44°48′	80°40′	600	43½°	39°	68°	45°	22°
Collingwood, "	44°30′	80°20′	576	44°	..	..	..	..
Goderich, "	43°44′	81°43′	576	45°	..	..	..	..
Toronto, "	43°39′	79°21′	342	44½°	41°	68°	47°	24°
Hamilton, "	43°15′	79°57′	275	48°	44°	72°	50°	27°
Niagara, "	43°18′	79°08′	250	47°	44°	69°	51°	27°
Sarnia, "	42°58′	82°25′	572	46°	43°	67°	49°	26°
Waterloo, "	42°55′	79°00′	565	47°	44°	68°	48°	27°
Windsor "	42°20′	82°57′	570	47°	46°	68°	48°	27°
Amherstburg, "	42°05′	82°58′	565	48°	47°	69°	49°	28°

PART IV.

CLIMATIC DIVISION OF CANADA.

Canada extends in length from the coast of Labrador 52° north latitude, westwardly, to the mouth of the Kaministiqua River, where stands Fort William, in latitude 48° 23′ near the western extremity of Lake Superior, about sixteen hundred miles; the average breadth is about two hundred and fifty miles. It contains an area of about three hundred and fifty thousand square miles, and is washed by the waters of the Gulf of St. Lawrence on the east, and by the waters of the Great Lakes of America on the west.

This important region of country lies between the 41st and 52d parallels of north latitude, and the meridians of 61°and 90° west from Greenwich. The range of mean annual temperature varies from 35° on the north, to 48° Fahr. on the south. The winters being excessively cold in the Lower or Eastern Province, and the summers warm; while in the Upper or Western Province the cold and heat are modified by the Great Lakes lying on its southern border.

The natural features of Upper and Lower Canada are, for the most part, very different. In Lower Canada, the scenery is of a far bolder character than in Upper. On the lower part of the St. Lawrence, both sides of the river are mountainous, and on the northern side the range which runs as far as the vicinity of Quebec presents the most sublime and picturesque beauties. On the southern side, the highlands or mountains which divide the waters flowing into the St. Lawrence from those running into the St. John's and other rivers of Maine, commence in the District of Gaspé, and about sixty miles below Quebec, turn off and enter the United States, forming, in part, the boundary between the two countries.

The District of Gaspé, facing the Gulf of St. Lawrence, being mostly surrounded by water, has a mild and favorable summer climate, while the cold of winter is perceptibly modified by the same cause. The lands in this district are composed of a light but fertile soil, producing most kinds of grain and vegetables in abundance; the potatoes in particular being highly esteemed for their fine quality and flavor. There are also found a variety of forest trees producing an abundance of timber for ship-building and other purposes. The great *pine region* of Canada, which may be said to extend from Saguenay to the St. Maurice and Ottawa Rivers, and even farther inland, afford the great wealth and distinctive feature of Canada in a commercial point of view.

The large extent of uninhabited country lying north of the isothermal line of 40° mean annual temperature, and along the 47th parallel of latitude, embracing the northern half of Canada, is almost entirely unfit for cultivation, owing to the absence of continued warm weather during the summer months; this whole region being subject to killing frosts every month of the year.* It is, however, valuable for its timber, minerals, and fur-bearing animals.

The valley of the St. Lawrence, below its junction with the Ottawa River, embracing the Eastern Townships, has a favorable and healthy climate; the soil producing all kinds of cereals and vegetables, as well as nutritious grasses. The Island of Orleans, below Quebec, and the Island of Montreal and its vicinity, may be called the "gardens of Lower Canada," where is produced wheat, rye, oats, barley, vegetables, and many kinds of fruit. The mean annual temperature varying from 40° to 45° Fahr., having a summer heat, for the most part, as high as Central New York or Northern Illinois.

The healthy influence of this great valley is proverbial—its inhabitants being a vigorous and long-lived class of people. Here thousands of seekers after health and pleasure resort during the summer months, enjoying alike the beautiful river and mountain scenery, with a healthy and invigorating climate. The cold winter weather continues from Dec. to March, inclusive.

* This depressing influence is caused, no doubt, from the cold, chilly winds coming off Hudson Bay and the more northern cold region.

UPPER, or WESTERN CANADA, is comprised within the parallels of 41° to 49° north, and the meridians of 74° to 90° west of Greenwich, and embraces an area of about one hundred thousand square miles. As compared with the Lower Province, Upper Canada is in general a level champaign country, with gentle undulating hills and rich valleys. At a distance of from fifty to one hundred miles north of Lake Ontario, there is a ridge of high rocky country running towards the Ottawa or Grand River, behind which there is a wide and rich valley of great extent, bounded on the north by a mountainous country, of still higher elevation. From the division line on Lake St. Francis, (near Fond du Lac) to Sandwich, along the shores of the St. Lawrence and Lakes Ontario and Erie, there is not an elevation of any consequence; and throughout this extent the soil is generally remarkably rich, and the climate salubrious.

The following Extracts, from a BRIEF OUTLINE OF CANADA, published by authority, will convey a correct idea of the climate of Canada West.

"The most erroneous opinions have prevailed abroad respecting the climate of Canada. The so-called rigor of Canadian winters is often advanced as a serious objection to the country by many who have not the courage to encounter them, who prefer sleet and fog to brilliant skies and bracing cold, and who have yet to learn the value and extent of the blessings conferred upon Canada by her world-renowned 'snows.'

"It will scarcely be believed by many who shudder at the idea of the thermometer falling to zero, that the gradual annual diminution in the fall of snow in certain localities, is a subject of lamentation to the farmer in Western Canada. Their desire is for the old-fashioned winters, with sleighing for four months, and spring bursting upon them with marvellous beauty at the beginning of April. A bountiful fall of snow, with hard frost, is equivalent to the construction of the best macadamized roads all over the country. The absence of a sufficient quantity of snow in winter for sleighing, is a calamity as much to be feared and deplored as the want of rain in spring. Happily neither of these deprivations is of frequent occurrence. The climate of Canada is in some measure exceptional, especially that of the Peninsular portion. The influence of the Great Lakes is very strikingly felt in the elevation of winter temperatures and in the reduction of summer heats.

"Perhaps the popular standard of the adaptation of climate

to the purposes of agriculture is more suitable for the present occasion than a reference to monthly and annual means of temperature. Much information is conveyed in the simple narration of facts bearing upon fruit culture. From the head of Lake Ontario, round by the Niagara frontier, and all along the Canadian shores of Lake Erie, the grape and peach grow with luxuriance, and ripen to perfection in the open air, without the slightest artificial aid. The Island of Montreal is distinguished everywhere for the fine quality of its apples, and the Island of Orleans, below Quebec, is equally celebrated for its plums. Over the whole of Canada the melon and tomato acquire large dimensions, and ripen fully in the open air, the seeds being planted in the soil towards the latter end of April, and the fruit gathered in September. Pumpkins and squashes attain gigantic dimensions. Indian corn, hops, and tobacco, are common crops, and yield fair returns. Hemp and flax are indigenous plants, and can be cultivated to any extent in many parts of the Province.

"The most striking illustration of the influence of the Great Lakes in ameliorating the climate of Canada, especially of the western peninsula, is to be found in the natural limits to which certain trees are restricted by climate. That valuable wood, the black walnut, for which Canada is so celebrated, ceases to grow north of latitude 41° on the Atlantic coast, but under the influence of the comparatively mild lake climate of Peninsular Canada it is found in the greatest profusion, and of the largest dimensions, as far north as latitude 43°."

Dr. Lillie, in his "Essay on Canada," remarks, that "Prof. Hind holds the climate of Canada West to be superior to those portions of the United States lying north of the 41st parallel of latitude, in mildness—in adaptation to the growth of cereals—in the uniformity of the distribution of rain over the agricultural months—in the humidity of the atmosphere—in comparative indemnity from spring frosts and summer droughts—in a very favorable distribution of clear and cloudy days for the purposes of agriculture—and in the distribution of rain over many days—as also in its salubrity. In the following points he regards it as differing favorably from that of Great Britain and Ireland, viz. :—in high summer means of temperature—in its comparative dryness—and in the serenity of the sky."

Climate of Canada, as described by Early French Authors.

"The first Europeans who came to Canada were surprised to observe the remarkable difference between the temperature of the Old World and of the New, under the same parallels of latitude. Thus Quebec, 46° 50′ north latitude, is hardly more northerly than La Rochelle, France, while it is more than two degrees to the south of Paris, and yet the winters of the ancient capital of Canada are much more rigorous than that of those two European cities. It had been thought that, according to a general rule, the intensity of cold and the rigor of climate increased in proportion as the Poles were approached, but in Canada these calculations were found to be at fault.

"Father Bressaui, an early writer on the subject, says: 'The first Frenchmen who inhabited the country believed that the immense forests which entirely covered it, were the cause of such an excessive cold. For my part, I think that if the forests, naked and leafless as they are in winter, can keep the sun from warming the earth and tempering the rigor of the cold, they should be a still greater shield in summer, when they are decked with thick foliage. They do not, however, produce that effect; for the heat, even in the middle of these woods, is then excessive, although it freezes during certain nights, just as in winter.'"

"The historian Charlevoix is of a different opinion, and thinks that, even in his time, the clearings had made some difference in the temperature, and rendered it less cold than in the first years of the colony's existence. In our day, the enquiry is sometimes made, whether the destruction of the great forests which there were on the banks of the St. Lawrence, has caused any improvement in the climate of the country—and the question is a very interesting one, which it is worth while examining. As reasoning could not possibly shed any light on the subject, we have thrown together some meteorological observations, of a nature to enable us to see whether there are grounds for hope that the severity of our climate will be abated.

"The oldest observations made as to the temperature of Canada are those of Jacques Cartier, during the winter of 1535–36, which he spent near the River St. Charles, in the vicinity of the present site of Quebec. 'From the middle of November,' says he, 'until the 18th of April, we were continuously shut up in the ice, which was more than two fathoms (brasses) thick, and, on land, the snow was four feet deep and more, so that it was above the bulwarks of our ship, and it lasted until the time above mentioned, so that all our drink was frozen in our casks, * * * and all the said river (St. Lawrence) was frozen, as much of it as is fresh water, as far as above Hochelaga,' now Montreal.

"So, during the winter that Cartier spent at Stadaconé (Quebec), more than three centuries ago, the ice took about the middle of November, and the thaw occurred about the middle of April (a period of five months); the earth was covered with a sheet of snow, four feet thick.

"About eight years later, Jean Alphonse, pilot to the Lord (Sieur) of Roberval, made the remarks which follow:—'The whole extent of these regions may well be called New France, for the air is as temperate as in France, and they are situated in the same latitude. The reason why they are so cold in winter is, that the fresh water river is naturally colder than the sea, and also because it is wide and deep, and in some places is more than half a league in width, and also because the land is not cultivated nor full of people, but altogether covered with forests, which is the cause of the cold. * * * * * If the country were cultivated and fully inhabited, it would be as warm as at La Rochelle; and the reason why it snows oftener than in France is because it rains less.'

"These remarks contain nothing precise; they only show that it was expected that the climate would become milder in proportion as settlements extended. Let us pass on to the observations of Champlain.

"1608, October 3d.—Some white frost appeared, and the leaves of trees began to fall on the 15th. 24th. I had some of the vines of the country planted. November 18, a quantity of snow fell, but only remained on the ground two days.

"1613.—When Champlain arrived at Quebec, on the 7th of May, the trees were putting forth leaves, and the fields were variegated with flowers. The winter had been mild and the river free from ice.

"1623, March 19th.—A violent storm, accompanied with wind, hail, thunder and lightning, although at this time the air is still cold, and the country full of snow and ice. April 16th, there was a foot of snow in some parts. 20th of said month, grain was sown behind the house, where the snow had melted sooner than in other places, being sheltered from the northwest wind.

"Towards the end of November, the River St. Charles was almost covered with ice. From the beginning to the end of November the weather was very variable, and the days were chiefly pretty cold mornings with frost, although it was usually fine the rest of the day: there was sometimes rain and snow, which often melted as it fell. We remarked that there is not a fortnight's difference between the commencement of winter temperature, one year with the other, that is to say, from the 20th of November until April, when the snow melts, and May is the spring-time; making six months of frost and six months of mild weather for this region. Some years the snow is deeper than others, the depth being from one and a half to three or four feet at most, on the level country.

"December 10th.—The River St. Lawrence was filled with ice, and the shore ice taking, navigation stopped.

"1624, April 18th.—This is the time for hunting game, which is very plentiful until the end of May, when the birds go back to lay their eggs, and only come again towards the 15th of September, when you can shoot until the ice takes along the shore, which is about the 20th of November.

"May 8th.—The cherry-trees begin to open their buds to let the leaves grow; at the same time little grey and white flowers shoot from the ground, which are the first fruits of spring in these regions. 9th. Strawberries began to bud, and all herbs to spring up out of the earth. 12th. White violets were seen in bloom. 15th. Trees were budding out, and cherry-trees covered with foliage; wheat had grown a span high, and in the fields the sorrel was two inches high. 18th. The birch-trees put forth leaves, the other forest trees following close after; the oak had its buds formed, and the apple-trees, which had been brought from France, as well as the plum-trees, began to blossom, and Indian corn was sown. 29th. Strawberries began to blossom, and the oaks to put out pretty large summer leaves. 30th. Strawberries were all in blossom; apple-trees began to open their leaf-buds; the oaks had their leaves about an inch long; the plum and cherry-trees were in flower, and Indian corn began to come up."

Now, then, here is what Father Lalemant, superior of the museum in Canada, wrote to his brother in the year 1626:—

"The place where the French have domiciled themselves called Kébec, is on the parallel of about 46 degrees and a half, on the banks of one of the most beautiful rivers in the world. But, although the latitude of the country where we are is nearly two degrees to the south of Paris, yet the winter is usually five months and a half in length, the snow three or four feet deep, and so lasting, that they do not generally melt until the middle of April, although they always begin in the month of November. During that time you cannot see the earth, and our French people have even told me that they have drawn a sleigh in May.

"The mildest winter ever known is that last past, so say the old inhabitants; yet the snow began to fall on the 16th of November, and to melt towards the end of March. The length of time the snow lasts is such, you would hardly think wheat and barley could grow very well here; I have, however, seen crops of them just as fine as in France.

"From the mouth of the St. Lawrence up to this there is no cleared land—all is forest. These people do not engage in husbandry; there are only three or four families who have cleared two or three acres, where they sow Indian corn."*

Well, this was the climate of the neighborhood of Quebec 240 years ago, at the time when the French had only cleared a few score of acres: the winter began in November, and terminated in the middle of April; three or four feet of snow covered the ground around Quebec.

To procure some other terms of comparison, we present extracts from the Jesuits' journals:—

1645, Nov. 15th.—"The snow began to stay."

1646, April.—"From the 17th to the 18th the river was clear, and they began to sow about this time."

Nov. 7th.—It began to freeze, so as to form ice, and the next day it snowed for the first time."

1647, March 11th.—"Then began the break up of a winter without winter, for it had not been cold up to that time."

Nov. 4th.—"The snow began."

1648, Nov. 18th.—"The snow began to stay."

1649.—Navigation opened between Quebec and Three Rivers on the 22d of April; the ice in the River St. Charles broke up on the 27th, and on the 28th they began to sow."

1650, April 25th.—The breaking up of ice took place in the River St. Charles. On the 23d of November, a vessel going to carry fish to Montreal came back to Quebec, because the ice began to form in Lake St. Peter."

Such are the observations we have been able to gather relative to the climate of Quebec, such as it was in the first days of the colony; and here are some made in our own times, after a lapse of upwards of 200 years:—

1855, Nov. 18th.—The snow fell in abundance. 25th. Steamboats, starting for Montreal, were obliged by the ice to come back, and go into winter quarters at Quebec.

1856, April.—The ice of Lake St. Peter was going down from the 23d to the 27th; 20th, the snow had mostly disappeared.

1857, April 20th.—The middle of the fields was uncovered. 24th. The ice in the River St. Charles broke up. 28th. The Lake St. Peter ice broke up and passed Quebec. May 30th. Plum and apple-trees are in bloom. October 28th. The first snows whiten the earth, but soon disappear. Nov. 23d. Good sleighing. Dec. 3d. Navigation terminates between Montreal and Quebec.

* The Island of Orleans, situated a few miles below Quebec, is now highly cultivated, and very fertile; also other portions of the country, on both sides of the St. Lawrence, and River St. Charles.

1858, April 14th and 15th.—Lake St. Peter ice passes Quebec. June 1st. Apple-trees and plum-trees in bloom. Nov. 8th. First snow fell; 30th, navigation closes between Quebec and Montreal.

1859, March 30th.—Summer vehicles are used. 18th April, Lake St. Peter ice broke up, and navigation resumed. May 16th. Plum and cherry-trees in blossom; October 21, snow whitens the ground; 29th, navigation stopped between Quebec and Montreal.

1860, April 17th.—Navigation opened in the River St. Lawrence; May 19th, plum-trees in blossom; Nov. 18th, first snow falls; navigation remained open until the 7th of December.

1861.—The Lake St. Peter ice broke up on the 25th of April, and navigation opened on the St. Lawrence between Quebec and Montreal.

Mean Temperatures.

Comparative Monthly Tables of Mean Temperature of Montreal and Quebec, from recent Observations.

Months.	Montreal. 45° 30′ N. L. ° Fahr.	Quebec. 46° 49′ N. L. ° Fahr.
March,	29.40	28.06
April	43.50	36.14
May,	58.00	49.03
Mean Spring Temp.,	43.65	37.74
June,	68.30	60.34
July,	73.00	68.86
August,	70.70	62.50
Mean Summer Temp.,	70.60	65.56
September,	60.60	55.15
October,	46.40	45.43
November,	25.70	26.75
Mean Autumn Temp.,	43.56	43.10
December	19.00	18.00
January,	14.90	12.50
February,	17.80	10.55
Mean Winter Temp.,	17.30	13.68
Yearly Mean,	44.60	40.02

Although the early observations on record do not give the exact range of the thermometer, as is now attainable, yet still it seems safe to infer that the mean annual temperature of Canada has not materially changed during the past three centuries.

"On comparing the meteorological observations made in the sixteenth and seventeenth centuries with those of the middle of the nineteenth, it is easily to be convinced that the climate of Canada—at least the neighborhood of Quebec—is about the same now as it was 300 years ago. Then, three or four feet of snow; now at least as much; the first snow falling the first fortnight of November, the break up of the River St. Charles from the 18th to the 27th April; navigation on the St. Lawrence, between Quebec and Montreal, interrupted by the ice in the last week of November, and opening toward the end of April—five months, closed on an average. The apple, cherry, and plum-trees blossoming the last of May and beginning of June; this is what we find at both epochs. In this respect nothing appears changed; and the *clearings* made until now have had very little influence in the present temperature of Canada.

"It is, then, to other causes besides the existence of forests, that the great cold of the winter of our country is due. They are to be looked for in the dryness of the northern atmosphere; the neighborhood of Hudson Bay, which is covered with ice during a great portion of the year; in the frequency of the northwest winds, which carry away from America the heated moisture produced by the warm current of the Gulf of Mexico; perhaps in the proximity of the Magnetic Pole, which, according to Captain Ross, is to the north of the Continent of America, in about the 70th degree of latitude, while the greatest cold is felt on the 72d and 73d degree. Indeed, the nearness of the Poles of cold and terrestrial magnetism would seem to show that some relation exists between the temperature and the magnetism of the globe."

Canada as It Is, in a Climatic Point of View.

HOGAN, in his *Prize Essay on Canada*, says: "The acknowledged influence of the atmosphere, not only upon the productiveness of the soil of a country, but upon the temper, habits, and industry of its inhabitants, renders an inquiry into the climate of Canada a subject of great importance.

"Her Inland Seas, covering an area of about 100,000 square miles, and a supposed contents of 11,000,000 cubic miles of water—far exceeding half the fresh water in all the lakes in the world—exercise a powerful influence in modifying the two extremes of heat and cold. The uniformity of temperature thus produced, although low, is found to be highly favorable to animal and vegetable life. It is therefore found, that in the neighborhood of the lakes, the most delicate fruits are reared without injury, whilst in places four or five degrees further south, they are destroyed by the early frosts. The quantity of rain, which, for the most part, falls in summer and early autumn, is, no doubt, greatly increased by evaporation from these immense bodies of water. The winds are most variable, and rarely continue for more than two or three days in the same quarter. This has the effect of preserving the equilibrium, and renders the occurrence of disastrous storms less frequent. The southwest, the most prevalent wind, is generally moderate, with clear skies. The northeast and east bring continued rains in summer and early autumn, and the northwest, springing from the region of ice, is invariably dry, elastic, and invigorating. Since 1818, the climate has greatly changed, owing principally, it is supposed, to the large clearings of the primeval forests.

"The salubrity of the province is sufficiently shown by its cloudless skies, its elastic air, and almost entire absence of fogs. The lightness of the atmosphere has a most invigorating effect upon the spirits. The winter frosts are severe and steady, and the summer suns are hot, and bring on vegetation with wonderful rapidity. It is true that the spring of Canada differs much from the spring of many parts of Europe; but after her long winter, the crops start up as if by magic, and reconcile her inhabitants to the loss of that which, elsewhere, is often the sweetest season of the year. If, however, Canada has but a short spring, she can boast of an autumn deliciously mild, and often lingering on, with its "Indian Summer" and golden sunsets, until the month of December.

"A Canadian winter, the mention of which, some years ago, in Europe, conveyed almost a sensation of misery, is hailed rather as a season of increased enjoyment than of privation and discomfort by the people. Instead of alternate rain, snow, sleet, and fog, with broken up and impassable roads, the Canadian has clear skies, a fine bracing atmosphere, with the rivers and many of the smaller lakes frozen, and the inequalities in the rude tracks through the woods made smooth by snow, the whole face of the country being literally macadamized by nature for a people as yet unable to macadamize for themselves.

"It must not be supposed that the length of this season is necessarily prejudicial to the farmer, for mild winters are generally found to be injurious to fall crops of wheat, and a serious hindrance to business and travelling. The summer, short and eminently fructifying, occupies the whole of the farmer's time. It is in winter that the land is cleared of timber, the firewood dragged home from the woods on sleighs, over ground impassable for wheel carriages, and that the farmer disposes of his produce, and lays in his supplies for the future. The snow forms a covering for his crops, and a road to his market. On the arrival of winter, the care of his fat stock ceases, for the whole is killed, freezes, and can be disposed of as the state of the market suggests.

"Comparing the two Provinces, it is admitted that the climate of Upper Canada is the most favorable for agricultural purposes, the winter being shorter and the temperature less severe; but the brilliant sky, the pure elastic air and uninterrupted frost of Lower Canada, though perhaps lingering too long, are far more exhilarating, and render out-door exercise much more agreeable. Few who have enjoyed the merry winter of Quebec and Montreal, with the noble hospitality and charming society of these cities, their sleigh-rides and their pic-nics, can ever forget the many attractions of a winter in Lower Canada."

Of the general salubrity of the Province, its vital statistics, as compared with those of other countries, afford satisfactory evidence; and the following table, communicated by Professor Guy, is not devoid of interest, as showing the proportion of deaths to the population in various countries:—

Austria, . . .	1 in 40	Prussia, . .	1 in 39
Belgium, . .	1 " 43	Russia in Europe, .	1 " 44
Denmark, . .	1 " 45	Spain, . . .	1 " 40
England, . .	1 " 46	Switzerland, . .	1 " 40
France, . .	1 " 42	Turkey, . . .	1 " 30
Norway and Sweden,	1 " 41	United States, .	1 " 74
Portugal, . .	1 " 40	Canada, . . .	1 " 98

Climate of Canada, New Brunswick, and Nova Scotia.

"The climate of Western Canada and Nova Scotia is warmer than that of Canada East, Prince Edward Island, or New Brunswick, although a large portion of the latter is similar to that of Nova Scotia. In Newfoundland winter is severe; yet snow does not lie long on the southern coast. It is generally said that winter in these colonies, lasts five months, which in one sense is true, but in another it is not. Winter, in reality, cannot be said to last longer than three months, commencing about the middle of December, and ending about the middle of March. During this period there are, in the coldest section of

Lower Canada, from twenty to twenty-five cold days, when the thermometer ranges from 15° to 20° degrees below zero. The cold is driven from the Arctic regions by northwest winds, passing over the country in waves, lasting for about three days at a time—familiarly known as 'cold snaps.' During the intervals between these periods of cold the thermometer ranges about zero.

"There are generally from four to seven snow storms during each winter, when the snow falls, in Canada West, to the depth of about one and a half feet in the aggregate; in Nova Scotia, from one to two feet; in New Brunswick, Canada East, and Prince Edward Island, from two to four feet. To these general rules, however, there are frequent exceptions. Some seasons the snow exceeds these depths, and very frequently, in Nova Scotia and a large part of New Brunswick, the snow does not average one foot in depth. The January thaw often sweeps the snow from the face of the country, leaving the ground, contrary to the interests of agriculture, uncovered for weeks. In Western Canada, where a large quantity of winter wheat is raised, these thaws are particularly injurious. During a large portion of winter, in the cold parts of the colonies, the thermometer ranges from 10° to 40° above zero.

"Deep snow adds to the fertility of the soil. The ground is so pulverized by the action of the frost as to be rendered friable and more easily ploughed. In Lower Canada the snow appears early in December, and disappears finally about the middle of April; in Western Canada, it disappears three weeks sooner.

"By a wise and economical division of time, all classes of the people may be, and generally are, as profitably employed during the winter months, as in summer. It is a great mistake to say that winter is necessarily a period of idleness and inactivity; the reverse is the fact. Our winters are pleasant, and their long evenings afford the student ample time for the acquisition of useful knowledge. There is no season of the year so well adapted to the cultivation of literary, domestic and social intercourse, as that of a North American winter. It is the lecturing season in the institutions and halls, with which nearly every community is supplied; it is the season when the several colonial legislatures sit, and the season when the press is doubly vigilant in supplying the public with useful information. Indeed the winter season in these colonies is very pleasant, affording enjoyment and profit to the inhabitants.

"The prevailing winter winds are the northwest, north, and northeast; in spring, south; and in the summer, west and southwest. In the interior of Canada East and New Brunswick, the heat of summer sometimes rises to 80° and even 90°;

while along the sea-board the climate is more equable, and the air wholesome and bracing. Vegetation progresses with great rapidity.*

"The autumn is the most delightful season in the year. In the language of J. V. Ellis: 'The summer still lingers, as if regretting to quit the scenes of beauty it has created'—and then is produced the 'Indian Summer,' a season of rare and exquisite loveliness, that unites the warmth of summer with the mellowness of autumn.

"The fogs which sometimes prevail along a part of the Atlantic coast line, seldom extend more than five miles inland. The Gulf and River St. Lawrence are more free from fogs than the Bay of Fundy and the Atlantic coasts; but in none of these places are they found to impede navigation, or produce effects detrimental to the general interests of the country.

"There are no endemical, and few epidemical diseases in the habitable part of British North America. The country is remarkably healthy, as the longevity of human life testifies. The frosts are less severe than in many of the populous countries of Christendom, and the summers are less calid than in many of the southern climes where civilization is making rapid progress. Indeed, the climate of one-third, at least, of British North America is highly adapted to the progress of civilization."

Indian Summer in Canada.

Indian Summer is a phenomenon of constant yearly occurrence and marked characteristics in th northewest of the United States and Canada. The following table, furnished from the private memoranda of an assistant at the Provincial Observatory, at Toronto, establishes the fact, that hazy, warm, mellow weather, termed Indian summer, is a periodical phenomenon in Canada. The characters of Indian Summer are still more decided in the far Northwest than in the neighborhood of Lake Ontario. Sounds are distinctly audible at great distances; objects are difficult to discern unless close at hand; the weather is warm and oppressive, the atmosphere hazy and calm, and every object appears to wear a tranquil and drowsy aspect.

* "By a reference to the mean temperature of the years in Montreal, from 1826 to 1852," says Dr. A. Hall, "the fact will be apparent, that a gradual decrease of temperature has marked the years, as they have successively passed away; a circumstance not very consonant with the almost universally received opinion, that countries become gradually warmer in the ratio of their cultivation, population, &c. The year 1830 was the warmest on record (47.8° Fahr.), and the year 1835 (42° Fahr.) was the coldest; the year 1852 (44.6°) being the mean temperature for a period of twenty-seven years."

Indian Summer at Toronto, C. W.

Year.	Commencement.	Termination.	No. of Days.
1845,	24th October,	29th October, .	. 6
1846,	4th November,	7th November,	. 4
1847,	28th October,	31st October, .	. 4
1848,	20th November,	23d November,	. 4
1849,	13th "	18th "	. 6
1850,	7th "	13th "	. 7
1851,	6th October,	11th October, .	. 6
1852,	16th November,	21st November,	. 6
1853,	12th October,	20th October, .	. 9
1854,	24th "	28th " .	. 5
1855,	16th "	26th " .	. 11
1856,	19th "	22d " .	. 4
1857,	5th "	12th " .	. 8
1858,	18th "	28th " .	. 11
1859,	2d November,	8th November,	. 7
Mean result,	27th October,	2d November,	6½ days.

Climatic Observations near Montreal.

The extreme nature of the Climate will be sufficiently exhibited by the accurate observations of the weather of 1854, made near Montreal by Dr. Smallwood.

Months.	Mean Temperature. ° Fahr.	Rain in Inches.	Snow in Inches.	Range of Thermometer. ° Fahr.
January, . .	10.92	1.067	17.98	78.8
February, . .	12.20	0.150	23.96	71.7
March, . .	26.84	0.910	28.16	60.4
April, . . .	37.75	7.886	4.03	52.2
May, . . .	57.17	3.418		60.7
June, . . .	63.80	8.384		46.6
July, . . .	76.20	0.174		48.5
August, . .	68.31	2.265		48.2
September, .	58.01	6.167		64.2
October, . . .	48.40	4.844	3.10	55.5
November, . .	32.99	5.130	1.10	50.6
December, . .	27.35	0.110	18.67	78.1
Total, . .		40.505	97.00	

Spring, . . .	38° Fahr.	Autumn, . .	47° Fahr.
Summer, . . .	70° "	Winter, . .	17° "

Yearly Mean, 44° Fahrenheit.

"The ISLAND OF MONTREAL has been called 'the Garden of Canada.' The soil, however, can only be regarded as of secondary quality. The Trenton limestone prevails over the whole island, as an isolated patch among the surrounding primary rocks, rendering the soil genial to the growth of grasses, though not of winter wheat. Oats, barley, and potatoes are the staple crops, with small quantities of turnips. Good orchards, producing different kinds of fruit, are met with throughout the Island of Montreal, where the soil rests upon the limestone, and is more friable. The apple, however, does not thrive on the clay soils of the flats of the St. Lawrence, in consequence of their tenacious nature."—*Russell's Agriculture and Climate.*

Monthly Mean Temperature at QUEBEC; North latitude, 46° 49′; West longitude, 71° 16′, from Greenwich.

Months.	Mean Temp. °Fahr.	Maximum. ° Fahr.	Minimum. °Fahr.	Variation. ° Fahr.
January, . .	16.70	46.0	—14.0	60.0
February, . .	14.55	36.8	—29.5	66.3
March, . .	21.06	47.3	2.4	44.9
April, . . .	34.14	59.8	5.9	53.9
May, . . .	49.03	83.0	32.0	51.0
June, . . .	58.34	88.0	43.2	44.8
July, . . .	68.86	90.3	51.9	38.4
August, . .	61.54	85.0	38.3	46.7
September, .	55.15	81.3	34.7	46.6
October, . .	45.43	60.4	28.4	32.0
November, . .	28.75	34.3	10.0	24.3
December, . .	20.09	40.1	—19.2	59.3

Extreme range of Temperature, 119° Fahrenheit.
Mean Annual Temperature, 40.64° Fahr.

Health Statistics of the Principal Cities in Canada. From the Official Census.

Cities.	Pop. 1861.	Deaths.	Pop. 1851.	Deaths.
Montreal, C. E., . .	90,323	2,038	57,715	1,725
Quebec, " . .	51,109	1,111	42,052	1,064
Three Rivers, C. E. .	6,058	106	4,936	
Ottawa, C. W., . .	14,669	172	7,760	90
Kingston, " . .	13,743	129	11,585	172
Toronto, " . .	44,821	727	30,775	474
Hamilton, " . .	19,096	217	14,112	185
London, " . .	11,555	102	7,035	100

Total Population and Deaths in Canada,—1861.

	Population.	Deaths.	Ratio of Deaths.
Lower Canada, . .	1,111,566	13,224	1 in 84
Upper Canada, . .	1,396,091	10,160	1 in 137
Grand Total, . .	2,507,657	23,384	1 in 107

From the above Table of Deaths, &c., it seems that Montreal and Quebec, situated on the St. Lawrence River, are quite unhealthy, compared to Hamilton and Toronto, situated on the shores of Lake Ontario. The same disparagement is apparent between the Upper and Lower Province.

Mean Temperatures.

Comparative Tables of Temperature of St. John, N. B.; Halifax, N. S.; and Charlottetown, P. E. Is.

Months and Seasons.	St. John. 45° 16′ N. L. ° Fahr.	Halifax. 44° 39′ N. L. ° Fahr.	Charlottetown. 46° 15′ N. L. ° Fahr.
March, ...	29.61	28.50	29.00
April . . .	38.97	37.80	36.00
May, ...	50.11	50.40	51.25
Spring, . .	39.56	38.90	38.75
June, ...	58.04	57.20	59.50
July, ...	64.87	67.10	62.25
August, ...	64.02	61.60	66.00
Summer, . . .	62.31	61.63	62.58
September, . .	57.70	55.70	56.00
October, ...	48.55	50.40	39.25
November, . . .	36.80	42.80	29.50
Autumn, .	47.68	49.30	41.58
December	23.15	32.10	17.00
January, ...	19.06	22.00	14.50
February, ...	21.13	20.30	17.30
Winter, ...	21.11	24.80	16.26
Yearly Mean, . . .	42.66	43.50	40.00

Agricultural Produce of Canada,—1861.

	Acres of Wheat.	Bush. of Wheat.	Acres Indian Corn.	Bush. Indian Corn.
Upper Canada, .	1,386,366	24,620,425	79,918	2,256,290
Lower Canada, .	244,769	2,654,354	15,012	334,861
Total, . .	1,631,135	27,274,879	94,930	2,591,151
	Acres of Oats.	Bush. of Oats.	Acres of Barley.	Bush. Barley.
Upper Canada, .	678,337	21,220,874	118,940	2,821,962
Lower Canada, .	955,553	17,551,296	139,442	2,281,674
Total, . .	1,633,890	38,772,170	258,382	5,103,636
	Acres of Rye.	Bush. of Rye.	Acres of Peas.	Bush. of Peas.
Upper Canada, .	70,376	973,181	460,595	9,601,396
Lower Canada, .	83,931	844,192	234,035	2,648,777
Total, .	154,307	1,817,373	694,630	12,250,173
	Acres Buckwheat	Bush. Buckwheat	Acres Potatoes.	Bush. Potatoes.
Upper Canada, .	74,565	1,248,637	137,266	15,325,920
Lower Canada, .	75,605	1,250,025	118,709	12,770,471
Total, . .	150,170	2,498,662	255,975	28,096,391
	Tons of Hay.	Lbs. Maple Sugar	Lbs. Wool.	Lbs. Hemp & Flax
Upper Canada, .	861,844	6,970,605	3,659,766	1,225,934
Lower Canada, .	689,977	9,325,147	1,967,388	975,827
Total, . .	1,551,821	16,295,752	5,627,154	2,201,761
	Lbs. Butter.	Lbs. Cheese.	No. of Cows.	No. of Sheep.
Upper Canada, .	26,828,264	2,687,172	451,640	1,170,225
Lower Canada, .	15,906,949	686,279	328,370	682,829
Total, . .	42,735,213	3,373,469	980,010	1,853,054

From the above Table it appears that the great agricultural products of Canada are wheat and oats; showing conclusively that the climate of Canada is well adapted to the hardier cereals.

Mean Meteorological Results at Toronto.

Latitude, 43° 39′ North; *Longitude*, 79° 21′ West.

BY G. T. KINGSTON, M.A.

DIRECTOR OF THE PROVINCIAL MAGNETIC OBSERVATORY, TORONTO.

THE mean temperature of the year 1864 was 44°.70, or 0°.53 in excess of the average of twenty-five years. The deviation of the monthly means above or below their respective averages, and irrespective of sign, had an average amplitude of 1°.36; thus indicating a year of unusually equable temperature, the average amplitude in twenty-five years being 2°.33.

The mean deviations of temperature in the four seasons, with their proper signs, were: −0°.33 in Winter; +0°.79 in Spring; +2°.27 in Summer; and −0°.60 in Autumn.

As regards rain and snow, there was, on the whole year, an excess amounting to 0.655 inches of water. An excess occurred in Winter, Spring, and Autumn—the total precipitation exceeding the average by 1.136 inches in Winter, 1.788 inches in Spring, and 0.186 inches in Autumn. In Summer, the rain was deficient as compared with the average by 2.405 inches. This deficiency was not much greater than that of the summer of 1863; but the distribution among the three summer months was very different in the two years, for while in the summer of 1863 there was a moderate deficiency in each month, the rain in June, 1864, was less than one-fifth, and in July little over one-third of the average fall; that of August being above the average in the ratio of 5 to 3 nearly.

In the following summary several of the results for the year 1864 are compared with the averages derived from a series of years as well as with extreme values of analogous results given by the same series.

TEMPERATURE.

	1864.	Average of 25 years.	Extremes.	
	°	°	°	°
Mean temperature of the year...	44.70	44.17	46.36 in '46.	42.16 in '56.
Warmest month..............	July.	July.	July, 1854.	Aug. 1860.
Mean temp. of the warm'st month	69.73	66.98	72.47	64.46
Coldest month..........	January.	February.	Jan. 1857.	Feb. 1848.
Mean temp. of the coldest month.	22.79	22.99	12.75	26.60
Diff. between the temp. of the warm'st and the cold'st m'ths	46.94	43.99	....	....
Warmest day..................	June 25.	....	July 12, '45.	July 31, '44.
Mean temp. of the warmest day.	81.77	77.45	82.32	72.75
Coldest day....................	Feb. 17.	....	Feb. 6, '55 / Jan. 22, '57	Dec. 22, 42.
Mean temp. of the coldest day...	—4.62	—1.02	—14.38	+9.57
Date of the highest temperature	Aug. 8.	....	Aug. 24, '54.	Aug. 19, '40.
Highest temperature..........	94.0	90.6	99.2	82.4
Date of lowest temperature.....	Feb. 17.	....	Jan. 26, '59.	Jan. 2, '42.
Lowest temperature...........	—15.0	—12.4	—26.5	+1.9
Range of the year.............	109.0	103.0	118.2	87.0

General Meteorological

PROVINCIAL MAGNETICAL OBSER

	JAN.	FEB.	MAR.	APR.	MAY.	JUNE.
	°	°	°	°	°	°
Mean temperature............	22.79	24.32	29.12	40.95	54.81	63.03
Difference from av. (25 years).	— 0.82	+ 1.33	— 0.74	— 0.01	+ 3.13	+ 1.69
Thermic anomaly (L. 43° 40′).	—10.01	—10.38	—10.98	— 9.25	— 3.29	— 1.57
Highest temperature.........	44.2	45.0	50.2	59.4	79.0	93.4
Lowest temperature.........	— 9.0	—15.0	3.0	28.1	32.2	34.8
Monthly and annual ranges..	53.2	60.0	47.2	31.3	46.8	58.6
Mean maximum temperature.	29.58	31.52	35.59	47.48	62.86	73.06
Mean minimum temperature.	17.51	18.94	22.44	34.61	46.20	52.87
Mean daily range............	12.07	12.58	13.16	12.87	16.67	20.19
Greatest daily range.........	26.9	37.4	28.4	24.4	26.2	31.7
Mean height of barometer.....	29.5887	29.4914	29.5082	29.5968	29.4721	29.6545
Difference from av. (18 years).	—.0447	—.1208	—.0741	+.0098	—.1125	+.0921
Highest barometer...........	30.102	30.124	30.067	29.964	29.788	29.961
Lowest barometer...........	28.910	29.009	28.829	29.301	29.166	29.007
Monthly and annual ranges..	1.192	1.115	1.238	0.663	0.622	0.954
Mean humidity of the air......	.82	.82	.80	.75	.75	.63
Mean elasticity of aqueous vapor.	.110	.119	.135	.194	.333	.380
Mean of cloudiness............	.67	.72	.66	.74	.68	.30
Difference from av. (12 years).	—.05	+.01	+.06	+.15	+.15	—.22
Resultant direction of the wind.	S 73 W	S 84 W	N 53 W	N 41 E	N 7 W	N 55 W
" velocity of the wind..	6.00	6.48	2.29	3.39	1.86	1.72
Mean velocity (miles per hour)..	10.22	10.11	8.41	7.77	5.64	4.53
Difference from av. (17 years)	+2.22	+ 1.77	—0.26	—0.29	—0.95	—0.74
Total amount of rain..........	1.165	0.397	1.620	3.633	4.070	0.570
Differ. from av. (24 & 25 yrs.)	—0.166	—0.603	+0.063	+1.200	+0.864	—2.297
Number of days rain.........	5	2	9	16	18	5
Total amount of snow.........	26.3	9.5	3.7	3.5	0.0	....
Difference from av. (22 years.)	+11.15	— 8.55	— 5.40	+ 1.10	— 0.09	....
Number of days snow........	14	14	12	3	0	....
Number of fair days..........	14	13	14	14	13	25
Number of auroras observed....	0	4	2	4	3	5
Possible to see auro. (No. nights).	11	11	15	10	12	24
Number of thunderstorms......	0	0	0	0	5	2

NOTE.—The TORONTO MAGNETICAL AND METEOROLOGICAL OBSERVATORY is situated in the grounds of the University of Toronto, in latitude 43° 39′ N.; longitude 5h. 17m. 33s. W.; 108 feet above Lake Ontario, and approximately 342 feet above the level of the sea.

The duties of the Observatory are carried on by the Director, G. T. KINGSTON, M.A., assisted by several competent assistants.

Register for the year 1864.

VATORY, TORONTO, CANADA WEST.

July.	Aug.	Sept.	Oct.	Nov.	Dec.	Year 1865.	Year 1864.	Year 1863.	Year 1862.
°	°	°	°	°	°	°	°	°	°
69.73	68.58	56.36	45.17	36.91	24.66	44.92	44.70	44.57	44.35
+ 2.75	+ 2.37	— 1.48	— 0.48	+ 0.16	— 1.50	+ 1.75	+ 0.53	+ 0.40	+ 0.18
+ 1.03	+ 0.08	— 5.14	— 8.63	— 6.29	—11.34	— 6.08	— 6.30	— 6.43	— 6.65
90.2	94.0	73.0	67.0	60.2	50.4	90.5	94.0	88.0	95.5
49.0	47.0	37.8	28.0	21.0	—10.4	—10.0	—15.0	—19.8	— 5.2
41.2	47.0	35.2	39.0	39.2	60.8	100.5	109.0	107.8	100.7
79.95	77.24	63.94	52.05	42.85	32.23	....	....	....	....
59.79	61.41	48.96	39.73	31.31	19.71	....	....	....	....
20.16	15.83	14.98	12.32	11.53	12.52	15.43	14.57	14.73	14.43
31.2	29.2	27.0	26.0	24.2	31.4	36.9	37.4	39.6	37.0
29.6289	29.5450	29.6097	29.5207	29.5790	29.5198	29.6330	29.5596	29.6536	29.6248
+.0275	—.0763	—.0532	—.1293	—.0349	—.1282	—.0197	—.0537	+.0403	+.0115
29.831	29.863	29.975	29.890	30.126	30.327	30.354	30.327	30.502	30.469
29.319	29.099	29.230	29.026	28.671	28.854	28.707	28.671	28.704	28.805
0.512	0.764	0.745	0.864	1.455	1.473	1.647	1.656	1.798	1.664
.66	.73	.75	.80	.78	.82	0.75	0.76	0.77	0.77
.473	.516	.347	.248	.182	.121	.259	.263	.266	.262
.44	.70	.58	.74	.75	.80	.61	0.65	0.61	0.63
—.04	+.23	+.08	+.11	+.01	+.05	+.01	+.05	+.01	+.03
N 61 W	N 70 W	N 38 W	N 60 W	S 72 W	S 82 W	N 66 W	N 76 W	N 41 W	N 48 W
2.23	1.38	1.89	3.17	3.82	4.94	1.98	2.49	1.34	2.03
6.00	4.75	7.06	6.66	7.64	9.98	6.78	7.40	7.13	7.33
+1.03	—0.43	+1.52	+0.52	+0.17	+1.66	—0.10	+0.54	+0.27	+0.47
1.332	5.060	2.508	3.321	3.765	2.045	26.599	29.486	26.483	25.529
+2.142	+2.034	—1.222	+0.791	+0.617	+0.404	—3.344	—0.469	—3.472	—4.426
8	16	11	22	11	9	111	132	130	118
....	....	....	Inap.	4.5	27.1	63.3	74.6	62.9	85.4
....	....	....	— 0.78	+ 1.38	+12.41	— 0.06	+11.24	— 0.46	+22.04
....	....	....	1	8	18	68	70	74	72
23	15	19	9	12	9	201	180	181	189
3	6	4	2	1	0	55	34	44	48
19	12	14	11	9	10	201	158	182	176
4	5	4	0	0	0	17	20	24	24

The instruments used in the Observatory and the system of observation are of the most approved description, showing reliable results. At the regular observation hours a record is made of the general appearance of the sky, including the form, distribution, and motion of the clouds.

The magnetical and meteorological results are published at intervals of a few years, forming valuable acquisitions to knowledge of the most useful and reliable character.

Climate of Vancouver's Island, B. A.

"The climate of this important Island, lying on the Pacific coast, between 48° and 51° north latitude, is rendered proverbially genial, productive, and salubrious, from an interesting variety of causes. The temperature of the Pacific coast generally is known to be much milder than that which obtains on the corresponding shores of the North American Continent in the Atlantic. The isothermal line belonging to latitude 40° on the latter ocean, passes through the parallel of 50° in the former. For lucid illustrations of this principle the reader is directed to consult the instructive work of Lieutenant Maury, entitled 'The Physical Geography of the Sea.'

"We have the authority of eminent meteorologists for the action of cold under-currents flowing from the Arctic Sea, which lave the rocky foundations of the island during the hot season, and exert their tempering influence far beyond high-water mark. The Olympian range of mountains, in Washington Territory, extending in an easterly and westerly direction, regale the eye in the rich sunshine. The proximity of their grateful summits, capped with eternal snows, tends to modify what must otherwise be the intense heat of mid-summer. The prevailing winds at this season come from the south, charged with warm moisture drawn from the sea, and oppress with sultriness the atmosphere of northern regions in most easterly longitudes. But, by contact with the neighboring snowy heights, the humid element of these winds is condensed, and their excess of caloric absorbed, so that they are transmuted, as by a magic touch into breezes,

'Mild as when Zephyrus on Flora breathes!'

A vast rush of warm water, supposed to originate at the Equator, and producing climatical effects resembling those which result from the agency of the *Gulf Stream* in the Atlantic, softens the rigors of winter as the boreal action, already described, is believed to cool the scorching heat of summer. The phenomenon referred to is called the *China Current,* from the fact of its sweeping, in part, that coast, on its curvilinear path across the ocean, to break upon the shores of Vancouver's Island.*

* Another of these currents makes its escape through the Straits of Malacca, and being joined by other warm streams from the Java and Chinese Seas, flows into the Pacific, like another Gulf Stream, between the Philippines and the shore of Asia. Thence it attempts the great circle route for the Aleutian Islands, tempering climate, and losing itself in the sea on its route towards the northwest coast of America—as with the Gulf Stream so with the China Current.—The climates of the Asiatic coast correspond with those of America along the Atlantic, *and those of Columbia, Washington, and Vancouver, are duplicates of those of Western Europe and the British Islands.—Physical Geography of the Sea.*

"From observations, taken daily in Victoria during the years 1860-61, at nine A. M., three P. M., and nine P. M., it appears that the lowest *mean* of the thermometer, in that period, occurred in the thirty-one days in December, 1860, when the range of that instrument averaged 41.22° Fahr. Twenty-nine days in July, 1861, indicated the highest *mean* to be 60.97°. At intervals of from seven to ten years, however, as in Great Britain, winters of unusual severity are experienced, when snow lies on the ground for a month or six weeks. But with the exception of these extraordinary periods, snow continues for little more than a week; and sharp frosts extend over about a fortnight of the year. So mild is the cold season generally, that cattle can find enough of food in the fields without special provision having to be made for their shelter and maintenance.

"The winter of 1863-64 was mild throughout. As this part of the subject is so important to intending settlers, with respect to consideration of health as well as to farming operations, let us take a past year at random to aid the reader in arriving at a satisfactory conclusion on the matter.

Dr. Rattray, R. N., attached to Her Majesty's ship "Topase," in Esquimalt Harbor, in 1860-61, carefully tabulated, for the use of the Admiralty, the state of the weather from the beginning of April to the end of March following in those years. Subjoined are the result of his labors :—

No. of fine days, ...	187
" wet days, ...	17
" showery days, ...	101
" foggy days, ...	17
" days with strong wind, ...	35
" days with thermometer below freezing,	11
" days in which snow fell, ...	12

Temperature, &c., of the Upper Lakes.

SITUATION and MEAN ANNUAL TEMPERATURE of the Principal PORTS on the UPPER LAKES.

POSTS, &c.	Latitude.		Longitude.		Altitude.	Mean Temp.
	°	′	°	′	Feet.	° Fahr.
Agate Harbor, L. S., Mich., .	47	30	88	10	600	41
Ashland, L. S., Wis., . .	46	33	91	00	600	41
Bayfield, " " . . .	46	45	91	00	600	40
Beaver Bay, L. S., Minn., . .	47	12	91	18	600	38
Buchanan, L. S., Minn., . .	47	33	92	00	600	37
Bruce Mines, Can., . . .	46	20	83	45	576	40
Chicago, L. M., Ill., . . .	41	53	87	37	578	47
Collingwood, Can., . .	44	30	80	20	574	43
Copper Harbor,* L. S., Mich., .	47	30	88	00	620	41
Detroit,* Mich., ...	42	20	83	00	600	47¼
Eagle Harbor,* L. S., Mich., .	47	28	88	08	600	41
Eagle River, L. S., Mich., . .	47	25	88	18	600	41
Escanaba, Green Bay, Mich., .					578	42
Forrestville, L. Huron, Mich., .	43	40	82	34	574	45
Fort Gratiot,* " " .	42	55	82	23	598	46
Fort William, L. S., Can., . .	48	23	89	22	600	36
Goderich, L. Huron, " . .	43	44	81	43	574	45
Grand Haven, L. M., Mich, .	43	05	86	12	578	46
Grand Portage, L. S., Minn., .	47	50	90	00	600	37
Green Bay,* Wis., . . .	44	30	88	05	620	44½
Houghton, L. S., Mich., . .	46	40	88	30	600	41
La Pointe, L. S., Wis., . .	46	45	90	57	600	40
Manitouwoc, L. M., Wis., . .	44	07	87	45	578	45
Mackinac,* Mich., . . .	45	51	84	33	700	40½
Marquette, L. S., Mich., . .	46	32	87	33	600	41
Michigan City, L. M., Ind., .	41	41	86	53	578	49
Michicipoten, L. S., Can., . .	47	56	85	06	600	38
Milwaukee, L. M., Wis., . .	43	04	87	55	578	46
Munising, L. S., Mich., . .	46	20	87	00	600	41
Neepigon, L. S., Can., . .	49	00	88	30	600	36
Ontonagon, L. S., Mich., . .	46	52	89	30	600	40
Penetanquishene, Can., . .	44	51	80	40	574	43
Port Huron, Mich, . . .	42	58	82	25	572	46
Racine, L. M., Wis., . . .	42	45	87	48	578	47
Rock Harbor, L. S., Mich., .	48	08	88	50	600	37
Saut Ste. Marie,* Mich., . .	46	30	84	43	600	40
Sheyboygan, L. M., Wis., . .					578	45
Superior City, L. S., Wis., . .	46	40	92	03	600	40

* United States Military Posts, giving the elevation of Forts, &c.

PART V.

GREAT LAKES, OR INLAND SEAS.

The magnitude of the Lakes of North America, together with the St. Lawrence River, is an interesting theme, being fully comprehended by only a few intelligent minds. They deserve to be described both in prose and verse, in order to have the mind fully impressed with their purity, extent, and importance, in a commercial point of view, as well as their climatic influence.

From the head of Lake Superior, passing through this inland sea and the St. Mary's River into Lake Huron, and thence through Lake Erie, for a distance of upwards of 1,000 miles, there is much to see and admire; while from the Falls of Niagara to the mouth of the St. Lawrence, for another thousand miles, the traveller witnesses the most grand and instructive lake and river scenery imaginable—passing through a healthy region of country, varying from 36° to 48° Fahr., mean annual temperature—the dark blue waters of Lake Superior changing to the green waters of Lakes Erie and Ontario, and the St. Lawrence River. All the impediments to navigation, for the distance of about 2,000 miles, through the Great Lakes and their outlets, are overcome by a succession of ship-canals, connecting with the tide waters below Montreal. Navigation is usually interrupted by ice from the beginning of December to the middle of April, on the Upper Lakes and St. Lawrence River. River steamers usually start from Montreal the latter part of April, and arrivals from sea, at Quebec, commence early in May, affording seven months of uninterrupted navigation.

Lake Superior, at the height of 600 feet above the sea, is 420 miles long, 160 miles broad, and about 900 feet deep. It discharges its surplus waters by the Strait or River St. Mary, fifty miles long, into Lake Huron, which lies twenty-six feet below. This lake is computed to be 260 miles long, 110 miles broad, and 800 feet deep. A world of waters of itself, to say nothing of the *Georgian Bay*, on the northeast, or the *Saginaw Bay*, on the southwest, both embracing a large expanse of waters. The former, however, is studded with innumerable islands and islets, forming a perfect labyrinth for about 100 miles along its northeast shore, being entirely within the confines of Canada.

LAKE MICHIGAN, 578 feet above the sea, is 320 miles long and 80 miles broad, and about 800 feet in depth. It discharges its waters, in connection with *Green Bay*, through the Strait of Mackinac, fifty miles in length, into Lake Huron, nearly on a level. *Lake Winnebago*, the extreme southwest tributary of the St. Lawrence, in connection with the Fox or Neenah River, is an interesting body of water, lying 170 feet above Green Bay, or Lake Michigan. Navigation is now extended to the head of this lake, and up the Fox River, until the level is obtained, where the waters of the St. Lawrence unite with the waters of the Mississippi, by means of a canal of only between one and two miles in length—the junction being made with the Wisconsin River at Portage City, near old Fort Winnebago—thus forming an inland navigation from the Gulf of Mexico to the Gulf of St. Lawrence.

The accumulated waters of Lakes Superior, Michigan, and Huron are immense, all of which find an outlet through the St. Clair River, into St. Clair Lake, and thence through Detroit River into LAKE ERIE, the fourth great lake of this immense chain. This latter lake, again, running nearly east and west, at an elevation above the sea of 565 feet, is 250 miles long, 60 miles broad, and 200 feet at its greatest depth, but, on an average, considerably less than 100 feet deep, discharges its surplus waters by the Niagara River and Falls, into Lake Ontario, 330 feet below. The river is thirty-five miles in length; 160 feet being the descent at the Falls, the remainder made up of rapids above and below the Falls. Here navigation ceases for about seven miles, from a short distance above Niagara Falls Village to Lewiston, New York.

LAKE ONTARIO, the fifth and last of the Great Lakes of America, is elevated 235 feet above tide-water, at Three Rivers, on the St. Lawrence; it is 180 miles long, 60 miles broad, and

600 feet deep. Lake Ontario is the safest body of water for navigation, and Lake Erie the most dangerous, owing to its elevation and low depth of water.

The lakes of greatest interest to the tourist or scientific traveller are Ontario and Huron, together with Georgian Bay and North Channel, and Lake Superior.

Lake Region of North America.

In treating of the climate and phenomena peculiar to the Great Lakes, or "Inland Seas" of North America, Foster and Whitney, in their Report of 1850, remark: "The meteorological influence should be described in its most extended sense, as comprehending, according to Humboldt, all the changes of the atmosphere which seriously affect our organs—as temperature, humidity, variations in the barometrical pressure, the calm state of the atmosphere, or the action of opposite currents of wind, the purity of the atmosphere, and, finally, the degree of ordinary transparency and clearness of the sky, which is not only important, with respect to the increased radiation of the earth, the organic development of plants, and the ripening of fruits, but also with reference to its influence on the feelings and mental condition of men.

"To this great student of nature, science is indebted for having first suggested a system of lines, called Isothermal, Isotheral, and Isochimenal (implying the year, summer, and winter), connecting those places where the mean summer, winter, and annual temperatures have been ascertained; — thus running round the globe, defining the frigid, cold, temperate, sub-tropical, and tropical zones, or belts of temperature. These lines are by no means parallel, or defined by lines of latitude; various causes conspiring to produce divergencies—such as altitude above the sea, the geographical configuration of the country, the presence or absence of large bodies of water and mountain chains, the purity of the sky, and the prevailing currents of the ocean and the direction of winds.' This latter influence is fully apparent on the opposite sides of the American Continent, as well as in that portion of country where the cold blasts from off Hudson Bay and Baffin Bay crowd down the isothermal line.

"The presence of so vast a body of fresh water as is afforded by the American lakes modifies the range of the thermometer, lessening the intensity of cold in the winter and of the heat in summer. By the freezing of the water, a great volume of heat is evolved, and the intense cold of the northern winds is somewhat mitigated in sweeping over the open lakes. In the summer when the sun, often with obscured lustre, shines for sixteen hours in twenty-four, the intensity of the heat is modified by the breezes which are cooled in their passage over the surface of the lakes, the water of which is always at a low temperature."*

In order to show the equalizing effects of the lakes on the climate, we need only refer to the mean temperature of the following stations, situated in nearly the same parallels of latitude:—

Stations.	Latitude.	Longitude.	Mean An. Temp.	Winter.	Summer.	Range of Thermometer.
Fort Howard, Wis., .	44° 30′	88° 05′	44.50°	20°	68°	—14 +92
Fort Snelling, Min., .	44° 53′	93° 10′	44.84°	16°	71°	—32 +95

Comparative Fall of Rain in Inches.

	Spring.	Summer.	Autumn.	Winter.	Year.
Fort Howard, Wis., .	9.00	14.45	7.84	3.36	34.65
Fort Snelling, Min. .	6.60	10.92	5.98	2.00	25.50

Thus, during the winter, the mean temperatures of the Lake Stations are higher, but during the summer months they are lower, while the mean annual temperature is nearly the same.

Forts Howard and Mackinac are both situated in the proximity of large bodies of water, which essentially modify the temperature; while Forts Snelling and Ripley, both in Minnesota, are in the midst of a vast plain, in the Valley of the Upper Mississippi, with no mountain chain to break the force of the winds.

Lake Superior forms the upper basin of the lake region, and is the largest expanse of fresh water on the globe; it is fed by more than eighty streams, none of which attain any considerable magnitude, and are adapted only to canoe navigation. Those which flow down the northern slope of the basin are longer than those of the southern, and the water being more exposed to the direct rays of the sun, possesses a higher temperature.

Extent of the Great Lakes.

	Length.	Breadth.	Depth.	Elevation above Sea.	Area in Square Miles.
Superior, .	460	170	800	600	31,500
Michigan, .	330	90	700	576	22,000
Huron, . .	260	110	700	574	21,000
Erie, . .	250	60	200	565	9,000
Ontario, . .	180	60	600	235	6,400
Total length,	1,480				90,500

* The water at the surface of Lake Superior is usually the same as the mean temperature, varying from 40° to 42° Fahr.; but at sixty feet below the surface the temperature is uniformly at 38° Fahr.

The St. Mary, St. Clair, Detroit, Niagara, and St. Lawrence Rivers extend this great water-course 940 miles further, before entering the Gulf of St. Lawrence, making the entire length of water communication, 2,420 miles—all of which distance is made navigable by means of canals and locks, having a descent of 600 feet before meeting tide waters, near Quebec.

The waters of the St. Clair, Detroit, Niagara, and St. Lawrence Rivers assume a light green appearance, which, no doubt, is caused by the admixture of coloring matter found along their banks, strongly contrasting with the dark green waters of the Upper Lakes, which, when agitated by the waves or ripple of the passing steamer, presents a brilliancy peculiar to these pure waters; it then being an admixture of white foam with a lively green tinge assuming a crystal-like appearance.

The evaporation from the surfaces of the lakes must be immense. The combined area of Lakes Superior, Michigan, Huron, and Erie is about 85,000 square miles, and of their basins not less than 300,000 square miles.

"It has been estimated that the quantity of water passing into Niagara River at Black Rock, is 22,440,000 cubic feet per minute, or about 80½ cubic miles per annum.* This is equivalent to fifteen inches perpendicular depth of water over the area of the whole country drained. The annual amount of rain which falls within this area is about thirty-four inches; nearly one-half, therefore, of the water which falls within the basin of the Upper St. Lawrence is taken up by evaporation.

"At the Saut Ste. Marie, the outlet of Lake Superior, the spectator beholds a river near a mile in width, and of sufficient depth to float the largest vessel. In its onward progress it winds among numerous islands, and ultimately discharges itself, by several mouths, into Lake Huron. At Fort Gratiot he sees the same river, under another name, after having received all the tributaries of Michigan and Huron, contracted to a width of about 1,000 feet, but of increased depth, and he finds it difficult to realize that it is the same river which he saw three hundred miles above. So, too, the voyager who has coasted around Lake Superior, and gauged the streams which pour their annual floods into the great reservoir, when he stands on the brink of Niagara, and witnesses the fearful plunge of the cataract is induced to inquire what has become of the superfluous water."

The Waters of Lake Superior.

From a series of careful observations continued through a period of six years, from 1854 to 1859, inclusive, by Dr. G. H. BLAKER, of Marquette, L. S.; it has been shown that the annual

* *Silliman's Journal*, January, 1844.

rise and fall of the surface of Lake Superior, ranges between twenty and twenty-eight inches. From the first of May, when the snow begins to melt freely, until the first of September, the surface of the lake level continues to rise constantly, about six inches a month, until it gains, on an average, two feet by the middle of August; and by the first of September it begins to fall and so continues, through the winter, until about the middle of April. The permanent rise, however, has been about *two inches* more than the fall, during the above period, so that in the six years just passed, there has been an actual rise of some twelve inches in the lake level.

The cause of this annual rise and fall has ever been, with the learned, a question of doubt, and the following hypothesis has been submitted to the public for inspection. The extreme length of this extensive body of pure fresh water, from east to west, is about 470 miles, and from north to south, 175 miles, with a mean breadth, however, of but 85 miles; giving an area of 31,540 square miles, or 97,698,304,000 square yards of water surface.

The distance from the extreme west end of the lake to the culminating line, dividing the waters from the Mississippi, or more western rivers, is about fifty miles, and that of the extreme eastern end is about eighty miles; while the distance on the southern side is but twenty-five miles, and that on the northern side, up to the head waters of the rivers falling into James' Bay, or Hudson Bay, is 100 miles, making the extreme length of the lake basin, from east to west, of about 500 miles; and the extreme breadth, from north to south, 300 miles—giving an area of 150,000 square miles, or 464,640,000,000 square yards.

It has been shown by observation, that the whole quantity of water that falls in one year in this latitude is about forty inches, or $1\frac{1}{9}$ yards in depth; therefore, no less than 516,266,666,666 cubical yards of water must fall in the lake basin in one year.

From the above figures it will be seen, that the proportion of water to land within the lake basin is as 1 to $4\frac{1}{2}$, or $4\frac{1}{2}$ times the water surface is equal to the land surface; when it follows, that of this 40 inches of water that falls on the $4\frac{1}{2}$ parts of land surface, would, if all of it were found on the one part of water surface, raise it $4\frac{1}{2}$ times 40 inches, or 180 inches, provided that the outlet at the Saut Ste. Marie was closed, and the *area* of the water surface continued the same. But, so long as the outlet at the Saut continues a constant quantity, and the supply a constant quantity, the level of the lake would be the same throughout the year, if no obstruction to the supply interfered.

Both from experiment and observation, it has been seen that the amount of water that passes off from the surface of the lake by evaporation in one year, is equal to 18 inches in depth. Also, that the ratio of evaporation between that of land and water, is as 1 to 5, very nearly; and since the land surface of the basin is 4½ times larger than the water surface, it is plain that it would give or lose, in cubical yards, about the same amount of water by evaporation that the water surface would; therefore, one yard, or 36 inches, must be deducted from the rise of 180 inches, leaving but 144 inches, or 12 feet. Now, that part of the lake basin which is covered by water is almost 1⅕ of the whole quantity, therefore, one-half of this 1⅕ must express the whole amount of water that passes off the surface of the lake by evaporation, equal to 43,849,152,000 cubical yards of water; add evaporation from land surface of an equal amount, make a total of 97,698,304,000 cubical yards of water to pass off by evaporation. Now, from the whole amount of water that falls in one year, 516,266,666,666, deduct the full amount of evaporation, and there will remain 418,568,362,666 cubical yards of water to pass down the St. Mary's River in one year, giving nearly 14,000 cubical yards per second that must run over the rapids at the Saut Ste. Marie.

Observation has further shown that 12 inches, or about $\frac{3}{10}$ of the 40 inches of water that falls in one year, comes in the form of snow,* and but about five inches of this amount is lost, or passes into the lake during the winter season; the 7 inches, about ⅙ of 40 inches, remains on the ground until spring, and is cut off from the winter supply, and must be added to the summer supply as a surplus, since it must find its way to the lake during the summer season with the amount that falls during that period. Thus, if ⅙ of the winter's supply is cut off, and the outlet remains the same, it is plain that the surface of the lake must fall ⅙ of 4½ fifths of 12 feet, equal to 1.8, or 22 inches nearly, and if the supply which was cut off during the winter season all returns to the lake during the summer, the rise must be about equal to the fall. Whence it follows, that while observations give a mean rise and fall in six years of 23.5 inches, the hypothesis advanced gives nearly 22 inches, making a discrepancy of but 1.5 inches, which may be accounted for in the excess of water that must pass down the river when it is high, over that when it is low.

* During the winter of 1858–59, snow fell at Marquette, Mich., to the depth of twenty-eight feet on the average, extending through a period of six months, from November to May, inclusive. In April, 1859, there was on the ground four feet eight inches of packed snow or ice, so compact as to bear a man or horse. This fall of snow, when melted, produced fifteen inches of water. Thus this immense quantity of moisture, lying on the ground, must of necessity pass off into the earth and streams flowing into the lake during the spring and summer months.

Low Water in the Lakes.—The *Detroit Advertiser* says the low stage of water in the Western Lakes is something remarkable. A fall of about two feet has recently taken place (1864), and the water is now four feet lower than in 1861. At some of the ports on Lake Huron, it is now difficult to make landings, where formerly there was water to spare.

Remarkable Phenomenon—Fluctuations on Lake Superior.

Professor Mather, who observed the barometer at Fort Wilkins, Copper Harbor, during the prevalence of one of these fluctuations, remarks: "As a general thing, fluctuations in the barometer accompanied the fluctuations in the level of the water, but sometimes the water level varied rapidly in the harbor, while no such variation occurred in the barometer at the place of observation. The variations in the level of the water may be caused by varied barometric pressure of the air on the water, either at the place of observation or at some distant point. A local increased pressure of the atmosphere at the place of observation would lower the water level, where there is a wide expanse of water; or a diminished pressure, under the same circumstances, would cause the water to rise above its usual level."

In the summer of 1834, according to the Report of Foster and Whitney, made to Congress in 1850, an extraordinary retrocession of the waters took place at Saut Ste. Marie. "The river here is nearly a mile in width, and the depth of the water over the sandstone rapids is about three feet. The phenomena occurred at noon; the day was calm but cloudy; the water retired suddenly, leaving the bed of the river bare, except for the distance of about twenty rods, where the channel is deepest, and remained so for the space of an hour. Persons went out and caught fish in the pools formed in the rocky cavities. The return of the waters was sudden, and presented a sublime spectacle. They came down like an immense surge, roaring and foaming, and those who had incautiously wandered into the river bed had barely time to escape being overwhelmed."

Auroras, even in midsummer, are of frequent occurrence on Lake Superior, and exhibit a brilliancy and extent rarely observed in lower latitudes. The phenomena of this kind which most frequently occur are the following: A dark cloud, tinged on the upper edge with a pale luminous haze, skirts the northern horizon. From this streaks of orange and blue colored light flash up, and often reach a point south of the zenith. They rapidly increase and decrease, giving to the whole hemisphere the appearance of luminous waves, and occasionally forming perfect corona. They commence shortly after sunset, and continue during the night. The voyageurs regard them as the precursors of storms and gales, and our own observations have confirmed the result. Occasionally broad belts of light are seen spanning the whole arc of the heavens, of sufficient brilliancy to enable one to read. In the autumn and winter months these phenomena are much more frequent, and the ground or snow appears tinged with a crimson hue.

MIRAGE.—The difference between the temperature of the air and the waters of the lake give rise to a variety of optical illusions known as *mirage*. Mountains are seen with inverted cones, headlands project from the shore, where none exist. Islands, clothed with verdure or girt with cliffs, rise up from the bosom of the lake, often presenting a strange appearance. On approaching Keweenaw Point, Mount Houghton is the first object to greet the eye of the mariner. In peculiar stages of the atmosphere, its summit is seen inverted in the sky long before the mountain itself is visible. On approaching the north shore, the *Mamelons*, or Paps, two elevated peaks near the entrance of Neepigon Bay, at one time appear like hour-glasses, and at another like craters, emitting long columns of smoke, which gradually settle around their cones.

Agricultural Products of Lake Superior.

On the south shore of Lake Superior, where is to be found an annual mean temperature of from 40° to 42° Fahrenheit, being about the same as at Quebec, Canada, the principal products of the soil, as found in the forest, are white and red pine, hemlock, spruce, balsam, and cedar of the evergreen species; sugar maple, soft maple, ash, birch, bass-wood, and mountain ash of the hard-wood species; wild plums, raspberries, whortleberries, and cranberries, are found in abundance, besides many berries of inferior quality, and wild flowers.

The cultivated fruits are the hardier varieties of apples, pears, plums, cherries, currants, and gooseberries. The cereals are wheat, rye, oats, and barley, while grasses and clover of different kinds flourish all along the lake shore, yielding a profitable return, usually producing from one to two tons of hay per acre. The vegetable productions are potatoes, turnips, beets. carrots, parsnips, cabbage, pumpkins, onions, squash, melons, cucumbers, lettuce, peas, beans, rhubarb or pie-plant.

The soil consists of sand, clay, loam, and gravel, the former predominating near the lake shore. The surface soil in many localities is rich and productive, particularly when having a southern exposure, or protected from the cold current of air blowing off the lake from the northwest; the land breeze or south wind being usually warm during the spring and summer months.

The streams along the south shore are usually free from ice about the middle of April, and planting commences during the early part of May. June, July, August, and September, are usually free from frost, the north or lake winds tending to modify the air during the early months of autumn. The foliage of the forest trees are affected by frost in October, when they assume a variegated and beautiful appearance.

Snow appears about the middle of November, and lays during the winter months, protecting the soil from frost in the coldest season. Potatoes and other root vegetables are often left in the ground and collected in the spring.

"The whole face of the lake country is covered with a dense forest, unbroken, save by the clearings of the settlers and a few natural meadows and open marshes, which are scattered here and there along the rivers, especially near their mouths. The constitution of this forest is such as is characteristic of so high a northern latitude. The peculiarities consist, not as much in the introduction of new and exclusively northern species—for there is hardly one of any importance that does not occur also in the highlands of the Middle States—as in the increased frequency and predominance of certain northern types, and the total absence or great rarity of many which are the most familiar to the eye of the dweller further south. Thus, all the trees that have esculent fruit, the oak, walnut, chestnut, beach, &c., are either quite wanting, or of very unfrequent occurrence; while the spruces, the fir, the cedar, the red pine, the birches, the maple, the aspen, poplar, &c., are the prevailing growth."—*Foster and Whitney's Report.*

PART VI.

FORESTS AND PRAIRIES OF NORTH AMERICA.

LIMIT OF THE WOODS.—"FOR a number of degrees around the North Pole the amount of heat is insufficient for the growth of those plants which tend to become trees. The line of limitation is remarkably distinct, and the commencement of forests usually so sudden and uniform in their position, as to indicate a fixed law of temperature as its cause. It does not form a circle with the Pole as its centre, but a three-sided area with rounded corners; the most northern side reaching to about latitude 70° in Norway and Siberia, near the level of the Arctic Sea, while on the Western Continent the limit descends from the Arctic Circle, near Behring Strait, in a nearly direct line, to about latititude 60°, near Fort Churchill, on Hudson Bay, and crossing Labrador, in about the same parallel, strikes over the Atlantic, south of Greenland, excluding that large country and Iceland from the regions of tree growth." The vegetation of Greenland mainly consists of grasses and lichens in the north, and of a few scattered birches, alders, and willows in the south, where are also raised small quantities of grain, potatoes, turnips, and other hardy vegetables.

Dr. Richardson, in his account of "Expeditions in Search of Sir John Franklin," tells us that at the mouth of Mackenzie's River, latitude 68°, the following trees occur, generally dwarfed, but many showing, by their annual rings, a very great age:—Paper birch, green alder, American aspen, poplar, white spruce, northern willow, balsam, poplar, and northern juniper.

Further south, on Mackenzie's River, the following are met with at the Arctic Circle, 67° 30′:—Choke cherry, gray scrub-pine, black spruce, and larch tamarock; at 62°, American balsam fir; at latitude 59°, he found black wild-cherry, and American mountain ash; at latitude 56°, the red or "Norway" pine. Crossing from the basin of the Mackenzie to that of the Saskatchewan, in about latitude 54°, the following trees make their appearance:—Smooth sumach, white ash, gray oak, Canadian yew, box elder, black water ash, Canadian arbor vitæ, ground hemlock, red wild cherry, white or weeping elm, and red cedar or juniper.

At the south of Lake Winnipeg, (British America), about

latitude 50°, are first met: Linden, basswood, post or iron oak, sugar maple, white oak, balm of Gilead, poplar, red maple, red beach, white pine, and mountain maple. Further east, in the basin of the Great Lakes and St. Lawrence, the following species are added:—Hemlock, spruce, fir, Canadian balsam, black birch, burr or overcup oak, cherry or sweet birch, yellow birch, and soft aspen, poplar, cranberry bushes, raspberry bushes, and whortleberry bushes, also abound in this region; being, however, destitute of any of the *nut* bearing species of trees.

All the forest trees found north of the Great Lakes and about Hudson Bay are also met with in the United States, as well as many others which do not reach to this limit in British America. The great belt or region of forest trees extends from the Rocky Mountains, east, to the coast of Labrador, continuing south along the Alleghanies to North Carolina and Georgia, as far as the thirty-fourth parallel of latitude.

Near latitude 44°, on the uplands, near the Atlantic coast, the following trees are first met with, viz.:—In Maine, great laurel, green ash, button-wood, or sycamore, butternut, chestnut, and northern pitch-pine. In New Hampshire, sassafras, shell-bark hickory, pignut or broom hickory, and rock chestnut oak. In Vermont, northern fox grape, winter or frost grape, flowering dogwood, red ash, American mulberry, black willow, and southern balsam fir. In New York, cucumber tree, paw-paw or custard apple, western crab-apple, yellow pine, locust tree, and thick shell-bark hickory.

The *Atlantic States*, south of latitude 45°, and *east* of the border of the prairies, which, commencing at the west end of Lake Erie, west longitude 83°, forms a curve nearly parallel to the Atlantic, terminating on the Gulf of Mexico, near the mouth of the Mississippi, embracing all the Appalachian region, was originally covered by a dense forest of remarkable richness, both in the variety and beauty of its trees—in this respect surpassing any other part of the temperate, and even many parts of the tropical zone. The quantity of rain falling on this region, running through sixteen degrees of latitude, averages from thirty-five to sixty inches annually, the four seasons for the most part being about equally divided.

The great "Central Plain" or Prairie Region of North America consists in their comparative destitution of forests and nearly uniform surface, gradually rising from the level of the Gulf of Mexico to the base of the Rocky Mountains, where they attain an elevation of from 4,000 to 5,000 feet. This immense tract of country is, for the most part, sparingly supplied with rain and snow; hence the forest trees are mostly found to skirt the numerous streams which flow onward toward the Gulf of Mexico, on the south, and Hudson Bay, on the north. The valley of the Rio Grande forms the southern boundary of this treeless or half-wooded region, lying *east* of the Rocky Mountains.

The Oregon region, extending northward to British America, west of the Cascade Mountains, and south, about latitude 40°, embracing Northern California, has its large and peculiar group of trees, which completely cover the surface from the level of the sea up to the regions of perpetual snow, excepting a small extent of prairie found in some of the valleys.

On the North Pacific coast, near Sitka, in latitude 57°, extending southward to Vancouver's Island, is found Sitka spruce, broad-leaved maple, Oregon crab-apple, Oregon alder, western hemlock, spruce, Nootka cypress, red or black fir, and Oregon white oak, and other species; making twenty new form of trees in this far northwestern region, north of latitude 49°, and others will doubtless be added, showing a very rich forest growth.

At latitude 48°, Straits of Fuca, are found Oregon ash, madrona laurel, scrub or twisted pine, Oregon yew, heavy yellow pine, and northwestern larch. At about latitude 47°, scrubby cherry, Oregon hawthorn, California green dogwood, coast willow and yellow fir. In Oregon and California are found, in addition to most of the above species, smooth manzanita, evergreen chestnut, downy-cone spruce, Cascade Mountain spruce, Oregon silver fir, California nut pine, western pitch pine, sugar pine, leafy-cone spruce, California grape, California buckeye, California plum, Mexican sycamore, California red oak, redwood, giant redwood, California cedar, California white oak, and long acorn live-oak.

Further south, between latitude 36° and 32°, west of the Coast mountains appear—Mexican walnut, laurel, sumach, sharp-toothed live-oak, oblong-leaved alder, hairy-pod poplar, Gouan's cypress, Mexican arbor vitæ, Mexican pistachio tree, and Torrey's pine; all of these may be supposed to occur in the Peninsula of Lower California, and many extend their range east of the coast mountains toward the Sierra Nevada.

Very little is known of the trees of Northern Mexico, or of their distribution, between latitude 32° and 20°, where abounds the mesquite, green acacia, New Mexican nut pine, Arizonian barberry, Arizonian live-oak, saguaro cactus, prickly pear, Wislizenis cactus, Chihuahuan pine, and different species of willows.

Many of these trees of the southern border, along the boundary line, scarcely rise above the growth of shrubs within our limits, and they are accompanied by a large number of tree shrubs, which cover large tracts with a kind of miniature forest, as is the case in portions of the Rocky Mountain range.

The successive tables of land with which the eastern slope of Mexico rises, from the Gulf to the *centre*, have each peculiarities of vegetation, &c., sufficient to indicate a division into the natural regions.

The northern limit of forests on this continent, no doubt, depends chiefly on the temperature, although careful observations having been made at only few and scattered localities along the line, it is not possible to say what isothermal lines correspond exactly with it; the soil and moisture forming an important element in the growth of particular species. The degree of cold to which the temperature sinks in winter must be duly considered, as it is well known that certain trees, particularly fruit-trees, may grow well for years in some localities, and then be killed by an unusually cold winter.

The main points established in regard to the northern limit of vegetation, on the American Continent, are, that it reaches its highest point near Behring's Straits and Mackenzie's River; its lowest, at or near Hudson Bay, about longitude 95°, west from Greenwich, following near the mean isothermal line of 20° Fahr. The first of these facts corresponds with the mildness of the winters towards the North Pacific coast, the second, with their severity near the 95th meridian, all the way south to the region of the Great Lakes, and eastward along the 45th parallel of latitude.

Prairie Region.

"The prairie regions form a great feature in the natural vegetation of the North American Continent. A line drawn from the centre of Southern Michigan to Cairo, Illinois, and extending southwestward to Texas, would form a rough boundary between the wooded and the treeless country. West of this line, the trees are generally stunted, unless along the margin of rivers, whereas the country eastward to the Atlantic coast was almost everywhere densely clothed with timber, when discovered by Europeans.

"The productive powers of the prairies are best brought out under cultivation, which renders the light and open mould absorbent to moisture. Indian corn, wheat, and oats are, therefore, relatively far more abundant in their produce than grasses or even trees. The dry prairie, which only yields annually a ton of hay to the acre, after it has been seeded with timothy, will produce from six to seven quarters of oats for twenty years in succession without manure, and still show little falling off in quantity. The same land, when well cultivated, will produce from forty to sixty bushels of Indian corn, with upwards of two tons of stalks and leaves. I do not know of any instance in which the cultivated produce of the soil exceeds that of the natural growth to such an extent as it does on the prairies.

"As illustrating the influence of climate on the growth of trees and other plants, it is worthy of remark that the banks of the streams and rivers which run through the prairies are invariably clothed with timber, and the surface of the ground is comparatively destitute of the dark mould that is found in the naked prairies, and forms their deep fertile soil. Like the large rivers, the smaller have also dug shallower beds out of the soft plateau, and their banks afford more moisture to the roots of trees. It is interesting to see how trees clothe the sides of the streams over such immense stretches of country in the prairie region, furnishing strong evidence in support of the opinion that the prairies arise from a deficiency of rains. Dr. Hooker's remarks on the climatic conditions which favor the growth of trees in different parts of the Himalaya Mountains are greatly in favor of this view. Indeed the llanos and pampas of South America are but extreme instances of the effects of a want of moisture at certain seasons of the year being averse to the growth of timber. The thinly timbered lands of the oak openings in Canada West are the first symptoms as we go westward of the climate becoming less favorable to the growth of trees; and as we approach the Mississippi, the natural grasses that clothed the surface of the ground, when the white man first took possession, indicate that their habits are better suited than those of trees to a scanty and less regular distribution of rain.

"More than three-fourths of the surface of Illinois consists of prairie. In many parts, not a tree or shrub is to be seen in the distance—a circumstance which has prevented its being cultivated. Wood for fuel and fencing is one of the first requisites to the working farmer, and he will rather hew for himself a farm out of the forest than sit down upon the treeless prairies. Immense beds of bituminous coal extend through the country, however, which will no doubt soon be made available, as they are often found very near the surface."

The prairie lands of Northern Indiana, south of Lake Michigan, are in the beginning of summer beautifully clothed with grass and flowers, presenting a most lovely appearance, in connection with the clumps of forest trees, in the distance looking like islands in the ocean.

"The Vegetable Kingdom of America," says a late English writer, "throughout all its regions, but especially in those of the Equatorial Zone, exceeds in the rarity, luxuriance, and multiplicity of its productions the botanical riches of any of the corresponding climates of the Old World. Its indigenous flora already number in its classifications upwards of 15,000 phanerogami. The most northernly latitude in which vegetation has been discovered is Melville Island, in 74° 30′ north latitude. In this desolate region such vegetable species as grasses, saxifrages, mosses, and lichens find existence, and the *Protococus nivalis*, which exists in even higher latitudes, tinges the snow with its crimson flowerets; but the only plant of woody structure it is capable of producing, is the Arctic willow, which here attains only six inches in height. The vegetation of Greenland and of the coast of Baffin and Hudson Bay, bears a close affinity to that of the High Alps or of Lapland, consisting of stunted willows, birches, poplars, pines, a few species of herbaceous plants, remarkable for the large size of their flowers and the rapidity of their development, and different species of cryptogami, which are exceedingly abundant, covering to a great extent the soil of these polar regions. Proceeding southwards, we meet with vast forests of spruce firs, beneath which the reindeer moss and other lichens overspread the soil, and various berry-bearing shrubs and papilionaceous plants. Next follow the majestic poplars of Canada, &c., pines, birches, various oaks, ashes, butternuts, and hickories. On the southern frontier of the British Possessions, the sugar-maple and azaleas abound, many kinds of asters stud the woods and meadows with their star-like flowers; and wheat, oats, maize, and even tobacco, form common field crops, marking the transitions to the flora of the United States. The American elm is properly a Canadian tree, as it is in the north, that this (the most magnificent of the Temperate Zone) attains its finest proportions. The botanical region of the United States, including the whole central district of North America, from about 50° to 25° north latitude, consists of the vast and originally uninterrupted forest tract, extending from Hudson Bay to the Mexican Gulf, and westward (but confined to the banks of the river), far beyond the Mississippi. Some prairies or unwooded tracts occur in Illinois, Indiana, Iowa, &c., and in the southern districts of Mississippi and Alabama. Of the 140 species of trees which are found in this forest, more than eighty attain a height of sixty feet and upwards. The most characteristic are the hickories, the tupelas, the lyriodendron or tulip-tree, the taxodium or American cypress, the locust and coffee-trees, and the negando. It likewise presents numerous species of oak, ash, and pine, and possesses several of the magnolia tribe, one of the Gordonia, the sycamore or buttonwood, liquid amber, and the tree-andromeda, two species of the walnut, three tilia, the red-bay, hackberry, &c. The shrubs and herbaceous plants which form the undergrowth of this forest, belong generally to the classes which require more or less protection from the sun. In the prairies—regions where the grasses usurp the domain of trees and shrubs, the northern district presents a strong analogy to the Tartarian Steppes, not only in their physical aspect and numerous salines, but in the gay profusion of their floral vegetation. In the southwest parts vegetation is very thinly diffused; and towards the Rocky Mountains it is so scanty, that the name of desert has been given to an extensive tract; but there is no district altogether destitute of streams, or where cactuses and yuccas may not occasionally be met with, or even some cucurbitaceous plants and grapevines spreading over the sands. The western district appears to be less extensive than the eastern, contains fewer but not less gigantic species. Spruces prevail in the northern; pines, maples, oaks, and poplars in the central district; and pines (white and yellow) in the southern region.

"The most characteristic feature of the North American flora are exhibited in the United States. Here the forests consists of pines and larches unknown in the old world, of many kinds of oaks, of locust-trees, black walnnt of enormous size, hickories and ashes, among which the noble tulip-tree rears its towering head. In the swamps grow the deciduous cypress, the white cedar, two species of fir, the rhododendron or rose bay, the glaucous kalmia, andromedas, sarracenias, and the glaucous magnolia. The sides of the mountains are covered with the arbor vitæ, with magnolias and hemlock-spruces, intermingled with the arborescent azaleas, the sorrel-tree, and the beautiful mountain laurel. The undergrowth of the woods and plains contains endless varieties of the aster, several species of azalea and asclepias, the dwarf pyrus, and the exclusively American genera of liatris, phlox, &c. Tobacco, maize, and wheat are the staple articles of cultivation. Of the aquatic plants (many of which are remarkable for their beauty) the principal are the hydropeltis, the orontium, various singular sagittarias or arrowheads, the white and the yellow water-lily, the *Vallisneria Americana*, used as food by the canvass-backed ducks, the pickerel-weed, &c. The gramineous species contain several grasses of peculiar forms, numerous rushes, the large and beautiful wild-rice, some carices, &c. Of the indigenous ferns (which are very numerous), the United States possess none in common with the Old World. In the more southern districts, from 35° to 25° north latititude, the vegetation becomes more varied and characteristic. In addition to the greater number of the above-mentioned plants, it includes many belonging to warmer temperatures. Amongst the climbing plants, which are very numerous, are clematis, vines, passifloras, tillandsias, &c. Among the herbaceous and smaller plants are several lupines, sarracenias, gentian, the fly-trap, common in swampy situations, but like the *Cabomba aquatica* indigenous likewise in tropical climates, sundews, asarums, amarylli, the superb yuccas, gerardias, parietaras, &c. Acquatic plants also abound, such as the magnificent, nelumbium, the nuphar, lobelias, &c.; and the cane (a gigantic grass) occupies extensive tracts. To the west of the Rocky Mountains we are presented with an entirely new botanical region, distinguished by characteristic productions, but possessing some species in common with the eastern district of the American Continent. A single pæonia, the only species of that plant indigenous to America, belongs to this region. The vegetation of the northwest coast bears considerable affinity to that of the United States and of the opposite shores of Asia. In the Southern States, the climate of which exceed in heat and humidity that of any other corresponding latitude, a botanical region is presented, in which the productions of Mexico are commingled with those of the north. Here, in addition to the principal productions of Virginia and Kentucky, cotton, rice, indigo, and the sugar-cane, are objects of cultivation, and the plane and deciduous cypress acquire gigantic dimensions. A solitary epidendron inhabits the branches of the Magnolia, near Savannah; and in the same locality is found the Pinckneya, a plant allied to the Peruvian bark. The parasitical, gigantic, long-moss is exceedingly abundant in all the forests of the Carolinas, Florida, Alabama, &c.

"Southern Mexico exhibits extreme diversity in vegetable productions; while its coasts, as well as the shores of Antilles, present those of tropical regions, the botany of its higher elevations bears considerable affinity to that of temperate latitudes, but exhibits some productions closely related to equatorial species. The principal productions belonging to the warm regions of Mexico are palms, bananas, plantains, yams, coffee, indigo, sugar-cane, maize, the cocoa-tree, the pine-apple, which grows wild in the woods, the American aloe, and various cactuses, which abound in localities where nothing else can find existence. The low forests of Honduras produce immense quantities of mahogany, logwood-trees, tamarinds, lignum vitæ, and vanilla. The *jalap* abounds near the city from which it derives its name. Of the numerous productions of the more elevated or temperate regions of Mexico, the principal are oaks. The higher elevations exhibit many species of plants belonging to European genera, such as valerians, roses, violets, salvias, &c. The *Cheirostemon platanoides*, a tree remarkable for the beauty and singularity of its organization, forms immense forests in the northern vicinity of Toluca. The caryophylaceous and rhodoraceous tribes common to northern climates here, forms the vegetable of altitudes verging on the regions of perpetual snow. The indigenous vegetation of the lower districts of the equatorial regions of America is characterized by extreme luxuriance and diversity."

Northwestern States—Healthy and Fruitful Region.

The great health-restoring region of the United States, embracing the Upper Peninsula of Michigan, Northern Wisconsin, Minnesota, Dacota, and Montana, east of the Rocky Mountains, lies between the forty-fourth and forty-ninth parallels of latitude. Starting from the Straits of Mackinac, on the east, it embraces the shores of Green Bay and Lake Superior, including the valleys of Upper Wisconsin, the St. Croix, the Upper Mississippi, the Minnesota, the Red River of the North, and the Upper Missouri, extending to the base of the Rocky Mountains.

The coldest part of this extensive region, considering its latitude, is on the southeast border; *Green Bay*, Wis., 44° 30′ north latitude, having a mean annual temperature of 44° Fahr., while *St. Paul*, Minn., 44° 50′ north, has a mean of 46° Fahr., situated five degrees of longitude westward. *Fort Benton*, Montana, in north latitude 47° 49′, and twenty-two degrees westward, has a mean annual temperature of 48° Fahrenheit, being one of the warmest stations on record, considering its position east of the Rocky Mountains.

Over this immense prairie region, which is destitute of mountain ranges, and for the most part of a large growth of forest trees, the land gradually rising to an elevation of 1,500 feet above the ocean, there seems to be a healthy influence prevailing throughout the entire year—particularly in regard to the absence of cases of consumption and malignant fevers.

It is difficult to explain satisfactorily this strange phenomenon, but the following influences are perceptible to an observing mind: Clear pure water, pure invigorating air, with serene calms and alternate high winds, and deluging rain-storms; an absence of fruit-trees, including chestnuts, hickory nuts, and sweet acorns, all of which seem to disappear in this otherwise favored latitude. The elm, the maple, the birch, the willow, the poplar, the mountain ash, the dwarf oak, the white and Norway pines, the firs, and the spruces prevail. The wild animals are mostly of the fur-bearing species, while the birds consist of the pigeon, the prairie-hen, the partridge, the hawk, the eagle, ducks, wild geese, and gulls on the large bodies of water. The agricultural products are principally wheat, oats, hay, potatoes, and other kinds of vegetables, which are produced in great abundance.

In the vicinity of St. Paul, and northwestward to the Rocky Mountains, a Siberian summer prevails, giving three or four months of warm weather, being ample time to ripen all the cereals, grasses, and vegetables. The summer temperature of this region is its most remarkable feature—it being as warm in St. Paul during the months of June, July, and August, as in the city of New York, situated about 4° southward—this same summer temperature extending northwestward to the valley of the Red River of the North, in 48° north latitude. Minnesota may be said to excel any of the Western States in men, women, horses, wheat, and vegetables, owing to its pure atmosphere and fruitful soil: and any portion of the Union in a healthy and invigorating climate.

When these known results are obtained, how can we reconcile them with the climate on the north shore of Lake Superior and on the coast of Labrador, where the growth of grain and vegetables ceases? The only reasonable solution is, that one is influenced by warm currents of air from the North Pacific Ocean, flowing over Russian and British America, and the other by the cold, chilling winds from Hudson Bay and Baffin's Bay, which come sweeping down from within the Arctic Circle, lowering the temperature in Canada, and the whole of the United States east of Lake Superior, including the Atlantic States to the coast of Florida.

While the warm summer weather does not prevail on the shores of Lake Superior, the same beneficial results are observable in regard to the prevalence of health. Here, during the summer months, the weather is very changeable, usually ranging from 50° to 70° Fahr., being a difference of about 10° from the summer temperature of St. Paul and its vicinity. It is, however, perceptibly warmer at Superior City, at the west end of the lake, than at Marquette, or the more eastern portion of this great inland sea.

On the south shore of Lake Superior the clouds usually run low and give out a chilling influence, which greatly retards vegetation—hence the agricultural products will never vie with those of the Upper Mississippi, or the Red River of the North. The mineral wealth, however, of this region is inexhaustibe, both on the American and Canadian shores, which, combined with its health-restoring influence, will always make Lake Superior a great place for business, and resort for invalids and pleasure-seekers.

The influence of the Great Lakes in raising the *winter* temperature and depressing the *summer* temperature of Wisconsin and surrounding country, is most singularly shown on a map drawn by J. A. Lapham, L L. D. From this exhibit of the mean temperature of Jannary (13° Fahr.), it appears that it is no colder on Keweenaw Point, Lake Superior, (47° north latitude) than at the Falls of St. Anthony, Minn., in 45°, being distant about 300 miles in a southwest direction. The mean temperature during the same month at Mackinac, Mich., 46° north latitude, (19° Fahr.), being the same as at Prairie du Chien, Wis., 43°, being distant about 400 miles in the same direction.

The difference in the *July* temperature (72° Fahr.) is equally singular, it being about the same in St. Anthony as in Chicago, Ill., situated about 3° southeast—thus reversing the influence of the January temperature—showing conclusively that the Great Lakes, or other controlling influences operate most strangely in producing this singular phenomena.

The July temperature again rises on the east side of Lake Michigan, which is continued through the summer and autumn months; thus, no doubt, producing a favorable effect on the growth of fruit of different species, for which Western Michigan is celebrated.

The summer temperature (70° Fahr.) is the same in the City of New York, Pittsburgh, Chicago, and St. Paul, Minn., continuing northwestward to the Red River of the North, where corn and most kinds of grain and vegetables alike flourish.

The rain or moisture which falls on this line or belt, extending for about 2,000 miles, is about the same during the summer months, but much less during the winter months in Minnesota and the adjacent regions—ten inches of moisture in the shape of rain or snow falling in the Middle States, while only two inches of moisture in the shape of snow falls in the vicinity of St. Paul, Minn., while the air is extremely cold, dry, and invigorating.

These well-established facts in regard to the extensive region drained by the Mississippi, the Missouri, and the Red Rivers of the North, are of the utmost importance to the American public, as the favorable climate and soil are not confined to Minnesota and Dacota, but extend north to the valleys of the Assiniboine and Saskatchewan Rivers in British America. A writer says: "I have seen Indian corn growing at Red Lake, in latitude 48° north, which produced thirty bushels to the acre. Further west, in Minnesota and Dacota, and north in the valley of the Red River, about Lake Winnipeg, and in the valley of the Saskatchewan, is a tract large enough for several States, where wheat flourishes as a certain and abundant crop. Those who consider this extreme region to be a cold barren waste, make a gross mistake."

Railroads are now in progress of construction, extending north and west from St. Paul toward Pembina, situated on the

Red River of the North, and toward Fort Union on the Upper Missouri, to be continued westward through Montana and Idaho to Puget's Sound. Soon the shrill whistle of the locomotive will be heard on these extensive prairies, giving life to this new region of country, destined ere long to contain millions of hardy freemen.

The Geography of Consumption.

"Consumption originates in all latitudes, from the Equator, where the mean temperature is 80° Fahrenheit, with slight variations, to the higher position of the Temperate Zone, where the mean temperature is 40°, with sudden and violent changes. The opinion long entertained that it is peculiar to cold and humid climates, is founded in error. Far from this being the case, the tables of mortality warrant the conclusion, that consumption is sometimes more prevalent in tropical than in temperate countries. Consumption is rare in the Arctic regions, in Siberia, Iceland, the Orkneys, and Hebrides, also, in the northwestern portion of the United States.

"In North America 'the disease of the respiratory organs, of which consumption is the chief, have their maximum in New England, in latitude about 42°, and diminish in all directions from this point inland. The diminution is quite as rapid westward as southward, and a large district near the fortieth parallel is quite uniform at twelve to fifteen per cent. of deaths from consumption, while Massachusetts varies from twenty to twenty-five. At the border of the dry climate of the plains, in Minnesota, a minimum is attained as low as that occurring in Florida, and not exceeding five per cent. of the entire mortality. It is still lower in Texas, and the absolute minimum for the continent in temperate latitudes is in Southern California.'

"The Upper Peninsula of Michigan, embracing the whole of the Lake Superior region, Minnesota, Nebraska, and Washington Territory, are all alike exempt, in a remarkable degree, from the above fatal disease. Invalids suffering from pulmonary complaints and throat disease are almost uniformly benefited by the climate of the above northern region, having a mean annual temperature of from 40° to 50° Fahrenheit.

Diseases of the Respiratory System.

With reference to diseases of the Respiratory System, Assistant Surgeon G. K. Wood, U. S. A., stationed at Fort Laramie, submits the following remarks:—

"The climate of these broad and elevated table-lands which skirt the base of the Rocky Mountains on the east, is especially beneficial to persons suffering from *pulmonary diseases*, or with a *scrofulous diathesis*. This has been known to the French inhabitants of the Upper Mississippi and Missouri for many years; and it has been their custom, since the settlement of that portion of the country, to send the younger members of their families, who showed any tendency to diseases of the lungs, to pass their youth among the trappers of the plains and mountains. The beneficial result of this course, no doubt, depends, in a great measure, upon the mode of life led by these persons —their regular habits, constant exercise in the open air, and the absence of the enervating influences incident to life in cities; but that more is due to the *climate* itself, is shown by the fact, that among the troops stationed in this region (whose habits are much the same everywhere), this class of disease is of very rare occurrence. The reports from the line of posts stretching from the Upper Platte, through New Mexico, to the Rio Grande, give a smaller proportion of cases of pulmonary disease than those from any other portion of the United States. The air in this region is almost devoid of moisture; there are no very sudden changes of temperature; the depressing heats of the eastern summers are never felt; and although in the north the winters are extremely cold, a stimulant and tonic effect is the only result of exposure in the open air.

He adds: "It is of great importance that the climate of this region should be generally known, that the present injudicious course of sending consumptives to the hot, low, and moist coast, and the islands of the Gulf of Mexico, should be abandoned. In diseases of debility, the remedies are tonics and stimulants. What is more debilitating than affections of the lungs? and what less tonic than heat and moisture combined, as is found in the climate of the Gulf coast? It is simply not cold, and has no other advantage over the Northern States. The towns of New Mexico (or still further north) should be selected as a refuge for those showing a tendency to disease of the lungs, or scrofula, anywhere east of the Rocky Mountains, and west of the region where 'northers' prevail.

Assistant Surgeon R. Bartholomew remarks: "A question well worthy of consideration, Is this climate adapted to the amelioration and cure of the tubercular diathesis? As phthisis is annually on the increase in the United States, and as the subject of its hygienic management proves to be more important than the treatment by medicaments, the consideration of the climate is, necessarily, of the first consequence. In my report I remarked the beneficial influence of the journey over the plains upon those in whom 'a phthisical tendency was marked and imminent.' The purity of the atmosphere and the equability and dryness of the climate are conditions highly favorable to such improvement. The entire immunity of the mountaineers from all forms of pulmonary disease indicates the healthfulness of the country in this particular. Moreover, the various commands stationed at Fort Laramie, have been remarkably free from all forms of pulmonary disease, and all such as came thither laboring under the incipient or well established symptoms of consumption speedily improved."

Meteorological Observations kept at Fort Laramie.

Latitude, 42° 12′; Longitude, 104° 47′; Altitude, 4,519 feet.

Month.	Mean Temp. ° Fahr.	Highest. ° Fahr.	Lowest. ° Fahr.	Fall of Rain. Inches.
January, . .	31.03	53	11	.27
February, . .	32.60	57	—4	.71
March, . .	36.81	70	9	1.37
April, . . .	47.60	71	32	1.93
May, . . .	56.11	69	33	5.39
June, . . .	67.34	88	49	2.95
July, . . .	74.70	87	59	1.83
August, . .	73.78	94	55	.94
September, . .	64.21	91	36	1.33
October, . . .	50.91	75	20	1.26
November, . .	35.83	66	22	1.37
December, . .	27.98	68	13	0.65
Yearly Mean, .	50.00			20.00

COMPARISON BETWEEN FORT LARAMIE AND FORT BENTON.

FORT LARAMIE, DACOTA TERRITORY.

Lat.	Long.	Alt. feet.	Temperature. Fahr. Spring.	Sum.	Autumn.	Win.	Year.
42° 12′	104° 47′	4,519	49°	73°	50°	30°	50°

FORT BENTON, MONTANA TERRITORY.

Lat.	Long.	Alt. feet.	Temperature. Fahr. Spring.	Sum.	Autumn.	Win.	Year.
47° 49′	110° 36′	2,780	49°	72°	44°	25°	48°

"Forts Benton and Laramie hold a similar position in relation to each other, and seem to be wholly influenced by the climate of the western part of the continent."—*Governor I. I. Stevens' Northern Pacific Railroad Report,* 1855.

PART VII.

CLIMATIC BOUNDARY OF THE UNITED STATES.

THE UNITED STATES OF AMERICA, extending from the Atlantic to the Pacific, is washed by two great oceans, one on the east and one on the west, each exercising a great and varied climatic influence. It is bounded on the north by the British Possessions, and on the south by the Gulf and Republic of Mexico. Its extremes of latitude are from 24½° to 49° north, and from 67° to 125° west longitude, from Greenwich.

Its *Northern* limit, on the Atlantic side, is 47° 15′ north latitude, where stands Fort Kent, Maine, having a mean annual temperature of 37° Fahrenheit; *centrally*, at Pembina, Minn., 49° north latitude, having a mean annual temperature of 38°. On the Pacific side (Puget's Sound), its northern limit is 48° 30′ north latitude, here having a mean annual temperature of 50° Fahr.; variation, **13**° Fahr.

Its *Eastern* limit, on the Atlantic, in north latitude 44° 54′; 66° 58′ west longitude, from Greenwich (Eastport, Me.), having a mean annual temperature of 43° Fahr.; *centrally*, Norfolk, Va., near the mouth of Chesapeake Bay, 36° 45′ north latitude, having a mean annual temperature of 60°. On the Atlantic coast, in 25° north latitude; 81° west longitude (the southern part of Florida), having a mean annual temperature of 76° Fahr.; variation, **33**° Fahr.

Its *Southern* limit on the Gulf of Mexico (Key West), 24° 32′ north latitude, having a mean annual temperature of 76° Fahr.; *centrally* at the mouth of the Rio Grande (Fort Brown, Texas), 25° 53′ north latitude, having a mean annual temperature of 74°. On the Pacific side, 32° 31′ north latitude; 117° 06′ west longitude (near San Diego, Cal.), having a mean annual temperature of 62° Fahr.; variation, **14**° Fahr.

Its *Western* limit on the Pacific coast, (San Diego), having a mean annual temperature of 62° Fahr.; *centrally*, San Francisco,

37° 48′ north latitude; 122° 26′ west longitude, having a mean annual temperature of 55°. On the Strait of San Juan de Fuca, (Cape Flattery), 48° 30′ north latitude; 124° 40′ west longitude, here having a mean annual temperature of 50° Fahr.; variation, **12°** Fahr. Its extremes of mean annual *temperatures* are from 37° to 77° Fahr., and running through 24½ degrees of latitude, and 40 degrees of temperature.*

Within this wide limit, embracing upwards of 3,000,000 square miles, are produced on the North all the more hardy products of the forest, furnishing the finest of timber, together with the cereals, the grasses, and the vegetables. On the South is produced cotton, rice, sugar, and most of the fruits peculiar to a warm or sub-tropical climate—thus placing the United States, in a climatic point of view, in the best possible position on the face of the globe.

The Atlantic slope, for the most part, is favored with a healthy climate, and rich in agricultural and mineral products. West of the Alleghany range of mountains, the great valley of the Mississippi, presents a virgin soil and favorable climate, which, combined with the basin of the Great Lakes, furnishes a cultivable field unequalled in the Eastern or Western Continent.

The northern boundary of the United States, for about half its distance across the continent, runs nearly parallel to the northern limit of the Temperate Zone (40° mean annual temperature), above which line killing frosts are liable to occur during each of the summer months—hence the uncertainty of raising grain or vegetables north of the Upper Lakes.

To the south of the United States boundary along the confines of Mexico, the climate assumes a tropical character. "The boundary," says Colonel Emory, "is embraced in the zone separating the tropical from the sub-tropical or temperate regions. It is indeed a *neutral region*, having peculiar characteristics so different as to stamp upon animal and vegetable life features of its own. The vegetation assumes a tropical character, and the margin of the Rio Grande, near its mouth, which is exposed to overflow, abounds in reed, cane-brake, palmetto, willow, and water-plants. That which, perhaps, creates as much

* On the Pacific coast, running through 16° of latitude (from 32° to 48°), there is only a variation of 12° of mean annual temperature.

as any other one cause the difference in its botanical and zoological productions, is the hygrometric state of the atmosphere."

Thus it may be perceived that the United States possesses a singular and marked *climatic* boundary, both on the North and on the South, although defined by no mountain ranges; giving to the Union nearly all of the temperate and sub-tropical zones on the Continent of North America, without any, or little, of the enervating influence of a purely tropical climate.

A portion of Canada, New Brunswick, and Nova Scotia are included in the above favorable Temperate Zone; also, the southern part of British America lying west of the head of Lake Superior.

Population, Health, and Agricultural Products.

The Area of the United States may be divided into *four* great Climatic Divisions, as follows:—

1. The most healthy region is the *Northern Division*, embracing the States of Maine, New Hampshire, Vermont, Northern New York, Michigan, Wisconsin, Minnesota, Dacota, Montana, Idaho, and Washington Territory, lying mostly between the 43d and 49th parallels of north latitude. 0.96 per cent., or 1 in 116 dying annually in the above States and Territories. This extensive region, embracing 825,000 square miles of territory, has three or four months of cold winter weather, with delightful summers; containing only five inhabitants to the square mile. It produces, wheat, oats, hay, potatoes, and other vegetables in great abundance. On the northern limit, the mean annual temperature varies from 37° to 44° Fahr., rising to 47° on the southern limit.

2. The *Middle Division*, lying mostly between the 39th and 43d parallels (the most favored zone), comprises the States of Massachusets, Rhode Island, Connecticut, New York, New Jersey, Pennsylvania, Ohio, Indiana, Illinois, Iowa, Nebraska, Southern Idaho, and Oregon, having an area of 453,000 square miles, and containing thirty inhabitants to the square mile. The mean annual temperature varies from 47° to 53° Fahr.; it being the centre of the Temperate Zone; 1.18 per cent., or 1 in 93 dying annually. It produces all the cereals, grasses, and fruit of different kinds, suitable for the sustenance of man and

beast. About one-half of the entire product of wheat, Indian corn, and oats, and more than half the hay, butter, and cheese being raised in this fertile region, according to the Census of 1860, while nearly half the population of the Union is found within the same belt of territory, extending from ocean to ocean.

3. The *Division* lying between the 36th and 40th parallels of latitude, embracing the Border States, including Delaware, Maryland, Virginia, West Virginia, Kentucky, Tennessee, Missouri, Kansas, Colorado, Utah, Nevada, and Northern California, forming an area of 676,000 square miles. It contains nine inhabitants to the square mile. The mean annual temperature varies from 50° to 60° Fahr., being greatly affected by altitude in different sections; 1.30 per cent., or 1 in 79 dying annually. Although rich in agricultural and mineral productions, and with a genial climate, this section of the Union has not increased as rapidly in population and wealth as the more northern divisions. It produces wheat, Indian corn, tobacco, and hemp, and is also favorable for grapes and other kinds of fruit.

4. The *Southern Division*, lying between the 24th and 36th parallels of latitude, includes the States of North Carolina, South Carolina, Georgia, Florida, Alabama, Mississippi, Arkansas, Louisiana, Texas, New Mexico, Arizona, and Southern California. The mean annual temperature varies from 60° to 77° Fahrenheit. It embraces an area of 1,091,000 square miles, being the largest division; containing six inhabitants to to the square mile; 1.48 per cent., or 1 in 69 dying annually. Here is produced cotton, rice, sugar, and Indian corn in great quantities; being for the most part a sub-tropical climate, where snow or ice are seldom to be found, and is subject to malignant fevers of different kinds.

Total Area, Population, &c., of the United States—1860.

Divisions.	Mean Temp. ° Fahr.	Sq. Miles.	Population.	No. to Sq. Mile.
Northern, . .	37 to 47	825,190	4,276,476	5
Middle, . . .	47 to 53	453,358	14,189,349	30
Border or Central,	50 to 60	676,706	6,208,583	9
Southern, . .	60 to 76	1,091,413	6,684,349	6
		3,046,667	31,358,757*	12½

* Of whom in 1860, 3,953,760 vere Slaves of African descent.

Habits and Character, as Influenced by Climate.

The habits and character of the people inhabiting these different sections are more or less influenced by *climate*—thus the inhabitants of the New England States, with a temperate, cool climate, are found to be generally intelligent, industrious, humane, and frugal—living in comfortable houses, encouraging education, religion, the arts, and all the helps which go to ameliorate and advance the human race. The same may be said of all the Northern and Northwestern States, where an agricultural community predominates. Here ship-building and manufactures, as well as agricultural pursuits of different kinds, are carried on very extensively.

The middle belt, having a mean annual temperature from 47° to 53° Fahr., is crowded with inhabitants, and full of enterprise—where the arts, commerce, institutions of learning and agricultural pursuits are alike encouraged. Here are the great cities and marts of trade—where are found steam-ships and railroads in rapid motion, communicating in a measure the same impetus to its citizens. Common schools, academies, colleges, scientific, benevolent and religious institutions, and manufacturing establishments are generally found to exist. Republican institutions here find firm supporters, while oppression of every kind is discouraged and opposed by the great mass of the people.

On this greatly favored belt or zone, commerce has its permanent and chief seat—here sailing vessels and steamers are enabled to run during the entire year with safety to passengers and freight, while in the more northern parts the harbors and rivers are generally closed for three or four months by ice—while to the south, during the summer months, excessive hot weather and sickness enfeebles and retards commerce. This may be further illustrated by comparing the navigation of the St. Lawrence River and ports south of the mouth of Chesapeake Bay, with Boston, New York, and Philadelphia. So on the Pacific side of the continent, nearly all the commerce being confined to ports having a temperate and healthy climate.

The inhabitants of the Border or Central States, including Maryland, Kentucky, &c., are of a mixed character in regard to many of the habits and traits enumerated above—often resorting to scenes of violence, in order to enforce their opinions.

Here, to a considerable extent, the arts, commerce, and manufactures are fostered, while agriculture is greatly encouraged.

The white population of the Southern or Cotton-growing States, as a whole, are more united and fixed in their character, than any other portion of the American people. Climate and the institution of slavery combined, has tended to render them haughty, domineering, and impatient of restraint—in a measure unfitting them for a republican form of government. These objectionable traits of character, however, are confined mostly to the vicinity of the sea-board, while in the mountainous regions of Virginia, North Carolina, and Tennessee are found good and loyal citizens, firmly maintaining republican principles. Here is a sub-tropical climate, the temperature varing from 60° to 76° Fahrenheit, mean annual temperature.

The inhabitants of the Pacific States and Territories are of a varied and mixed character, where may be found the descendants of the European, Asiatic, Mexican and African races, as well as the native Indian. Their habits of life and social relations differing according to their different nationalities. The white American race predominates in the government and business affairs of the country, being sincere in their support of republican principles; while a fine climate, fruitful soil, and mountain ranges engenders noble sentiments in the breast of man.

Further observations in North America are necessary in order to show the influences operating, in a climatic point of view, on its inhabitants, now making rapid progress in civilization and liberal forms of government, giving equal privileges to all races of men; thus elevating all classes of the human family, residing on this continent, to the standard designed by an overruling Providence.

I.—Agricultural Products of the United States—1860.

Giving the AREA of the STATES and TERRITORIES.

1. NORTHERN DIVISION,

Having a Mean Annual Temperature between 40° and 47° Fahrenheit.

STATES, &c.	Area, Sq. Miles.	Wheat, Bushels.	Indian Corn, Bushels.	Oats, Bushels.	Potatoes, Bushels.	Hay, Tons.
Maine,.........	35,000	233,876	1,546,071	2,988,939	6,376,052	975,803
New Hampshire,.	9,280	238,965	1,414,628	1,329,233	4,137,704	642,741
Vermont, ...	10,212	437,037	1,525,411	3,630,267	5,254,121	940,178
Nth'rn New York	16,000	2,893,701	6,687,016	11,725,044	8,818,307	1,188,264
Michigan,.......	56,243	8,336,368	12,444,676	4,036,980	5,300,737	768,256
Wisconsin,......	53,924	15,657,458	7,517,300	11,059,260	3,820,705	855,037
Minnesota, ...	83,531	2,186,993	2,941,952	2,176,002	2,566,277	179,482
Dakota Ter.,.....	230,000	945	20,269	2,540	9,489	855
Montana Ter.,...	160,000					
Idaho Ter.,......	100,000					
Washington Ter.,	71,000	86,219	4,712	134,334	163,612	4,580
Total,.........	825,190	30,071,562	34,102,035	37,082,599	36,407,004	5,555,196

2. MIDDLE DIVISION.

Having a Mean Annual Temperature between 47° and 53° Fahr.

STATES, &c.	Area, Sq. Miles.	Wheat, Bushels.	Indian Corn, Bushels.	Oats, Bushels.	Potatoes, Bushels.	Hay, Tons.
Massachusetts,...	7,800	119,783	2,157,063	1,180,075	3,202,517	665,331
Rhode Island,....	1,306	1,131	461,497	244,453	543,855	82,722
Connecticut., ...	4,750	52,401	2,059,835	1,522,218	1,835,858	562,425
Sth'rn New York,	30,000	5,787,404	13,374,033	23,450,090	17,636,616	2,376,529
New Jersey,.....	8,320	1,763,218	9,723,336	4,539,132	5,206,522	508,726
Pennsylvania, ...	47,000	13,042,165	28,196,821	27,387,147	11,696,154	2,245,413
Ohio,...........	39,964	15,119,047	73,543,190	15,409,234	8,999,546	1,564,502
Indiana, ...	33,809	16,848,267	71,588,919	5,317,831	4,165,163	622,426
Illinois,.........	55,409	23,837,023	115,174,777	15,220,029	5,846,544	1,774,554
Iowa,...........	55,000	8,449,403	42,410,686	5,887,645	2,858,082	813,173
Nebraska Ter.,..	70,000	147,867	1,482,080	74,502	162,356	24,458
Oregon,.........	100,000	826,776	76,122	885,673	303,654	27,986
Total,.........	453,358	85,994,485	360,168,279	101,118,029	62,456,867	11,269,245

3. BORDER SOUTHERN STATES.

Having a Mean Annual Temperature between 53° and 60° Fahr.

STATES, &c.	Area, Sq. Miles.	Wheat, Bushels.	Indian Corn, Bushels.	Oats, Bushels.	Potatoes, Bushels.	Hay, Tons.
Delaware,.......	2,120	912,941	3,892,337	1,046,910	520,144	36,973
Maryland,.......	11,124	6,103,480	13,444,922	3,959,298	1,501,169	191,744
Dis. of Columbia,	60	12,760	80,840	29,548	37,299	3,180
Virginia,...... } West Virginia, }	37,352 24,000	13,130,977	38,319,999	10,186,720	4,253,215	445,133
Kentucky,.......	37,680	7,394,809	64,043,633	4,617,029	2,814,088	158,476
Tennessee,.......	45,000	5,459,268	52,089,926	2,267,814	3,786,677	143,499
Missouri, ...	65,000	4,227,586	72,892,157	3,680,870	2,325,952	401,070
Kansas,.........	83,000	194,173	6,150,727	88,325	306,300	56,232
Colorado Ter.,..	104,000					
Utah Ter.,......	121,000	384,892	90,482	63,211	141,001	19,235
Nevada,........	90,000	3,631	460	1,082	5,886	2,213
Nth'rn California,	56,333	1,976,156	170,236	347,668	667,923	101,884
Total,........	676,706	39,800,673	251,175,719	26,288,475	16,359,654	1,569,639

4. Southern Division.

Having a Mean Annual Temperature between 60° and 76° Fahr.

States, &c.	Area. Sq. Miles.	Wheat, Bushels.	Indian Corn, Bushels.	Oats, Bushels.	Potatoes, Bushels.	Hay, Tons.
North Carolina,..	50,700	4,743,706	30,078,564	2,781,860	6,970,604	181,365
South Carolina,..	34,000	1,285,631	15,065,606	936,974	4,342,723	87,587
Georgia,........	58,000	2,544,913	30,776,293	1,231,817	6,812,330	46,448
Florida,.........	59,268	2,808	2,834,391	46,899	1,148,525	11,478
Alabama,.......	50,722	1,218,444	33,226,282	682,179	5,931,563	62,211
Mississippi, ...	47,156	587,925	29,057,682	221,235	4,978,193	32,901
Arkansas,.......	52,198	957,601	17,823,588	475,268	1,984,550	9,356
Louisiana, ...	41,346	32,208	16,853,745	89,377	2,355,636	52,721
Texas,..........	274,356	1,478,345	16,500,702	985,889	2,020,794	11,865
Indian Territory,	70,000					
N. Mexico Ter., Arizona Ter.,..	110,000 131,000	434,309	709,304	7,246	5,403	1,113
Sth'rn California,	112,667	2,952,314	340,472	695,338	1,335,847	203,771
Total,.........	1,091,413	16,238,204	193,266,629	8,154,082	37,885,768	700,816

Recapitulation, by Climatic Divisions.

Divisions.	Wheat, Bushels.	Indian Corn, Bushels.	Oats, Bushels.	*Potatoes, Bushels.	Hay, Tons.
Northern Division, ..	30,071,552	34,102,035	37,082,599	36,407,004	5,555,196
Middle Division,.....	85,994,485	360,168,279	101,118,029	62,456,867	11,269,245
Border South'n States,	39,800,673	251,175,719	26,288,475	16,359,654	1,569,639
Southern Division,,...	16,238,204	193,266,629	8,154,082	37,885,768	700,816
Grand Total,......	172,104,924	838,712,662	172,643,185	153,109,293	19,094,896

* 42,095,026 were sweet potatoes.

II.—Agricultural Products of the United States—1860.

Also, the Average Mean Annual Temperature of the Several States and Territories.

1. Northern Division.

States, &c.	Yearly Temp.	Butter, Lbs.	Cheese, Lbs.	Tobacco, Lbs.	Wool, Lbs.	Flax, Lbs.
	° Fahr.					
Maine,..........	37 to 46	11,687,781	1,799,862	1,583	1,495,060	2,997
New Hampshire,...	40 to 46	6,956,764	2,232,092	18,581	1,160,222	1,347
Vermont, ...	42 to 47	15,900,359	8,215,030	12,245	3,118,950	7,007
Nth'rn New York,	44 to 47	34,097,280	16,548,289	1,921,537	3,151,491	506,008
Michigan,........	40 to 49	15,503,482	1,641,897	121,099	3,960,888	4,128
Wisconsin.,.......	40 to 47	13,611,328	1,104,300	87,340	1,011,933	21,644
Minnesota,.......	38 to 46	2,957,673	199,314	38,938	20,388	1,983
Dakota Territory,..	38 to 48	2,170		10		
Montana Ter.,....	38 to 48					
Idaho Ter.,.......	40 to 50					
Washington Ter.,..	44 to 52	153,092	12,146	10	19,819	
Total,........		100,870,729	31,752,930	2,201,343	13,938,751	545,114

2. Middle Division.

States, &c.	Yearly Temp.	Butter, Lbs.	Cheese, Lbs.	Tobacco, Lbs.	Wool, Lbs.	Flax, Lbs.
	° Fahr.					
Massachusetts,...	45 to 50	8,297,936	5,294,090	3,233,198	377,267	0,165
Rhode Island,....	48 to 50	1,021,767	181,511	0,705	90,699	
Connecticut.	46 to 50	7,620,912	3,898,411	6,000,133	335,896	1,187
Sth'rn New York,	47 to 51	69,000,000	32,000,000	3,843,045	6,302,983	1,012,017
New Jersey,.....	48 to 53	10,714,447	182,172	149,485	349,250	48,651
Pennsylvania,...	46 to 53	58,635,511	2,508,556	3,181,586	4,752,522	312,368
Ohio,...........	47 to 54	48,543,162	21,618,893	25,092,581	10,608,927	882,423
Indiana,........	48 to 54	18,306,651	605,795	7,993,378	2,552,318	97,119
Illinois,.........	47 to 54	28,052,551	1,848,557	6,885,262	1,989,567	48,235
Iowa,...........	46 to 52	11,953,666	918,635	303,168	660,858	30,226
Nebraska Ter.,...	46 to 50	342,541	12,342	3,636	3,302	
Oregon,.........	48 to 54	1,000,157	105,379	0,405	219,012	162
Total,.........		260,456,680	69,174,343	56,686,581	28,242,701	2,432,553

3. Border and Western States.

States, &c.	Yearly Temp.	Butter, Lbs.	Cheese, Lbs.	Tobacco, Lbs.	Wool, Lbs.	Flax, Lbs.
Delaware,.......	53 to 55	1,430,502	6,579	9,699	50,201	
Maryland,.......	50 to 58	5,265,295	8,342	38,410,965	491,511	8,112
Dis. of Columbia,	55 to 56	18,835		15,200	100	14,481
Virginia,...... } West Virginia, }	52 to 60 50 to 56	13,464,722	280,852	123,968,312	2,510,019	487,808
Kentucky,.......	53 to 60	11,716,609	190,400	108,126,840	2,329,105	728,234
Tennessee,.......	54 to 60	10,017,787	135,575	43,448,097	1,405,236	296,464
Missouri, ...	52 to 60	12,704,837	259,633	25,086,196	2,069,778	109,837
Kansas,.........	50 to 56	1,093,497	29,045	20,349	24,746	1,335
Colorado Ter.,...	46 to 54					
Utah Ter.,.......	48 to 60	316,046	53,331		74,765	4,343
Nevada,........	48 to 58					
Nth'rn California,	48 to 58	1,547,517	671,844	1,575	1,341,554	
Total,.........		57,575,647	1,635,598	339,087,233	10,297,015	1,550,584

4. Southern Division.

States, &c.	Yearly Temp.	Butter, Lbs.	Cheese, Lbs.	Tobacco, Lbs.	Wool, Lbs.	Flax, Lbs.
North Carolina,..	54 to 66	4,735,495	51,119	32,853,250	883,473	216.490
South Carolina,..	56 to 68	3,177,934	1,543	104,412	427,102	344
Georgia,........	58 to 70	5,439,765	15,587	919,318	946,227	3,303
Florida,.........	60 to 77	408,855	5,280	828,815	59,171	
Alabama,.......	60 to 70	6,028,487	15,923	232,914	775,117	111
Mississippi, ...	60 to 70	5,006,610	4,427	159,141	665,959	50
Arkansas,.......	56 to 66	4,067,556	16,810	989,980	410,382	3,821
Louisiana, ...	62 to 72	1,444,762	6,153	39,940	290,847	
Texas,..........	60 to 74	5,850,583	275,128	97,914	1,493,738	115
Indian Territory,	56 to 64					
N. Mexico Ter., } Arizona Ter.,.. }	48 to 72 56 to 74	13,259	37,240	7,044	492,645	
Sth'rn California,	56 to 74	1,547,517	671,844	1,575	1,341,554	
Total,........		37,719,801	1,101,034	36,234,303	7,786,315	224,234

Recapitulation, by Climatic Divisions.

Divisions.	Butter. Lbs.	Cheese, Lbs.	Tobacco, Lbs.	Wool, Lbs.	Flax, Lbs.
Northern Division, ..	100,870,729	31,752,930	2,201,343	13,938,751	545,114
Middle Division,.....	260,456,680	69,174,343	56,686,581	28,242,701	2,432,553
Border States, etc....	57,575,647	1,635,598	339,087,233	10,297,015	1,550,584
Southern Division,,...	37,719,801	1,101,034	36,234,303	7,786,315	224,234
Grand Total,.......	456,622,857	103,683,905	434,209,460	60,264,782	4,752,485

III.—Agricultural Products of the Sub-Tropical States—1860.

States, &c.	Cotton, Bales, 400 lbs.	Rice, Lbs.	Cane Sugar, Hogsheads.	Molasses, Gallons.
North Carolina,..........	145,514	7,593,976	38	12,494
South Carolina,..........	353,412	119,100,528	198	
Georgia,...............	701,840	52,507,652	1,167	546,749
Florida,................	65,153	223,704	1,669	436,357
Alabama,..............	989,955	493,465	175	85,115
Mississippi,.............	1,202,507	809,082	506	10,016
Tennessee (Southern),....	228,194	40,372	2	2,830
Arkansas,...............	367,393	16,813	402	22,305
Louisiana,..............	777,738	6,331,257	221,726	13,439,772
Texas,..................	431,463	26,031	5,099	408,358
Other States,...........	50,473	24,134	—	—
Total,.................	5,318,782	187,167,032	*230,982	14,963,996

* 230,982,000 pounds cane sugar; *maple sugar*, 40,120,083 pounds, mostly raised in the Middle and Northern States.

IV.—Comparative Agricultural Statistics of the United States, 1850—1860.

Products, &c.	Quantities.		Proportion to Each Inhabitant.		Increase and Decrease.	
	1850.	1860.	1850.	1860.		
Inhabitants, Number,	23,191,876	31,443,222			In.	8,251,446
Wheat,.... Bushels,	100,485,944	173,104,924	4.33	5.50	"	72,618,980
Indian Corn, "	592,141,230	838,792,740	25.	27.	"	256,651,510
Oats,....... "	146,584,179	172,643,185			"	26,059,006
Potatoes, .. "	104,037,562	153,109,293	6.34	5.	"	49,071,731
Hay,.......... Tons,	13,838,642	19,083,896			"	5,245,254
Butter,........ Lbs.	312,948,915	459,681,372	13.	15.	"	146,733,457
Cheese,........ "	105,535,599	103,663,927	4.50	3.50	De.	1,871,672
Tobacco, ... "	199,752,746	434,209,461	8.50	14.	In.	234,446,715
Wool,......... "	52,518,143	60,264,913	2.25	2.	"	7,746,770
Cotton,........ "	978,317,200	2,127,512,800	42.50	70.	"	1,149,195,600
Rice,.......... "	215,312,710	187,167,032	9.25	6.	De.	28,145,678
Sugar, ... "	236,814,000	230,982,000	12.	9.	"	5,832,000
Wine,...... Gallons,	221,219	1,627,242			In.	1,405,993

States in the Order of their Indian Corn Product in 1860.

States.	Bushels.	States.	Bushels.
1. Illinois,	115,174,777	18. South Carolina,	15,065,606
2. Ohio,	73,543,190	19. Maryland,	13,444,922
3. Missouri,	72,802,157	20. Michigan,	12,444,676
4. Indiana,	71,588,919	21. New Jersey,	9,723,336
5. Kentucky,	64,043,633	22. Wisconsin,	7,517,300
6. Tennessee,	52,089,926	23. Kansas,	6,150,727
7. Iowa,	42,410,686	24. Delaware,	3,829,337
8. Virginia,	38,319,999	25. Minnesota,	2,941,952
9. Alabama,	33,226,282	26. Florida,	2,834,391
10. Georgia,	30,776,293	27. Massachusetts,	2,157,063
11. North Carolina,	30,078,564	28. Connecticut,	2,059,835
12. Mississippi,	29,057,682	29. Maine,	1,546,071
13. Pennsylvania,	28,196,821	30. Vermont,	1,525,411
14. New York,	20,061,049	31. N. Hampshire,	1,414,628
15. Arkansas,	17,823,588	32. California,	510,708
16. Louisiana,	16,853,745	33. Rhode Island,	461,497
17. Texas,	16,500,702	34. Oregon,	76,122
Seven Territories,			2,388,147
Total Bushels,			838,792,740

Being to each inhabitant in the U. States, 27 Bushels.

States in the Order of their Wheat Product in 1860.

States.	Bushels.	States.	Bushels.
1. Illinois,	23,837,023	18. New Jersey,	1,763,218
2. Indiana,	16,848,267	19. Texas,	1,478,345
3. Wisconsin,	15,657,458	20. South Carolina,	1,285,631
4. Ohio,	15,119,047	21. Alabama,	1,218,444
5. Virginia,	13,130,977	22. Arkansas,	957,601
6. Pennsylvania,	13,042,165	23. Delaware,	912,941
7. New York,	8,681,105	24. Oregon,	826,776
8. Iowa,	8,449,403	25. Mississippi,	587,925
9. Michigan,	8,336,368	26. Vermont,	437,037
10. Kentucky,	7,394,809	27. N. Hampshire,	238,965
11. Maryland,	6,103,480	28. Maine,	233,876
12. California,	5,928,470	29. Kansas,	194,173
13. Tennessee,	5,459,268	30. Massachusetts,	119,783
14. North Carolina,	4,743,700	31. Connecticut,	52,401
15. Missouri,	4,227,586	32. Louisiana,	32,208
16. Georgia,	2,544,913	33. Florida,	2,808
17. Minnesota,	2,186,993	34. Rhode Island,	1,131
Seven Territories,			1,070,623
Total,			173,104,924

Being to each inhabitant in the U. States, $5\frac{1}{2}$ bushels.

Total Deaths in the United States—1860.

Showing the Number Dying from Consumption and Fevers.

STATES, &c.	Consumption.	Ratio.	Fevers.	Ratio.	Total. Deaths.
Maine,	2,169	29.5	616	8.3	7,614
New Hampshire,	1,163	26.6	340	7.7	4,469
Vermont,	779	24.4	253	7.6	3,355
Massachusets,	4,845	24.0	965	4.6	21,304
Rhode Island,	567	23.4	81	3.4	2,479
Connecticut,	1,269	21.7	341	5.8	6,139
Total, New England States,	10,792	24.9	2,596	6.2	45,361
New York,	8,199	18.4	1,663	3.7	46,941
New Jersey,	1,350	17.5	314	4.3	7,525
Pennsylvania,	5,011	17.6	1,932	5.0	30,241
Delaware,	201	18.0	77	6.0	1,246
Maryland,	1,197	17.2	393	5.9	7,374
District of Columbia,	255	22.0	60	5.1	1,285
Total, Middle States,	16,213	18.4	4,439	5.0	94,612
Virginia,	2,109	11.3	1,453	7.7	22,474
North Carolina,	761	7.2	1,503	14.2	12,617
South Carolina,	390	4.5	1,120	13.9	9,749
Georgia,	491	4.5	1,455	13.4	12,816
Florida,	97	6.2	235	9.5	1,769
Alabama,	596	5.3	1,466	13.1	12,760
Mississippi,	554	5.1	1,710	15.9	12,214
Louisiana,	843	7.5	1,384	13.6	12,324
Total, Southern States,	5,851	6.4	10,326	12.6	96,723
Texas,	420	5.1	1,346	21.7	9,377
Arkansas,	329	4.2	1,510	19.5	8,856
Tennessee,	1,440	10.9	1,745	13.2	15,156
Kentucky,	1,742	14.2	1,669	13.7	16,467
Missouri,	1,302	8.3	2,462	15.8	17,654
Kansas,	117	8.0	373	25.7	1,567
Total, Southwestern States,	5,350	8.4	9,105	18.2	69,077

Total Deaths in the United States, &c., continued.

STATES, &c.	Consumption.	Ratio.	Fevers.	Ratio.	Total Deaths.
Ohio,	3,495	14.1	1,650	6.6	24,726
Indiana,	1,805	12.8	1,746	12.6	15,326
Illinois,	1,948	10.9	2,339	13.1	19,300
Iowa,	784	11.1	878	13.2	7,259
Nebraska Ter.,	28	8.2	66	19.4	381
Total, Western States,	8,060	11.4	6,679	13.0	66,992
Michigan,	1,187	17.0	636	9.1	7,401
Wisconsin,	910	13.8	484	7.3	7,141
Minnesota,	151	15.2	83	8.3	1,109
Total Northwestern States,	2,248	15.0	1,203	8.2	15,651
New Mexico,	34	3.4	207	20.7	1,305
Utah,	18	5.5	16	4.9	374
California,	524	15.1	301	8.7	3,705
Oregon,	30	11.1	26	9.6	300
Washington Ter.,	8	16.0	—	—	50
Total Pacific States,	614	10.2	550	8.8	5,734

RECAPITULATION.

	Consumption.	Ratio.	Fevers.	Ratio.	Total Deaths.
Eastern States,	10,792	24.9	2,596	6.2	45,361
Middle States,	16,213	18.4	4,439	5.0	94,612
Southern States,	5,851	6.4	10,326	12.6	96,723
Southwestern States,	5,350	8.4	9,105	18.2	69,077
Western States,	8,060	11.4	6,679	13.0	66,992
Northwestern States,	2,248	15.0	1,203	8.2	15,651
Pacific States,	614	10.2	550	8.8	5,734
Grand Total,	49,118		35,898		394,150

In 1860, the deaths by Consumption were 13.79 per cent. of the whole number of deaths, and by Fevers, 9.79 per cent.; making 23.58 per cent. of deaths caused by the above diseases.

NOTE.—Northern Michigan, Wisconsin, and Minnesota, in the vicinity of Lake Superior, is one of the healthiest regions in the United States.

Rain in the United States.

[Extract from "Agriculture and Climate of North America," by R. RUSSELL.]

"In consequence of the Continent of North America being powerfully heated by the rays of the sun in summer, the southerly winds are more prevalent during that season. This fact is well established by the researches of Professor Coffin. The United States and Canada thus owe their fertility to the abnormal course of the tropical winds. These aerial currents, hot and moist from the Equatorial Zone, after crossing the Caribbean Sea and Gulf of Mexico, flow northward in summer over the continent almost with the regularity of a monsoon. Indeed, from the Gulf of Mexico to the British Possessions in America, the country is liberally watered by summer rains. There is no break about latitude 30° in this vast rainy region. Unlike the climate of corresponding latitudes in Europe and Africa, the West India Islands have their hurricanes and their luxuriant cane-fields; the Mississippi Valley, its summer tornadoes and its cotton and maize fields. The following observations even show that latitude 30°, along the Gulf coast, is the most rainy on the Atlantic side of the American Continent:—

FALL OF RAIN IN DIFFERENT LOCALITIES.

STATIONS, &c.	Lat.		Rain in Inches.				
	°	′	Spring.	Sum.	Autumn.	Win.	Year.
New Orleans, Lou.	30	00	11.29	17.28	9.62	12.71	50.90
Mobile, Ala. .	30	42	12.60	19,30	12.10	16.90	60.90
Pensacola, Flor.,	30	18	12.86	18.69	13.71	11.72	56.98
Savannah, Geo., .	32	4	11.90	23.00	9.70	8.40	53.00
Wilmington, N. C.,	34	20	6.83	15.52	16.32	7.34	46.00
Norfolk, Va., .	36	50	9.77	15.08	10.16	10.17	45.18
Washington, D. C.	38	53	10.45	10.43	10.15	10.07	41.20
Cincinnati, Ohio,	39	6	11.13	9.80	8.50	13.40	42.83
City of New York,	40	42	11.55	11.33	10.30	9.63	43.23
Buffalo, N. Y., .	42	53	8.59	9.23	13.54	7.53	38.80
Boston, Mass., .	42	21	8.60	8.42	9.27	9,01	35.30
Plattsburgh, N. Y.	44	41	8.36	10.03	10.05	4.95	33.39
Portsmouth, N. H.	43	4	9.03	9.21	8.95	8.38	35.57
Eastport, Me., .	44	54	8.88	10.05	9.85	10.61	39.39
Fort Kent., Me., .	47	15	5.46	11.65	9.64	9.71	36.46
St. Paul, Min., .	44	52	6.61	10.92	6.00	2.00	25.44

"The summer rain usually falls in thunder showers during sultry weather. The autumn rains are sometimes protracted for two or three days. The winter rains and snows are accompanied with violent winds and great fluctuations in the barometer.

The winter storms sweep the whole continent east of the Rocky Mountains. At all seasons the rains and snows are preceded by southerly winds from the Gulf of Mexico. But in winter these warm and moist winds are invariably succeeded by cold winds from the west, which render the fluctuations in the temperature of the most extreme character. I think that the principal phenomena of the American storms can be accounted for by the action of these two winds—the *south* and the *west*. The cold wind in Canada and the Northern States is usually from the north of west; in the latitude of Washington from the west, and on the Gulf of Mexico often due north. The manner in which the two winds—the south and the west—of the most opposite characters, alternately displace each other, *involves the whole theory of North American storms.*"

It appears on the Continent of North America, that about the same diminution of rain occurs, from south to north, as is found to exist on the Eastern Continent. The annual quantity falling at Rome, north latitude, 41° 53′, is thirty-nine inches; at London, north latitude, 51° 30′, twenty-four inches; at St. Petersburg, north latitude, 60°, sixteen inches.

Changes in the Climate of the United States.

"In the United States," says Professor Lovering, "and perhaps in the whole of North America, it has been observed that the temperature of January and July, have approached each other, the extremes not being so great, since European settlements began. The rivers do not freeze so thick, or so long, as they once did. When Philadelphia was first settled, the Delaware was covered with ice as soon as the 1st of November. Now it is rarely frozen at all. Hudson River is open a month longer in the winter than formerly, now usually closing the latter part of December, and opening about the middle of March, being an average of about three months.

"When New England was first settled, the winters set in regularly, and continued for three or four months, without interruption, and broke up at nearly the same time, as is now the case in Canada. The snow is diminished, and the period of sleighing is less. The changes of the seasons are all of them less sudden and uniform. It is also affirmed that there has been a great alteration in the prevalent winds. The force of the west winds has abated, while the east winds are increasing in frequency and extent. A century ago, they did not penetrate more than thirty or forty miles into the country; now they reach eighty miles from the sea-shore or upwards.

Notwithstanding these alleged facts, Dr. Enoch Hale gives the result of his careful discussion of the Meterological Journal of Dr. Holyoke, kept at Salem, from 1786 to 1821, as follows: It thus appears that this journal does not support the opinion that there has been a progressive increase of the temperature of our climate *in regard to the whole year.* If we compare the spring months of the different years, we find the results nearly the same, both in respect to the whole spring, and to the months of March and April; thus showing that the opinion is equally unfounded, which has been maintained, that the spring advances more rapidly in proportion to the temperature of the whole year, than it did formerly. The mean temperature of the first ten years is 48.77 Fahr., of the last ten years 47.85; the highest year of all was 1793, the mean annual temperature being 50.96; the lowest, 1812, the mean annual temperature being 45.28.

It appears from the published observations at Boston, by Mr. Jon. P. Hall, from 1821 to 1856, that the average temperature of the whole year, during the period of thirty-six years, was 48.66 Fahr. The warmest year was 1828, the temperature being 51.78; the coldest year was 1836, the temperature being 45.34 Fahr.

As the observations of Mr. Hall are not strictly comparable with those of Dr. Holyoke, because the places were ten miles apart, and the hours and instruments also different, we may compare the observations of Mr. Hall with each other; and we find the mean temperature for the first period of nine years 49.36; for the second, 47.76; for the third, 49.00; and for the fourth, 48.54 Fahr.

The mean annual temperature derived from the above observations, during forty-three years, are as follows:

Mean yearly temperature of Boston, 48.86 Fahrenheit.
Mean yearly temperature of Salem, 48.66 "

The mean annual temperature of Salem appears to have fluctuated irregularly to the extent of 5.68°; and that of Boston to the extent of 6.44° Fahr.

The average annual fall of snow in Boston and its vicinity, during the past twenty years, as ascertained by actual observation, amounts to four feet three inches.

In the vicinity of Lake Superior, the average annual fall of snow amounts to twenty-four feet, while at St. Paul, Minnesota, the annual fall is only two feet, or two inches of water, showing that the annual fall of snow is owing to local causes more than to difference of latitude.

"Dr. Hugh Williamson attributes the change of climate which he thinks has taken place in the United States, at least in the neighborhood of Philadelphia, to the settlement upon the soil and its cultivation. When the settler enters the new country, the trees disappear, the sun strikes down to the surface, and penetrates the upturned soil, the drainage is perfected, and evaporation and cold diminish in the winter. The land becomes more heated than the water, and the sea breeze, which before scarcely passed the edge of the coast inland, now makes farther and farther inroads. The summers will be less overheated and the winters will not be so excessively cold as before man began his cultivation."

The mean annual temperature of Philadelphia, for the last sixty years, at periods of ten years, are as follows:

From 1800 to 1809,	average,	. . 51.80	° Fahr.
" 1810 to 1819,	"	. . 51.20	"
" 1820 to 1829,	"	. . 52.70	"
" 1830 to 1839,	"	. . 52.00	"
" 1840 to 1849,	"	. . 52.70	"
" 1850 to 1859,	"	. . 53.00	"

The coldest year was 1816, being a mean of 49° Fahr., and the warmest, 1858, being 54°, making a variation of 5° Fahr.

I.—Meteorological Table.

SHOWING THE PRINCIPAL CITIES AND MILITARY STATIONS IN THE UNITED STATES HAVING A MEAN ANNUAL TEMPERATURE BETWEEN 37° & 47° FAHR.

Stations, etc.	Latitude.	Longitude.	Altitude.	Yearly Mean.	Four Seasons.			
					Spring.	Summer.	Autumn.	Winter.
			Feet.	° Fahr.	° Fahr.	° Fahr.	° Fahr.	° Fahr.
Eastport, Maine.........	44°54′	66°58′	70	43.00	40.15	60.50	47.52	23.90
Fort Fairfield, Maine....	46°46′	67°49′	415	38.10	36.29	61.58	40.30	14.28
Hancock Barracks, Maine.	46°07′	67°49′	620	40.50	39.15	63.33	43.15	16.41
Fort Kent, "	47°15′	68°35′	575	37.00	35.22	61.68	39.88	11.36
Bath, "	43°50′	69°52′	20	44.50	41.70	64.80	47.60	24.00
Portland, "	43°39′	70°20′	20	45.22	42.77	65.24	48.16	24.70
Portsmouth, N. H.......	43°04′	70°49′	40	45.80	43.22	64.38	49.00	26.60
Manchester, "	42°59′	71°28′	300	46.14	.....	.....	.....	.....
Concord, "	43°13′	71°29′	300	44.50	42.60	65.40	47.30	22.70
Hanover, "	43°42′	72°17′	530	42.29	.....	.....	.....	.....
Windsor, Vt............	43°30′	72°27′	.....	45.46	41.80	66.40	49.27	24.37
Burlington, Vt..........	44°29′	73°11′	350	45.00	42.70	67.90	47.80	21.60
Newburyport, Mass.....	42°49′	70°50′	.....	47.34	.....	.....	.....	.....
Andover, "	42°40′	71°08′	150	47.40	44.70	68.70	49.30	26.90
Lawrence, "	42°42′	71°11′	133	45.89	.....	.....	.....	.....
Worcester, "	42°16′	71°48′	536	47.60	.....	.....	.....	.....
Amherst, "	42°22′	72°31′	260	46.70	45.00	68.60	48.70	24.70
Williamstown, "	42°43′	73°13′	930	45.90	43.60	67.90	47.90	24.20
Troy, N. Y............	42°43′	73°40′	50	47.80	46.10	70.00	50.25	24.60
Salem, "	43°15′	73°30′	600	46.56	45.00	68.26	48.41	24.50
Plattsburgh, N. Y.......	44°71′	73°25′	180	44.00	42.32	66.76	46.67	20.22
Malone, "	44°50′	74°23′	700	43.40	43.16	64.19	45.00	21.30
Ogdensburgh, "	44°41′	75°32′	280	43.50	42.80	66.84	48.00	22.66
Utica, "	43°06′	75°13′	470	46.66	45.33	67.94	48.12	25.24
Sacket's Harbor, N. Y...	43°55′	76°00′	260	45.00	42.52	66.84	48.00	22.66
Oswego, " ..	43°20′	76°40′	250	46.44	43.70	66.92	50.40	24.72
Rochester, " ..	43°07′	77°51′	500	47.00	44.60	67.60	48.90	27.00
Buffalo, " ..	42°53′	78°56′	650	46.25	42.73	66.93	47.92	27.42
Detroit, Mich..........	42°20′	83°00′	580	47.25	45.89	67.60	48.67	26.84
Port Huron, Mich.......	42°53′	82°24′	600	47.00	43.68	66.80	49.00	25.60
Grand Rapids, "	43°00′	86°00′	852	45.95	.....	.....	.....	.....
Fort Mackinac, "	45°51′	84°33′	728	41.00	38.73	62.00	43.85	20.00
Saut Ste. Marie, Mich..	46°30′	84°43′	600	40.37	37.60	62.00	43.54	18.30
Marquette, L. S., " ..	46°32′	87°41′	630	41.00	41.26	61.10	43.00	17.53
Copper Harb'r, L.S., Mich.	47°30′	88°00′	620	41.00	38.47	60.80	42.96	21.78
Ontonagon, " "	46°52′	89°30′	600	40.00	40.00	61.00	42.00	17.00
Manitowoc, Wis........	44°07′	87°37′	600	45.00	.....	.....	.....	.....
Milwaukee, "	43°03′	87°55′	600	46.40	42.30	67.30	50.10	26.10
Green Bay, "	44°30′	88°05′	620	44.50	43.52	68.50	46.00	19.92
Fort Winnebago, Wis...	43°31′	89°28′	770	45.00	45.50	68.00	46.00	20.00
Fort Crawford, " ...	43°05′	91°00′	640	47.68	48.66	72.28	48.53	21.25
Bayfield, L. S., " ...	46°45′	91°00′	620	40.00	38.00	62.00	43.00	15.60
Superior, " " ...	46°38′	92°03′	600	41.00	49.00	63.00	42.00	15.00
Fort Atkinson, Iowa.....	43°00′	92°00′	700	46.00	46.63	68.00	46.13	20.62
Fort Snelling, Min......	44°53′	93°10′	820	44.54	45.57	70.64	45.90	16.07
Fort Ridgely, "	44°15′	94°48′	.....	43.82	44.13	70.00	44.21	17.00
Fort Ripley, "	46°19′	94°19′	1,130	39.30	39.33	65.00	42.90	10.00
Pembina, "	48°56′	97°00′	900	39.00	34.30	70.00	42.00	12.00
Fort Randall, Dakota....	43°01′	98°12′	1,245	46.80	45.13	75.40	48.32	18.36
Fort Union, "	48°00′	104°00′	2,000	46.00	.....	.....	.....	.....

PART VIII.

MILITARY POSTS AND CITIES ON THE NORTHERN FRONTIER.

Climatic Features.

THIS belt of temperature, having an annual mean varying from 37° to 47° Fahrenheit, lies mostly between 43° and 49° north latitude, but does not extend to the Pacific Ocean within the bounds of the United States, running northward into the British and Russian Possessions.

Nova Scotia, New Brunswick, Canada, and the southern portion of the Hudson Bay Company's Territory, west of Lake Superior, also possess the same climatic influence, varying about 10° mean annual temperature. On the Pacific slope of the continent, the above range of climatic extends from the 49th to the 57th degree of north latitude, terminating on the north near Sitka, or New Archangel, in Russian America.

This immense region of country, extending through 80° of longitude, produces the cereals, grasses, and vegetables of the more hardy kind, sufficient to sustain a dense population. Wheat, rye, oats, barley, potatoes, beans and peas being produced in great abundance. The chief articles of export are fish and lumber, the former affording profitable employment to a large amount of tonnage, and thousands of seamen of different nations. The lumber trade is extensively and profitably pursued, both in the British Possessions and in the northern portion of the United States, where ship-building is carried on to a very large extent. The pine, the oak and the maple are the most valuable trees of the forest. The trade in furs and maple sugar are also important items of home consumption and export, the former slowly decreasing in amount, and the latter increasing. The ice crop is also rising into importance, both for home consumption and export; nearly all being collected north of the mean annual temperature of 50° Fahrenheit, where the thermometer occasionally falls to zero. The streams are usually closed by ice for four months of the year, from December to March, while more or less snow covers the face of the earth, affording both warmth and moisture.

The great geographical feature of this region is the Lakes or "*Inland Seas*," which are immense basins, containing the largest deposit of fresh water on the globe, their surplus waters flowing northeast through the St. Lawrence River into the Gulf and Atlantic Ocean; other large streams flow eastward and northward into Lake Winnipeg, and the Hudson Bay, draining altogether an immense section of country, lying east of the Rocky Mountains. The mouths of the principal navigable rivers on the Atlantic slope are the St. John, Penobscot, Kennebec, Connecticut and Hudson Rivers, while many other streams take their rise in this region, and flow southward, falling into the Atlantic Ocean or Gulf of Mexico.

Military Posts and Cities.

FORT KENT, situated in the most northern part of the State of Maine, at the junction of the Fish River with the St. John's, in latitude 47° 15′ north, has a mean annual temperature of 37° Fahr. The coldest winter month (February) had a mean of 10°, and the warmest summer month (August) had a mean of 68° Fahr. The greatest extremes being from 96° above to 36° below zero, showing a variation of 132° Fahrenheit. "The region adjacent to Fort Kent is probably one of the healthiest within the limits of the United States, and though rigorous, the climate seems to be productive of the most robust health. Fevers and other diseases of a malarious origin are unknown, and other acute diseases are by no means of common occurrence. The soil is a light loam, which rests upon a stratum of gravel and pebbles. In consequence of its geological formation, the drainage of the land is excellent, and numerous springs of fine water are found in every direction. With the exception of the immediate banks of the St. John's River, the whole country is still covered by a dense, unbroken forest. The hardier woods, different varieties of the maple, beech, birch, and ash, are found on the more elevated and rocky soil, while the lower grounds are occupied by the spruce, fir, larch, and cypress. The white and yellow pines, which produce the fine lumber, the staple of the country, are found scattered through the forest, generally more or less isolated and distant from each other. Large elms are generally seen on the interval lands; the generality of the forest trees, however, with the exception of the pines, are of a rather diminutive size.

"The climate of Fort Kent, like that of the colder regions of Northern Europe, does not seem favorable for the production of pulmonary phthisis. During my sojourn at the post," says Assistant Surgeon Wotherspoon, "I have neither seen or heard of a case of this disease among the French or American settlers. Assistant Surgeon Isaacs, who, during the two years he was resident at the fort, had a much better opportunity than myself of becoming acquainted with the diseases of the country, informs me, not only that he never saw a case of consumption in the country, but that some of the inmates of the garrison, who were affected with suspicious symptoms, recovered from them entirely. The present revenue officer at the post—a man of decidedly scrofulous temperament—had suffered a slight attack of hæmoptysis, and other symptoms of incipient pulmonary disease, when he was ordered to this post. Though liable to catch cold when exposed, his cough no longer troubles him; he has gained flesh and strength, and considers himself free from the disease. The children in and near the garrison have generally enjoyed the best of health, and have been afflicted with none of those complaints so common in warmer climates." —*Medical Statistics U. S. Army.*

Fort Sullivan, the most northern military post on the Atlantic coast, is situated on a rocky eminence on Moose Island, Passamaquody Bay, in the immediate vicinity of the town of Eastport, Maine. Owing to its situation, and its proximity to the Bay of Fundy, the climate is damp, and fogs are frequent in the earlier summer months. The winters are cold, and in this season, the thermal variations are often sudden; yet more extreme cold is felt in the interior, on the main land, than on the island.

Plattsburgh Barracks.—This station is on the west shore of Lake Champlain, about a mile from the village of Plattsburgh, N. Y., in north latitude 44° 41′. A range of mountains borders the lake on the west, rising into the Adirondack range, and on the east the Green Mountains of Vermont are to be seen throughout their whole extent. The weather is very variable; sudden and great changes frequently occur. The thermometer has an extreme range of 124°, being 100° in summer and —24° in winter; the mean annual temperature being 44° Fahr. The mean annual precipitation in rain and snow is 33.40 inches. The prevailing winds are from the south and southwest; those from the south are often very cold, and frequently accompanied with snow or rain.

MADISON BARRACKS.—This station is at Sacket's Harbor, N. Y., in latitude 43° 50′ north. It is situated on the southern side of the bay formed by the entrance of Black River into Lake Ontario. The river is the third in size that is wholly in the State of New York. The color of the water is quite dark—a feature not uncommon in this region, and not readily accounted for. The water is not drunk by those who live near it, being thought unwholesome. The forest trees are maple, beech, birch, walnut, bass, ash, elm, and hemlock. Esculent vegetables are produced in great abundance and variety. The staple agricultural product is wheat; the soil in general being rich. This post is ranked as very healthy, the troops usually stationed here suffering but very little from disease of late years.

FORT NIAGARA, one of the oldest fortifications in the United States, is situated at the mouth of the River Niagara, on the south shore of Lake Ontario, in north latitude 43° 18′. It is fourteen miles below the Falls of Niagara, and thirty-two miles from Lake Erie. Mean annual temperature, 47° 90′, the greatest extremes being from 95° above to 5° below zero; variations 100°. The general character of the climate, being modified by the surrounding large bodies of water, is very favorable for health and longevity. The fruit and vegetable productions of most kinds flourish here luxuriantly.

DETROIT BARRACKS, at the city of Detroit, Mich., is situated in north latitude 42° 20′; having a mean annual temperature of 47° Fahrenheit. The surrounding country is flat. The soil is a stiff clay, combined with the carbonate of lime; hence, in the rainy season, the land is in a great degree saturated with water, and to a certain extent submerged. The smaller streams emptying into the Detroit River and Lake St. Clair are sluggish, bordered with extensive marshes, and in the autumn abounding with decayed vegetable matter. As may be supposed, from this brief outline of its topography, intermittent and remittent fevers, diarrhœa, and dysentry prevail among the troops, and also among the inhabitants of the city at certain seasons of the year."

FORT MACKINAC, located on the Island of Mackinac, in the straits connecting Lakes Huron and Michigan, in north latitude 45° 51′, has a mean annual temperature of 40.65° Fahr.; the temperature being modified by the surrounding waters of the great lakes Huron, Michigan, and Superior. This post is one of the most healthy in the United States, and the town is a great resort for invalids and seekers of pleasure during the summer months. This romantic island is about nine miles in circumference, and rises on its eastern and southern shore in abrupt rocky cliffs, the highest point, old Fort Holmes, being 318 feet above the lake, while the present fortress stands elevated 150 feet, overlooking the village and the surrounding waters.

FORT BRADY, situated at the Saut Ste. Marie, Mich., in north latitude 46° 30′, lies on the southern bank of the river or strait which connects Lake Superior and Huron. The river at this point is twenty feet below Lake Superior, and 580 feet above the ocean level. Here is a ship-canal with two locks, through which vessels of 1,000 tons and upwards can pass with safety. The mean annual temperature of this post is 40° Fahr. The coldest winter month (February) had a mean of 4° Fahr., and the warmest summer month (August) had a mean of 63°. The greatest extremes being from 80° above to 32° below zero, showing an extreme of 112° Fahr.

This old and important post and settlement, lying on the northern confines of the United States, is the limit of settled country toward the north. The Hudson Bay Company have a post on the opposite side of the river, surrounded by a few dwellings, but to the northward, except near the river, are no dwellings to be found, a wild expanse of country extending north to Hudson Bay, some four or five hundred miles distant. The Chippewa tribe and other Indians, however, are to be found in this region, and far to the north and west. Many kinds of grain and vegetables come to perfection in this latitude, but the early frosts often disappoint the husbandman, rendering most crops very uncertain.

FORT SNELLING, near St. Paul, Min., is situated in north latitude 44° 53′, west longitude 93° 10′, on the west bank of the Mississippi, 2,050 miles from the Gulf of Mexico, by the course of the river. The mean annual temperature, deduced from a continuous series of observations for thirty years, is 44° 54′ Fahrenheit, with a maximum of 100°, a minimum of —30°, and an extreme of 130°, the mean annual range being 120°. The average annual fall of rain and snow is 25.43 inches; of which 6.61 fell in spring, 10.92 in summer, 5.98 in autumn, and only 1.92 inches in winter. The *summer* temperature of St. Paul is about the same as New York and Chicago, while the winter months are much colder.

"Fort Snelling, situated on the angle formed by the confluence of the St. Peter's and Mississippi Rivers, is elevated ninety-four feet above those waters, and 820 feet above the level of the Gulf of Mexico. The St. Peter's, a navigable stream, at its mouth, is 150 yards wide and sixteen feet deep; and the Mississippi, at this point, is about 400 yards wide, but is much less deep than the former, navigation being here interrupted by rapids and falls. The banks of the latter, up to the Falls of St. Anthony, a distance of eight miles, are about 200 feet high, the upper strata of which consist of limestone, and the lower of sandstone. Beyond the falls, the banks are less high, and the immediate valley of the river becomes more extended; navigation being resumed for upwards of 100 miles. The surface of the surrounding country presents an undulating prairie, studded here and there with 'islands' of timber. Large lakes, plentifully supplied with fish, are occasionally found. The soil, although sandy, is productive, producing the cereals and vegetables in great abundance. The climate is bracing and healthy; Minnesota being celebrated as a health-restoring region."

Fort Ripley, Min., is situated in north latitude 46° 10′; west longitude 94° 18′, upon the west bank of the Mississippi, elevated twenty feet above the river, and about 1,100 feet above the Gulf of Mexico, being the most northern post on the Mississippi River. The climate is subject to great variation, as will be seen by reference to the meteorological register. The coldest month, January, had a mean of 7° Fahr., and the hottest month, July, 67°. The extremes of temperature observed are 96° in August and —39° in January, 1852, showing a variation of 135 degrees of temperature, being the greatest of any recorded locality within the United States. This section of Minnesota, no doubt, being influenced by cold currents of air descending from Hudson Bay and the Arctic regions; while the western and more southern portions of the State are, no doubt, favored by a climatic influence proceeding from the Pacific coast, across the Rocky Mountains, in British America—hence the favorable climate of the Red River country of the north.

"Different kinds of oak and pines constitute the prevailing forest growth of this region. The sugar maple abounds in some places. The chestnut, walnut, and beach are unknown, as is every species of fruit tree, wild or cultivated. The soil is generally a sandy alluvium. The land, at least when first cultivated, is more productive than might be supposed, being what farmers term 'warm,' and adapted to the short summers. Wheat, oats, potatoes, and other hardy vegetables flourish, while maize is considered a very uncertain crop, owing to the shortness of the season. The average depth of snow, during winter, is from two to three feet, which lies for about five months, from November to April.

"The phenomena of spring, when once begun, often progresses with great rapidity; and from the climate of winter, the region sometimes seems to pass at once into that of midsummer. Wild strawberries, which are found here in great abundance, ripen from the 20th to the last of June. Green peas are ready for use about the second or third week in July. During the months of September and October the weather is generally clear and delightful."

Dakota Territory.

[Sanitary Report by Surgeon T. C. MADISON, U. S. A.]

FORT RANDALL, an important military post, is situated on the right bank of the Missouri River in the territory of Dakota, north latitude 43° 01′; west longitude 98° 12′; altitude above the sea 1,245 feet. "The country in the immediate vicinity of the fort is very hilly; but, after you ascend these eminences or bluffs two miles to the southwest, presents an expansive level prairie. From the summit of any of these hills you have a most picturesque view of the surrounding country, the Missouri River and adjacent territory. The plateau on which the post is built is about one fourth of a mile from the river, the width of which, at this point, is about the same distance. The river bottom is not extensive, and the only timber to be found is along it and the Dry Ravine (which latter extends about 15 miles), and is suitable for fuel only. There are several cedar islands from 15 to 30 miles above, from which the best lumber can be procured, and either floated down in summer or hauled down upon the ice in winter. The soil is chiefly selicious, and productive only in small localities. Our gardens would have succeeded better but for the inadequate supply of rain and the myriads of grasshoppers, which made their appearance about the first of August, continued throughout the month, and almost annihilated everything possessing verdure.

"The forest trees are cottonwood, elm, ash, cedar, scrub oak, hickory, and box-elder. Fruit—plums, choke-cherries, wild raspberries, gooseberries, buffalo and service berries, and wild grapes. Plants—sunflower, wild artichoke, wild onion, polar or magnetic plant, and a great variety of others. The wild rose is also most abundant, and a weed, called rattlesnake weed."

METEOROLOGY.—"The climate is uniformly cold in winter and not unpleasantly hot in summer. The lowest thermometrical observation was —26°, on the 17th January, and the highest 104°, on the 11th of August, 1857. The latest frost was on the 15th of May, and the earliest was on the 28th of September, when the thermometer was 34° Fahrenheit. There was no rain from the 24th of October, 1856, to the 4th of March, 1857. The annual precipitation of rain from October 1, 1856, when meteorogical observations were first commenced, to September 30, 1857, was 11.64 inches; and that of melted snow, 4.23 inches; total, 15.87 inches. It rarely rains during the winter months of any year. The wind blows almost incessantly, and most frequently from the north in winter, and south and southeast in summer. The snow storms are quite frequent, and usually most violent. The Missouri River was frozen over on the 2nd of December, and remained blocked until the 28th of March. It, however, requires much time after the river opens before it is safe navigation, owing to floating ice."

DISEASES AND DEATHS.—"The ratio of mortality per 1,000 of mean strength, exclusive of two deaths by cholera, was only 12, being lower than any other known military station on record. I do not believe that a single case of genuine intermittent fever has originated at the post. We have, at present, one, and the only case of febris typhoides, which might have occurred in the healthiest parts of Virginia. The climate is certainly unfavorable to the development of *phthisis* and the affections of the chest generally. Consumption, from obvious reasons, must be more frequent among the Indians in the vicinity, who are more exposed to atmospherical vicissitudes and not as well clothed and fed as soldiers.

"The only disease about which we need feel the slightest apprehension is *scorbutus*, the chief disease from which the troops have suffered from the commencement of the Sioux expedition up to the present time; but, after an experience of more than two years in the treatment of scurvy and its complications, I am compelled to believe that the use of too much salt meat is the true cause. The Indians eat nothing save fresh game or dried buffalo meat, and put up for winter quantities of dried plums, buffalo berries, &c.; hence their immunity.

"The whole district of country, east of the Rocky Mountain range, extending to the Missouri River, and running through several degrees of latitude, may be considered as a remarkably healthy region, where fevers, consumption, and throat diseases are seldom known, or prove fatal."

II.—Meteorological Table.

SHOWING THE PRINCIPAL CITIES AND MILITARY STATIONS IN THE UNITED STATES HAVING A MEAN ANNUAL TEMPERATURE BETWEEN 47° & 53° FAHR.

Stations, etc.	Latitude.	Longitude.	Altitude.	Yearly Mean.	Four Seasons.			
					Spring.	Summer.	Autumn.	Winter.
			Feet.	° Fahr.	° Fahr.	° Fahr.	° Fahr.	° Fahr.
Boston, Mass...........	42°21′	71°03′	50	48.90	46.30	69.10	51.60	28.90
New Bedford, Mass......	41°31′	70°56′	40	48.40	47.70	67.10	52.00	29.80
Nantucket, "	41°17′	70°06′	30	50.40	44.60	67.80	55.30	33.70
Springfield, "	42°06′	72°35′	200	48.10	45.44	70.43	51.72	24.50
Newport, R. I...........	41°30′	71°20′	30	49.90	45.90	68.80	53.50	31.30
New London, Conn......	41°21′	72°06′	25	49.60	46.40	69.30	52.90	29.90
New Haven, "	41°18′	73°38′	60	50.82	47.54	69.78	52.51	33.44
Jamaica, Long Is.........	40°41′	73°56′	50	50.00	47.30	68.90	51.80	30.40
New York,..............	40°42′	74°00′	25	51.00	48.70	70.10	54.50	31.40
West Point, N. Y........	41°23′	74°00′	167	50.50	48.70	71.30	53.20	29.70
Albany, "	42°31′	73°44′	130	48.20	46.70	70.00	50.00	26.00
Ithaca, "	42°27′	76°30′	417	48.12	46.37	68.12	49.35	28.62
Newark, N. J...........	40°45′	74°10′	30	50.50	46.74	71.25	52.24	31.80
Trenton, "	40°13′	74°45′	50	51.10	49.40	70.70	52.10	32.00
Philadelphia............	39°57′	75°13′	60	52.00	50.60	71.50	52.20	32.80
Lancaster, Penn.........	40°02′	76°21′	300	51.40	50.90	71.20	52.10	31.60
Harrisburg, "	40°16′	76°50′	300	49.50	49.60	60.66	50.00	28.66
Carlisle Barracks, Penn..	40°12′	77°14′	500	51.10	49.80	72.10	52.10	30.40
Pittsburgh, " ..	40°32′	80°02′	700	50.80	50.00	71.40	51.40	30.60
Meadville, " ..	41°38′	80°08′	1,000	50.26	48.59	71.32	51.84	29.27
Columbus, Ohio.........	39°57′	83°03′	740	52.00	47.60	72.70	54.20	35.00
Cleveland, "	41°30′	81°47′	640	49.76	.....	.....	.....	.....
Steubenville, Ohio.......	40°25′	80°41′	670	51.70	51.10	72.20	52.70	30.90
Toledo, "	41°45′	83°36′	565	50.00	.....	.....	.....	.....
Marietta, "	39°25′	81°31′	630	52.00	51.50	71.00	53.60	32.00
Hillsboro', "	39°15′	83°30′	1,130	50.70	50.76	69.66	51.20	31.20
Chicago, Ill.............	41°52′	87°35′	590	47.00	45.00	68.00	48.85	26.00
Fort Armstrong. Ill.....	41°30′	90°40′	530	50.00	50.50	74.00	51.00	24.88
Augusta, "	40°12′	90°58′	.....	50.50	.....	.....	.....	.....
Dubuque, Iowa..........	42°30′	90°50′	680	48.38	47.38	72.56	49.48	23.76
Keokuk, "	40°25′	91°21′	.....	51.00	.....	.....	.....	.....
Muscatine, "	41°25′	91°05′	586	49.30	46.38	78.43	50.44	27.15
Fort des Moines, Iowa...	41°32′	93°38′	780	49.70	51.50	71.55	47.00	28.90
Council Bluffs, Neb......	41°30′	95°48′	1,250	49.30	49.28	74.76	51.36	21.73
Fort Kearny, "	40°38′	95°57′	2,360	48.50	46.39	71.64	48.70	25.65
Fort Laramie, Dakota...	42°12′	104°47′	4,520	50.00	46.84	71.90	50.30	31.00
Camp Floyd, Utah......	40°13′	112°08′	4 860	48.65	47.17	75.65	48.44	23.32
Santa Fe, N. M.........	35°03′	107°14′	6,846	50.50	49.68	70.46	50.50	31.60
Las Vegas, "	35°35′	105°16′	6,418	49.14	48.30	67.35	48.34	32.56
Fort Jones, Cal.........	41°36′	122°52′	2,570	51.40	49.00	67.30	52.12	33.78
Fort Humboldt, Cal.....	40°46′	124°09′	50	51.40	51.80	57.55	52.88	43.35
Fort Umpqua, Or.......	43°52′	124°09′	.....	51.38	49.18	60.00	52.45	44.00
Fort Dallas, "	45°36′	120°55′	350	52.50	53.00	70.30	52.20	35.50
Fort Yamhill, "	43°37′	123°32′	.....	49.20	46.26	60.41	49.38	36.74
Fort Vancouver, "	45°40′	122°30′	50	52.50	51.80	65.60	53.50	39.50
Astoria, "	46°11′	123°48′	50	52.00	51.00	61.58	53.70	42.00
Fort Cascades, W. T.....	45°30′	121°30′	.....	49.68	48.30	64.92	51.58	33.90
Fort Steilacom, "	47°10′	122°25′	300	49.80	47.20	62.89	50.69	39.50
Olympia, "	47°00′	122°30′	.....	51.00	.....	.....	.....	.....
Victoria, Vancouver Is...	48°30′	123°00′	.....	50.00	.....	.....	.....	.....

Climatic Features.

The belt of temperature having a mean varying from 47° to 53° Fahrenheit, lies mostly between 40° and 43° north latitude, on the Atlantic coast, deflecting southward on crossing the Rocky Mountains, and then rising on the Pacific coast from 40° to 48° north, extending from Cape Mendocino to Puget Sound. In passing from the Atlantic to the Pacific Ocean it extends through 54 degrees of longitude.

This region stands unsurpassed as regards a favorable climate, fruitful soil, and rich mineral productions. On this favored belt is to be found the most dense and active population of any part of the Union. Here the cereals and grasses are produced in the greatest abundance, constituting the principal articles of export. Indian corn, wheat and hay form the principal items. The forest and mines also yield a rich return, as well as the agricultural products.

The four principal cities on the seaboard, Boston, New York, Brooklyn, and Philadelphia, with their two million of inhabitants, possess and exercise a preponderating influence from the Atlantic to the Pacific Ocean. New York alone stands unrivalled as a commercial mart, where the products of every clime are to be found in abundance; her commerce whitens every sea and seeks every port of the habitable world. The cities of the interior lying in the Valley of the Mississippi alike are alive with industry, while the broad spread country teems with rich agricultural products, particularly between the base of the Alleghany Mountains and the Rocky Mountains, embracing the Mississippi Valley north of the Ohio River.

On the western slope of the Rocky Mountains new States are springing into existence, no doubt soon destined to contain a dense population, who will possess all the energy and facilities of the inhabitants of the Atlantic cities. Here the mines and the soil will yield their rich reward.

The North Pacific, possessing a favorable climatic influence, with the adjacent shores of America and Asia, affords a new field for commerce. From this part of the republic, China, Japan and the long-sought East Indies are open to the American flag—that, too, by important treaty stipulations of late date. Puget Sound and adjacent waters afford ample accommodation for all the mercantile fleets of the world, being surrounded by immense quantities of timber valuable for ship-building.

The climate on the seaboard, from Massachusetts Bay to Delaware Bay, is well understood and appreciated by thousands of invalids who annually seek the favorite health-restoring resorts scattered along the coasts of Massachusetts, Rhode Island, New York and New Jersey. The interior of the country, east of the Atlantic range of mountains, is equally well understood and acknowledged to have a favorable climate. The Valley of the Mississippi is still subject to fevers of an intermittent type, but less subject to some other diseases, altogether comparing favorably with other parts of the Union. Iowa and the northern part of Missouri, situated between the Mississippi and Missouri rivers, partakes of the same character as the more eastern portion of this great valley, situated on the same parallel of latitude.

Military Posts.

WEST POINT, one of the most favored locations in regard to climate and healthy influences, is situated on the west bank of Hudson River, in north latitude 41° 23′, west longitude 74°, about midway in that part of the river called the "Highlands," 52 miles distant from the city of New York. The public buildings are on a plain about a mile square, having in its rear a range of hills of from 700 to 1,400 feet in height. On each side of this plain there are ravines that serve to carry off the great floods of water, which descend from the adjacent hills after heavy rains or spring freshets. The soil is gravelly, with frequent ledges of rock, either just below the surface, or rising above it in the form of boulders.

The mean annual temperature of this post, as determined by observations continued for thirty-one years, is 50° 50′ Fahr., with an extreme range of 110°, rising in summer to 100°, and falling in winter to 10° below zero. The prevailing winds are from the N. W. and S. The annual quantity of rain is about 52 inches. There are no diseases which can be considered peculiar to this station; acute inflammatory diseases are rare. The spring and autumn are most productive of severe catarrhal affections and rheumatism; the summer, of disorders of the digestive organs; and the winter is decidedly the most healthy period of the year. In addition to the officers, cadets and soldiers permanently residing at this post, with their families, altogether numbering about 800 souls, the hotel, during the summer months, is thronged with visitors from every section of the Union, enjoying the salubrity of the climate.

FORT LARAMIE, Dakota Ter., situated in north latitude 42° 12′, longitude 104° 31′, is a post of much importance, being on the most favored line of travel across the continent, where emigration flows westward toward the Pacific States and Territories. Its altitude is 4,519 feet above the level of the sea. The mean annual temperature is 50 degrees Fahr., rising in summer to 100 degrees and falling in winter to 20 degrees below zero. The mean annual precipitation of rain and snow is 20 inches. The soil in the vicinity appears to be sterile, owing, no doubt, to the extreme dryness of the air and almost total absence of dews. The mean annual temperature is 50° 32′ Fahrenheit. The maximum temperature during the year was 92° in July, and minimum 22° in December, showing an extreme of 70 degrees, being only about half as much variation as occurs in the same parallel of latitude on the upper Mississippi and Atlantic side of the continent. The annual quantity of rain that falls varies from 34 to 66 inches; average, for a number of years, 50 inches.

CAMP SCOTT, or BRIDGER'S FORT, the wintering place of the army of Utah in 1857-8, is situated in latitude 41° 18′ N., longitude 110° 32′ W. from Greenwich; altitude 7,800 feet. Fort Bridger, an Indian trading-post, lies on Black's Fork, a tributary of Green River. This mountain stream is of crystal clearness and purity, and is immensely valuable in this arid and thirsty region. The valley has an average width of about one mile and is separated from the higher table land by a range of irregular sand hills. During spring and summer the valley is covered with an abundant herbage, and offers a most striking contrast to the barren waste on either side.

Assistant Surgeon Barthalow, in his Sanitary Report, remarks:—"This region, as well as the Great Plains, like the steppes of Tartary, is adapted only to herds and grazing, and a nomadic population of savages or Indian traders, with their squaws and cattle. It can never become a nursery of civilized heroes; and thus, in the New World, may be revived, in somewhat the same form, the ancient patriarchal life, now almost extinct in the old.

"If we form an opinion of the mountain men from the reports of poetic explorers, we would probably accord them many virtues—integrity, steady friendship, a noble sense of justice, and high personal bearing. I did not find the original of this description in real life. They have some of the good qualities of the Bedouin Arab, many vices to which he is a stranger, but not many of the virtues of a good citizen. A country like the Great Plains, which has its analoque in the deserts of the East, would be incomplete without that other characteristic—a wandering people having a strong thirst for plunder, and acknowledging no law but the *lex talionis*.

"My observations on the climatology of this country have had but a limited scope, extending through the fall to midwinter. I have been very agreeably impressed, thus far, with the comparative mildness of the climate. Minus 18° Fahrenheit is the lowest degree to which mercury has yet fallen, and that was during the month of November, a degree of cold not since experienced.

"One distinguishing feature of this climate is its equability and dryness. No sudden transitions have been observed, and during the winter proper, whilst the cold has at no time been severe, the thermometer has rarely risen above the freezing point. The absence of moisture is well shown by the dryness and contraction of all kinds of wood-work, and the freedom of surgical instruments and arms from the slightest traces of rust.

"However deficient this region may be in the more humanizing influences, it has at least the great merit of being extremely favorable to health and longevity. There are two diseases which occasionally prevail—erysipelas, in an epidemic form, and mountain fever. Besides these I know of no diseases which may be said to have characters peculiar to this country.

"A question well worthy of consideration: Is this climate adapted to the amelioration and cure of the tubercular diathesis? As phthisis is annually on the increase in the United States, and as the subject of its hygenic management proves to be more important than the treatment by medicaments, the consideration of the climate is, necessarily, of the first consequence. In my late report I stated the beneficial influence of the journey over the plains upon those in whom a phthisical tendency was marked and imminent. The purity of the atmosphere and the equability and dryness of the climate are conditions highly favorable to such improvement."

Assistant Surgeon Wood, in a report upon the above subject, remarks:—"The climate of those broad and elevated tablelands, which skirt the base of the Rocky Mountains on the east, is especially beneficial to persons suffering from pulmonary disease, or with a scrofulous diathesis; that more is due to the climate itself, is shown by the fact that among the troops stationed in this region (whose habits are much the same everywhere) this class of disease is of very rare occurrence.

"From these facts it appears to me evident that to the subject of an hereditary or acquired predisposition to consumption, the Great Plains and the mountains offer more certain relief than any other climate in our country."

FORT STEILACOOM, Washington Territory, situated in lat. 47° 10′ north, and long. 122° 23′ west from Greenwich, is one mile east from Puget Sound, and about 300 feet above the level of the sea. "The Cascade range of mountains, running north and south, is east distant about thirty miles, and one of its snow-capped peaks, having an altitude of 14,000 feet, is directly in view; while the snow-peaks of the Olympian range, distant about forty miles on the west, are also visible, along the sound, varying from one to two miles in breadth, and near the mountains are dense and lofty forests. The country immediately around is composed of beautiful undulating prairies, intersected by numerous small streams, which have their sources in the fresh-water lakes with which the prairies are interspersed. The prairies are separated from each other, and surrounded by dense and almost impenetrable forests, while they are interspersed with numerous groves of oak, which give them a most beautiful and park-like appearance. Springs of pure water are abundant, both in the prairies and woodlands. The soil in this vicinity, particularly of the prairies, is composed of a mixture of sand and gravel, and is almost entirely unfit for agricultural purposes, except on the margins of the streams and in low places near the lakes. The soil of the woodlands is of a different nature, being a kind of loam; but so dense are the forests that years will elapse before it is brought into requisition."

"The forests are composed of pine, hemlock, fir, cedar, oak, maple, ash, cottonwood, yew, dogwood, alder, aspen, crab-apple, hazel, &c. The pine, cedar and fir growing on the high-lands; the oak on the prairies; and the maple, ash, cotton-wood, &c., on the bottom-lands near the streams. Blackberries, raspberries, gooseberries, cranberries, whortleberries, strawberries, dewberries, and currants, are very abundant. A species of fern is very common in every section of the country, and the uva ursi covers the ground on the margin of all the prairies. The country abounds with animals, which afford excellent amusement to the sportsman, and a principal article of food to the Indians. The birds of different kinds are numerous. During the latter part of autumn great numbers of swans, geese, ducks and cranes, make their appearance on their way to more southern latitudes, and are not generally seen again until the opening of spring, when they are returning north.

"The climate of this country, as regards temperature, possesses a medium between hyperborean, cold and intertropical heat. The seasons may be said to be divided into the rainy and dry. From the middle of October to the first of April is the rainy season. During April and May there are frequent showers, after which it rains occasionally, but seldom sufficient to thoroughly wet the ground. Snow falls to a greater or less extent every winter, but seldom remains on the ground over two or three days. Ice seldom forms over an inch thick.

"The prevailing winds during the rainy season are southerly; and during the dry, northerly. Southerly winds are always indicative of rainy weather, and northerly of dry. The country generally being high and dry, the lakes, all of pure fresh water, no marshes or alluvial bottoms being in the vicinity, diseases of a malarious origin are almost entirely unknown. Catarrhs, rheumatism, and diseases incident to exposure to cold, combined with moisture, are quite common during the rainy season."

Fort Dalles, Oregon, is situated in north latitude 45° 36′, west longitude 120° 55′, being elevated 350 feet above the ocean; mean annual temperature, 52° 79′ Fahr.; average annual fall of rain 15 inches. "The post at the Dalles of the Columbia, so called from the river being compressed by the encroaching rocky cliffs into a narrow *cut*, through which the whole volume of water rushes, is a few miles above the entrance of the river into the mountain ridges, jutting out from the Cascade range, and two hundred miles from the ocean. Like all of middle Oregon, this is an admirable grazing region; but, owing to the long dry season, is scarcely susceptible of cultivation. This is eminently a volcanic region, basalt and basaltic conglomerate abounding. The position may be considered perfectly salubrious. Within the experience of the residents in the vicinity, fevers of every description, or any local diseases, are entirely unknown."

Astoria, Oregon, situated on the south bank of the Columbia River, near its entrance into the Pacific, in north latitude 40° 11′, west longitude 123° 48′, has a mean annual temperature of 52° Fahrenheit. The coldest winter month (January) had a mean of 40° Fahr., and the warmest summer month (August) had a mean of 64°. "The most noticeable feature in the climate of Astoria is its equability. The summers are cold, dry and healthy; the winters stormy, rainy and disagreeable, but mild. The aurora is frequent during the spring, and intensely brilliant. Thunder storms are not frequent nor severe. Astoria, and the shores of the ocean southward, afford pleasant places of resort from the hot, dusty and malarial summer atmosphere of Portland and other places situated in the Willamectte Valley; and in the future growth of the country, Clatsop Plains will be on these western shores what Newport and Cape May are on the Atlantic.

"The soil in the vicinity of Astoria is, for the most part, a heavy red and black clay, mixed with some gravel, which becomes, during the rainy season, soft and sticky; and in the summer dry and fissured; the beach is covered with pebbles and conglomerate of clay and lime, enclosing petrified shells and marine animals.

"Pre-eminent among the forest trees are those of the pine tribe; three varieties found, including the *abis Douglasic*, often attaining incredible height and circumference; also, yew-leaved hemlock, red cedar, and a large-leaved maple, not found east of the Rocky Mountains. There are innumerable varieties of bushes and creeping plants, many of them producing delicious fruit in abundance. The potato, turnip, beet, and cabbage, are largely cultivated, and attain an enormous size and great perfection. The grasses, growing on the tide-lands, are tender, and afford a nutritious food to animals, who are able to keep in good condition throughout the year by grazing."

FORT CASCADES, Washington Territory, "is situated," says Surgeon J. K. Barnes, "on the north bank of the Columbia River, at the lower terminus of the portage around the rapids, in latitude 45° 35′ north, longitude 121° 30′ west. The immediate site of the post is a small plateau on the western slope of the Cascade range, surrounded on all sides by precipitous mountains, open only to the east and west by the river gorge, elevated but a few feet above the highest water level, and bearing unmistakeable marks of having at some remote period been a portion of the river bed. A line of isolated volcanic peaks, whose summits are covered with perpetual snow, extending in a direction nearly north and south, marks the western border of the elevated plateau between Pitt River and the Des Chutes Valley.

"Experience has shown a great difference in the seasons at points on the east and west side of the Cascade range upon nearly the same latitude. The spring is three weeks earlier at Fort Dalles than at Fort Vancouver, and five weeks earlier than at the mouth of the river. Some fruits and vegetables that come to great perfection at the Dalles, scarcely mature at Fort Vancouver, and cannot be successfully cultivated at Astoria, on the Pacific coast."

The difference in the amount of rain that falls in different localities in Oregon, and Washington Ter., is very remarkable, by far the greatest quantity falling near the sea-coast. The annual fall at Fort Orford is 68 inches; at Astoria, 60 inches;

Fort Steilacoom, 52 inches; Fort Vancouver, 45 inches, and at the Dalles, east of the coast range, 15 inches: average, 50 inches.

Ascent of Mount Hood.

This gigantic mountain of the Cascade range of mountains, situated in Oregon, about 45° 20′ north latitude, is supposed to be the highest peak in the United States, if not higher than Mt. St. Elias in Russian America.

The summit was reached in August, 1866, by a party of explorers, who give the following reliable information. "We have reached the summit of Mount Hood, and here succeeded in melting snow and boiling the water with the spirit lamp. Water here boils at 180° Fahrenheit. According to Prof. Porter's rule (given in his Chemistry)—and also by those in the Encyclopedia Brittanica—550 feet should be allowed for every degree. 32° by 550 gives the height of Mount Hood to be 17,600 feet above the sea. The highest point of vegetation is 11,000 feet (where commences the snow line.) The highest of trees, of stunted pine, is 9,400 feet. The ascent was difficult and hazardous."

Prof. A. Wood, one of the party, pronounces the Alpine flora of Mount Hood to be of a very interesting character. He finds at least thirty plants peculiar to this mountain, many of which are undoubtedly new. A considerable attention has been given to the geological, mineralogical and volcanic character of the mountain. The crater is about 1,000 feet below the summit, on the south side. Although at present not active, it is continually emitting a column of sulphurous steam and smoke, the odor of which is very nauseating.

The extreme summit is described as a circular ridge of three or four hundred yards in length, having its outward curve to the north. On this ridge are three or four eminences rising a few feet above the average of the ridge. The highest of these is the one to the east, though it is only a few feet higher than the others. The snow upon them was from six to ten feet in depth, and only in one place did a rock project through it. That was the extreme summit of the highest point of the ridge.

The scene around was overpoweringly indescribable. It would require the canvas and brush, and years of toil, to give an idea to the eye—yet a few general observations may be taken. The first is the Cascade range itself. From south to north its whole line is at once under the eye, from Diamond Peak to Mount Rainer, a distance of not less than 400 miles. Within that distance are to be seen Mount St. Helens, Adams, Jefferson, and the Three Sisters, making, with Mount Hood, eight snowy mountains.

III.—Meteorological Table.

SHOWING THE PRINCIPAL CITIES AND MILITARY STATIONS IN THE UNITED STATES HAVING A MEAN ANNUAL TEMPERATURE BETWEEN 53° & 60° FAHR.

Stations, etc.	Latitude.	Longitude.	Altitude.	Yearly Mean.	Four Seasons.			
					Spring.	Summer.	Autumn.	Winter.
			Feet.	° Fahr.	° Fahr.	° Fahr.	° Fahr.	° Fahr.
Fort Delaware..........	39°25′	75°34′	10	55.00	53.50	75.90	58.50	36.28
Baltimore (Ft. McHenry)	39°17″	76°35′	36	54.36	52.70	74.32	56.20	34.24
Frederick City, Md......	39°24′	77°18′	.....	53.34	50.37	78.32	53.13	34.81
Annapolis " ..[Obs.	38°58′	76°27′	20	55.40	53.78	75.30	57.76	34.82
Washington, D. C., Nat.	38°53′	77°02′	60	56.00	55.70	76.30	56.40	36.00
Fort Washington, Md...	38°43′	77°06′	60	57.80	57.40	77.76	58.90	37.36
Ashland, Va............	38°38′	81°57′	.....	58.00	.....	.....	.....	.....
Richmond, Va..........	37°20′	77°25′	120	59.27	57.39	77.90	60.40	41.38
Fort Monroe, Va........	37°00′	76°18′	10	59.14	56.87	76.57	61.68	41.45
Smithfield "	36°50′	76°41′	.....	57.70	55.91	76.11	56.64	42.12
Lynchburg, "	37°30′	79°07′	575	57.00	.....	.....	.....	.....
Kanawha, "	38°53′	81°25′	.....	53.47	.....	.....	.....	.....
Gaston, N. C............	36°32′	77°45′	.....	59.00	56.05	77.60	57.39	40.96
Murfreesborough, N. C..	36°30′	77°06′	.....	60.00	.....	.....	.....	.....
Chapel Hill, "	35°54′	79°17′	.....	59.30	57.81	78.14	58.57	42.64
Knoxville, Tenn........	35°56′	83°58′	960	55.70	55.80	70.80	56.70	39.30
Nashville, "	36°10′	86°49′	530	58.50	59.80	77.40	57.10	39.50
Glenwood, "	36°28′	87°13′	480	56.63	.....	.....	.....	.....
Memphis, "	35°08′	88°00′	400	60.50	61.00	78.00	60.00	42.60
Huntsville, Ala.........	34°45′	86°40′	550	60.00	61.00	76.00	60.80	42.20
Bardstown, Ken........	37°42′	85°18′	.....	56.23	55.55	76.82	55.10	40.72
Springdale, "	38°07′	85°34′	570	53.00	.....	.....	.....	.....
Newport, "	39°00′	84°29′	500	53.70	53 80	73.70	53.60	33.80
Paris, "	38°16′	84°07′	810	54.00	.....	.....	.....	.....
Louisville, "	38°08′	85°25′	600	54.50	55.00	73.10	54.70	36.30
Cannelton, Ind..........	37°58′	86°40′	450	55.88	.....	.....	.....	.....
New Harmony, Ind.....	38°08′	87°50′	320	55.73	.....	.....	.....	.....
Cincinnati, Ohio........	39°06′	84°29′	543	53.80	53.70	74.00	53.90	33.70
Portsmouth, "	38°45′	82°56′	540	55.00	54.80	74.20	55.00	36.00
St. Louis, Mo...........	38°40′	90°05′	450	54.50	54.15	76.19	55.44	32.27
Fort Scott, "	37°45′	94°35′	1,000	54.50	54.78	74.95	55.27	33.00
Fort Smith, Mo.........	35°23′	94°29′	460	60.00	61.29	77.60	60.00	41.12
Lawrence, Kan.........	38°58′	95°12′	800	54.50	55.54	77.06	53.11	32.38
Fort Leavenworth, Kan..	39°21′	94°44′	900	52.78	53.78	74.00	53.66	29.64
Fort Riley, " ..	39°00′	96°30′	1,000	52.70	55.00	80.36	55.46	20.00
Fort Gibson, Ind. Ter....	35°47′	95°10′	560	60.00	61.00	79.00	61.50	41.00
Fort Arbuckle, "	34°27′	97°09′	1,000	60.00	61.00	78.00	62.00	40.50
Albuquerque, N. M......	35°06′	106°38′	5,000	56.32	55.90	74.90	57.33	37.15
Fort Stanton, "	33°30′	105°38′	.....	54.00	52.52	71.00	55.65	36.00
Fort Conrad, "	33°34′	107°09′	4,576	59.40	59.80	77.46	60.80	39.50
Laguna, "	35°03′	107°14′	6,000	55.00	52.50	75.20	56.80	35.98
Fort Webster, "	32°47′	108°04′	6,350	54.80	52.90	71.70	53.48	41.29
Fort Buchanan, Ark.....	31°40′	111°35′	5,330	58.00	55.47	75.59	60.48	41.50
Great Salt Lake, Utah...	40°46′	112°06′	4,350	53.24	51.73	75.92	.....	32.00
Fort Tejon, Cal.........	34°55′	118°53′	.....	57.90	53.00	73.68	62.55	42.38
Monterey, "	36°36′	121°52′	140	56.30	54.00	60.64	57.30	51.20
Sacramento, "	38°33′	121°20′	50	59.80	59.16	72.85	61.27	46.29
Benecia, "	38°03′	122°08′	64	58.29	56.54	67.00	60.57	49.00
San Francisco, Cal......	37°48′	122°26′	150	55.80	54.50	60.30	56.83	50.80
Fort Orford, Oregon.....	42°44′	124°29′	50	53.60	51.82	60.00	55.22	47.48

Climatic Features.

The belt of territory having a mean, varying from 53° to 60° Fahrenheit, lies mostly between 40° and 36° north latitude on the Atlantic slope, and passes westward across the Valley of the Mississippi towards the Rocky Mountains. It then deflects southward, and again rises as it approaches the Pacific Ocean, being unsurpassed for fertility of soil and rich mineral productions. Here is produced Indian corn, wheat, tobacco, hemp, and almost every variety of the grape, producing large quantities of wine.

As regards climate and health it is very much varied according to altitude. Fevers exist on the Atlantic sea-board, in different forms, while for the most part the country to the westward is salubrious and invigorating. It is well watered, being favored with many navigable streams, that of the Ohio River passing near one thousand miles from east to west, before entering into the Mississippi. The population is also dense and fast increasing in all the elements of wealth. The inexhaustible beds of bituminous coal here deposited is alone a source of immense profit, while the richest gold fields of California, also silver mines on the eastern slope of the Sierra Nevada, lie within this belt of territory, extending from the Atlantic to the Pacific Ocean.

Military Posts.

FORT MONROE, situated a few miles north of Norfolk, Va., in lat. 37° 2′ north, long. 76° 12′ west, occupies the extremity of a level sandy beach, known as *Old Point Comfort*, standing on the western shore of the Chesapeake Bay, having an open exposure of water surface and within reach of the sea influence. "The geological formation of this peninsula is that of ocean sand resting upon clay. The general aspect of the country, both proximate and distant, is uniformly low and flat. The very limited extent of barren sandy ground, and the consequent almost entire absence of soil immediately around the fort, are necessarily productive of a circumscribed vegetation in the shape of trees, shrubs, plants or grasses. Within the enclosure of the work, however, and growing sparsely on the grounds immediately contiguous to it, the live oak, in its evergreen foliage, is quite conspicuous, having been here retained and preserved, for purposes of ornament, in the original clearing of the grounds; this point being the extreme northern limit in which it is found on the Atlantic coast.

"The geographical locality of Old Point Comfort would naturally denote the climate of this position to be that of an intermediate, mild, or temperate one; and such the leading meteorological phenomena of the different seasons impress upon it. The winters are open and mild, but seldom with such depressions of temperature as give rise to snow. Although the entire district of country contiguous to this important military post is annually subjected to the calamity of having rife every form of malarial fever, the immediate locality of Fort Monroe may justly claim exemption from this evil; this in fact being a favorite summer resort for invalids." The coldest winter month, January, had a mean temp. of 40° Fahr., and the warmest summer month, July, had a mean of 78°. Average annual temperature 59°. Annual fall of rain, 45 inches, the quantity being remarkably even during the different months of the year."

The temperature of Norfolk, Va., may be considered very similar to Old Point Comfort, although somewhat warmer and subject to malignant fevers.

"JEFFERSON BARRACKS," says Surgeon De Camp, "is situated on the right bank of the Mississippi River, ten miles below the City of St. Louis, upon a sloping ridge, elevated about 100 feet above the river, and distant from it about 150 yards. The ground continues to rise gently for one mile west of the barracks, attaining an elevation of about 200 feet above high water mark. The surface of the earth for many miles south and west, and for four or five miles north, is undulating; and as it frequently rises into abrupt hills with deep ravines, the drainage is perfect. The soil is a rich loam, based upon clay, with a substratum of limestone. The country around, with the exception of the public grounds, remains (1839) covered with a heavy growth of timber. Indications of lead are common, and stone-coal is found in abundance within a few miles of the post.

"The river is about one mile wide, and upon the opposite side, in Illinois, is the great 'American bottom,' which is said to be sixty miles long, and, on an average, seven miles wide. On the river it is skirted with forests, varying in breadth from a half to one mile, whilst the remaining space to the high ground consists principally of prairie, covered with a luxuriant growth of grass. This prairie is chequered with numerous lakes; and as the evaporation of the water during the latter part of the summer exposes the surface of the subjacent soil, a fruitful source of disease is engendered, the influence of which is sensibly felt at the barracks. The water used at the post is usually that of the river; but in summer it has been common to resort to wells and springs, the waters of which are prejudicial to health, causing, in many persons, bowel complaints.

"The buildings used as barracks are built of stone, and occupy three sides of a square. The position, with regard to health, is as good as any which could have been selected upon the river bank; but, from an acquaintance with diseases of this country far more than twenty-two years, I am able to state that fewer cases occur, and, when they do, they are much milder in their character, when removed from the river."

St. Louis Arsenal is situated within the incorporated limits of the city, in latitude 38° 37′, longitude 90° 15′. "It is elevated above the Mississippi, at an ordinary stage of water, about 12 feet. The main channel of the river runs east of the arsenal grounds about three quarters of a mile, an island intervening, and a small channel, at times, has but little water in it, leaving exposed a broad surface of a muddy deposit, covered to a considerable extent with decayed wood, brought by each rise of the upper rivers from the wood-drifts. Being situated at the lower end of a large city, where great numbers of dead animals are thrown into the river, not a few of them are deposited in the vicinity of the arsenal when the river is low. In addition to the above causes of disease, a little below the arsenal, on the opposite side of the river, there is a chain of lakes, which in midsummer become very low, leaving tracts of muddy ground exposed to the action of the sun. All these causes are fruitful sources of malaria, producing fevers of an intermittent type. From long residence in this vicinity," says Surgeon De Camp, "I am convinced that when the south and southeast winds prevail, those causes are made operative to a considerable extent. Elevation above the river, especially if a little removed from it, tends, to a great extent, to render the above causes inoperative. Persons residing in the rear of the arsenal, on high ground, are seldom attacked with chills and fevers; but, those who have recently come to reside at the arsenal, from distant parts of the country, are subject to this disease."

Fort Leavenworth, Kansas, situated in north latitude 39° 11′, west longitude 94° 44′, is elevated 896 feet above the Gulf of Mexico. The mean annual temperature is 53° Fahrenheit; the coldest month, January, having a mean of 28°, and the hottest month, July, 77° Fahr. Average annual fall of rain and snow, 30 inches. "This important military post is located on the right bank of the Missouri River about 500 miles above its confluence with the Mississippi. As the Missouri here is not more than 300 yards wide, being one of its narrowest points, the water is deep and current rapid. This mighty river is at times navigable for steamboats 1,750 miles above the fort, and always, unless obstructed by ice, to its mouth. "The soil, which is quite productive, consists of a sandy loam, covered with a rich vegetable deposit, the whole based on a stratum of clay and limestone. The forest abounds in trees valuable for timber or fuel. With the exception of pine, almost all kinds are to be found."

FORT SCOTT, Kansas, is situated in north latitude 38°, west longitude 94° 30′, four miles west of the Missouri line, and upon the military road from Fort Leavenworth to Fort Gibson, Indian Ter. The mean annual temperature is 54½° Fahrenheit. The coldest month, January, had a mean of 33°, and the warmest month, July, 77°. Average annual fall of rain, 42 inches. "Owing to the physical conformation of the country, the climate is one of extremes of heat and cold, of dryness and moisture. After a long and debilitating summer, the winter, most frequently commencing abruptly with cold storms from the northeast, a succession of alternations, the mercury falling or rising 30° to 40° in a few hours. Springs and wells supply an abundance of good water, which rarely fails, even in the dryest seasons. An accurate examination of the country, for several miles in each direction, has failed to discover any local feature which may be considered objectionable, or as remotely the cause of disease. The record of the post, however, showing so great a proportion of malarious fevers, an explanation is required of the statement that no appreciable *local* cause for them can be said to exist in this vicinity. This explanation may be found in the history of the occupation, habits and exposures of the troops; the meteorological conditions of the seasons when most prevalent; and in what I conceive to be the general characteristics of a rich prairie country."—*Medical Statistics U. States Army.*

BENECIA, Cal., is a military post situated in latitude 38° 8′ north, and 122° 4′ west, on the Straits of Carquenez, connecting the bays of San Pablo and Suisun, being about thirty miles east, in a direct line from the Pacific Ocean. The town of Benecia lies in the immediate vicinity of the barracks and is a place of importance. The coldest winter month, January, had a mean temp. of 47° Fahr., and the warmest summer month, July, had a mean of 67°. Average annual temperature, 58° 29′ Fahrenheit. Annual fall of rain, 17 inches.

The climate is mild, divided into two seasons—the wet and dry. The winter, a wet season, usually commences in November, and continues through March. A few weeks after the first rains, the grass springs up, and in a short time the country presents the appearance of spring. The fruit trees bloom in February and March; the vine and olive grow in great perfection. Snow seldom falls on the plains; occasionally the higher hills in the vicinity are covered for a short time; ice sometimes, but rarely, forms; the hills and valleys continue green until the last of May, when the oat and other grasses begin to ripen and turn yellow; and, by the middle of July, the ground is baked and cracked, and the whole country presents the appearance of the greatest aridity. At this, and other points near the coast, the sea-breeze blows regularly, commencing about 9 A. M., and continuing till sundown; the nights are cool and pleasant.

The most prevalent diseases at this post, and in the vicinity, are fevers and affections of the respiratory and digestive organs. The fevers are not severe, the remittent form being mild and easily managed; the intermittent is apt to return frequently and continue for a long time. The diseases of the respiratory organs are generally mild catarrhs, usually cured in a few days. The diseases of the digestive organs are diarrhœa and dysentery, both frequently proving extremely obstinate and difficult of cure.—*Medical Statistics U. States Army.*

The climate, &c., of San Francisco may be considered similar to Benecia, although somewhat cooler, the mean annual temperature being 55° Fahr.; the average annual fall of rain being 23 inches, mostly falling during the winter months.

Fort Jones, Cal.—"In latitude 41° 35′ N., longitude 122° 52′ W., and on the eastern slope of the 'Coast Range' of mountains, extending eastwardly towards the Sierra Nevada, is an oval basin thirty miles long by six wide, known as Scott's Valley. The southern as well as the northern extremity of this valley are bounded by two high ranges of mountains, the connecting links between the coast range and the Sierra Nevada. That range, bounding the northern extremity, is a spur of the coast range; that bounding the southern extremity is called Scott's Mountain, being some 8,000 feet above the level of the ocean, while the valley has an altitude of nearly 3,000 above the level of the sea.

"From a bird's-eye view of the geological characteristics of the surrounding country, one is led to believe that its origin is of rather recent date. On some of the mountains there are fresh appearances of scoriæ, and in the crater of one of the neighboring mountains, sulphurous ebullitions are distinctly visible. The pedrigal at the base of the mountains and in the small valleys appears to have undergone but little change, and is generally covered with an exceedingly young growth of timber. The character of the soil of this valley is principally argillaceous and arenaceous. The former presents a reddish appearance, and holds in combination, minute pebbles, and the latter is composed chiefly of micaceous matter. With plenty of moisture and sufficient heat, this peculiar soil would bring forth as delicious and luxuriant vegetables and fruit as the richest vegetable mould of the prairie lands of Iowa.

"Fort Jones is situated at the northern extremity of the valley, on a gentle slope of the mountain, and in a pine grove. The mean annual temperature of the post is 52° Fahr. Spring 52°, summer 73°, autumn 52°, winter 33°, the highest being 100°, and the lowest 3° below zero. Altitude, 2,570 feet.

"The climate of Fort Jones may be regarded, on the whole, as salubrious. The line of demarkation between each of the four seasons of the year is conspicuously drawn. The fall sets in about the middle of September, and continues, with cool nights and warm days, until the first of December, which is ushered in either by heavy rains or deep snows. This continues, at intervals, until the middle of February, when the hills and valleys are clad in verdure. March and April come and go with warm days and cool nights, and not unfrequently accompanied with frosts. June, July, and August, bring hot days, and, occasionally, a sultry night. The summer is not always attended with a drought as in the southern part of the State. A rain storm, accompanied with thunder and lightning, in July and August, is not an unfrequent occurrence.

"The most prevalent disease among the troops, as well as the citizens, is intermitting fever in some form. The overflowing of the river banks, with the rank vegetable matter that the water holds in combination, sufficiently accounts for this form of disease.

"It is a noticeable fact that when females from the Atlantic States arrive on this coast, those who have been barren for years, and those who have never borne children at all, no sooner become acclimated than the uterine organs assume a new tone, and conception immediately follows. This change of the functions is not temporary, but continues, and the once sterile female may calculate with the greatest certainty that the end of every eighteen months will bring an offspring. The American cow, after reaching this coast from the plains, will bring a calf every eleven or twelve months, and this calf will bring forth young when two years old. Sheep breed twice a year, and more frequently bring forth two at each birth than one. If I were to advance an opinion of my own on this great procreative tendency of both man and beast on this coast, I would attribute it, in a great measure, to this bland and stimulating climate. The climate has certainly the effect, on females who come here, of producing an immoderate action of the catamenial functions. But I will not attempt to offer any further cause for this, but will leave it for the more scientific."—*Extract from Sanitary Report of Asst. Surgeon C. C. Keeney*, 1856.

FORT READING, Cal., situated in latitude 40° 28′ N.; longitude 122° 7′ W., from Greenwich, is elevated about 700 feet above the level of the sea, lying in the northern part of the valley of the Sacramento River. Mean annual temperature, 52° Fahr., showing a great degree of heat for the latitude. The country around is, in a general view, an irregular prairie, bounded on the east by a range of mountains—Lassen's Mountains—running north and south, sixty miles distant, and one-fourth of the way to the range of the Sierra Nevada; on the west by the coast range, twenty-five miles distant; on the north by Shasta Butte, ninety miles distant, which appears to spread out east and west and connect with Lassen's and the coast range; and on the south it is continuous with a plain that follows the course of the Sacramento River. Mountain peaks covered with snow are here to be seen for most parts of the year in the distance. Shasta Butte, which is immediately under the 122d parallel, is estimated to rise 16,000 feet above the sea. The prairie is studded here and there with mots of white oak; and white oak, the nut pine, and willows, with long grass and dense undergrowth, skirt the water courses. Elsewhere the country is bare of everything that would intercept the winds. The cultivation of the soil is of such little extent that it cannot effect in any degree its healthfulness; yet, intermittent fever occurs here at all seasons. In point of climate and salubrity, the description of this part of the country is applicable to much of the country lying between the range of the Sierra Nevada Mountain and the coast range.

MONTEREY, California, situated forty leagues south from San Francisco, on the shore of Monterey Bay, in north latitude 36° 36′, is probably the most beautiful town on all the coast of California. "In all that constitutes beauty of scenery, derived from a proper proportion of woodland, water, hills, and distant mountains, Monterey will bear a comparison with other places of more celebrity. The atmosphere is humid, the temperature agreeably warm and equable; the prevalent winds are sea breezes from the west and north; the land winds from the east and south are much less frequent, blow less strongly, and may frequently be detected alone by the uncomfortable feelings they produce, without reference to the weather-vane. There is one rainy season, from November till April. This is about the average time the rains begin and terminate, although sometimes considerable rain will fall as early as October and continue until May. During this period there are frequent intervals of fine weather of such extraordinary beauty and balmy temperature, that the traveller arriving on the coast might well imagine, with Col. Fremont, that it resembled the climate of southern Italy. During the dry season the fogs rise from the sea late in the afternoon, float over the town, and disperse usually about 9 P. M. There is also a fog generally in the mornings until 10 A. M. These fogs are found on the entire coast of California as far south as Point Conception. In the rainy season, at which time the winds are from the south and east, there are no fogs; the sky, when not rainy, being clear and cloudless. There is a difference between the mean temperature of the summer and winter months of only from 6° to 7°; and hence the annual temperature (60° Fahr.) is very uniform, although the diurnal changes may be very considerable.

"This post is represented as being remarkably healthy, no particular disease could said to be endemic to this locality. The diseases from which the inhabitants are entirely free are contagious or infectious fevers; those from which they are nearly exempt, are consumption, dyspepsia, aneurism, and malignant tumors; and those which are mild, and of rare occurrence, are diarrhœa and dysentery.

IV.—Meteorological Table,

SHOWING THE PRINCIPAL CITIES AND MILITARY STATIONS IN THE UNITED STATES, HAVING A MEAN ANNUAL TEMPERATURE ABOVE 60° FAHR.

STATIONS, ETC.	Latitude.	Longitude.	Altitude.	Yearly Mean.	FOUR SEASONS.			
					Spring.	Summer.	Autumn.	Winter.
			Feet.	° Fahr.	° Fahr.	° Fahr.	° Fahr.	° Fahr.
Raleigh, N. C............	35°47′	78°48′	.....	61.00	.....	.....	.....	.....
Fort Macon, N. C.......	34°41′	76°40′	20	62.23	59.46	78.51	65.19	45.75
Fort Johnston, "	34°00′	78°05′	20	65.68	64.46	80.19	67.46	50.60
Columbia, S. C..........	34°00′	81°00′	200	60.00	59.00	78.00	58.00	40.00
Camden, "	34°17′	80°33′	275	61.00	58.20	80.20	59.40	41.00
Charleston (Ft. Moultrie).	32°45′	79°51′	25	66.60	65.85	80.59	68.11	51.88
Savannah, Geo..........	32°05′	81°07′	40	67.44	67.08	80.70	67.94	54.06
Augusta, "	33°28′	81°53′	600	64.00	64.37	80.21	63.37	48.07
Sparta, "	33°17′	83°09′	800	64.40	63.90	79.20	68.20	46.50
Fernandina, Fla.........	30°35′	81°30′	.....	69.80	69.40	79.20	70.80	59.80
St. Augustine, Fla.......	29°48′	81°35′	25	69.63	68.54	80.37	71.53	58.08
Fort King, "	29°10′	82°10′	50	70.00	70.72	80.22	70.64	58.41
Pilatka, "	29°34′	81°48′	25	69.64	70.60	80.56	70.20	57.18
New Smyrna, "	28°54′	81°02′	20	71.60	71.80	79.14	72.43	63.22
Fort Pierce, "	27°30′	80°20′	30	73.20	73.14	81.30	74.80	63.27
Fort Dallas, "	25°55′	80°20′	20	74.75	74.16	81.50	76.27	66.58
Key West, "	21°32′	81°48′	10	76.00	75.79	82.50	78.00	69.00
Jacksonville, "	30°15′	82°00′	14	69.50	.....	.....	.....	.....
Fort Brooke, "	28°00′	82°28′	20	72.48	72.08	80.20	73.44	62.35
Cedar Keys, "	29°07′	83°03′	35	69.60	70.08	79.67	71.04	58.22
Pensacola "	30°18′	87°27′	20	68.74	68.59	81.57	69.86	54.92
Mobile, Ala.............	30°42′	87°59′	25	70.00	70.00	82.70	71.00	57.00
Fort Morgan, Ala.......	30°14′	88°00′	20	67.00	64.81	80.33	69.17	53.68
Mt. Vernon Arsenal, Miss.	31°12′	88°02′	200	65.81	67.02	78.68	65.81	51.72
Natchez, Miss...........	31°34′	91°28′	240	67.10	68.00	81.00	67.10	52.00
Vicksburg, "	32°24′	91°00′	350	65.00	66.60	78.40	64.70	50.30
Little Rock, Ark........	34°40′	92°12′	150	62.30	63.60	78.60	62.00	45.00
Fort Pike, La...........	30°10′	89°38′	10	69.86	69.97	82.84	70.83	55.80
New Orleans, La........	29°57′	90°00′	10	69.86	69.94	82.27	70.71	56.53
Baton Rouge, "	30°26′	91°18′	40	68.14	69.84	81.21	68.21	54.21
Fort Jesup, "	31°33′	93°32′	80	66.39	67.00	81.27	66.22	51.00
Fort Towson, Ind. Ter...	34°00′	95°33′	300	61.69	62.39	79.16	61.27	43.92
Fort Washita, " " ...	34°14′	96°38′	645	62.21	62.16	79.29	63.25	44.14
Austin, Texas...........	30°20′	97°46′	650	67.50	68.46	82.92	67.00	30.00
Fort Belknap, Texas.....	33°08′	98°48′	1,600	64.00	64.90	80.95	65.15	44.90
Fort Worth, "	32°40′	97°25′	1,100	63.54	63.11	80.43	65.37	45.25
Fort Gates, "	31°26′	97°49′	1,000	66.12	64.79	82.30	67.80	49.58
Galveston, "	29°18′	95°01′	.....	74.00	73.20	87.50	70.60	60.40
San Antonio, "	29°58′	98°25′	600	69.25	69.68	82.16	71.26	53.90
Corpus Christi, "	28°05′	97°27′	20	70.95	71.45	82.53	73.11	56.72
Fort Brown, "	25°54′	97°26′	50	74.00	74.85	83.37	74.77	62.28
Fort Chadbourne, Texas.	32°02′	100°05′	2,120	62.38	64.35	76.77	62.55	45.87
Fort Duncan, Texas.....	28°42′	100°30′	2,842	70.00	72.75	84.48	72.25	53.92
Fort Quitman, "	30°40′	105°00′	3,700	61.89	61.92	82.78	62.42	40.46
Fort Fillmore, N. M.....	32°13′	106°62′	3,937	63.98	63.72	81,32	64.27	46.62
Fort Yuma, Cal.........	32°43′	114°36′	120	74.00	72.10	88.00	75.69	56.80
San Diego, "	32°42′	117°14′	150	62.00	60.00	71.26	64.40	52.29
Jurupa, "	34°00′	117°25′	1,000	63.28	61.00	72.44	65.78	53.89
Fort Miller, "	37°00′	119°40′	400	66.00	62.78	85.48	66.36	49.35
Stockton, "	37°57′	121°14′	.....	61.00	59.46	73.26	64.00	46.68

Climatic Features.

The southern portion of the Union having a mean temperature varying from 60° to 76° Fahrenheit, lies between 36 and 24 degrees north latitude. Key West, which is the most southern post in the United States, is also the hottest. It runs through 37 degrees of longitude on a line from Charleston, S. C., to San Diego, Cal., including the States of North Carolina, South Carolina, Georgia, Florida, Alabama, Mississippi, Louisiana, Arkansas, Texas, Southern California, the Territory of Arizona, and part of New Mexico.

This large section of the United States is mostly sub-tropical in its climatic character, producing cotton, rice and sugar in great abundance; the former product being annually exported to an immense amount. Indian corn, wheat and sweet potatoes are raised in most parts, while Texas furnishes pasturage for large numbers of cattle and sheep. Yellow or pitch pine, live oak, the cyprus, and many other kinds of forest trees abound in the lowlands, while a still greater variety flourish in the more hilly or mountainous sections. The orange, and other fruits peculiar to a tropical region, flourish in Florida, on the Atlantic and Gulf coast, while the grape and other kinds of fruit abound in Southern California on the Pacific coast.

The minerals, although not numerous, are very valuable; gold being found in North Carolina and South Carolina in considerable quantities, while Arizona is found to produce gold, silver, and copper, although as yet but partially explored. Southern California also yields gold, quicksilver, and other valuable minerals.

Military Posts and Cities.

CHARLESTON, S. C., situated in latitude 32° 46′; longitude 79° 57′; lies six miles in land from *Fort Moultrie*, on Sullivan's Island. The fort is surrounded in part by the village of Moultrieville, which is a fashionable resort during warm weather. Mean annual temperature 66° Fahrenheit; the mean of the winter months being 51°, and the summer months 81°. The average annual fall of rain is 45 inches; the greatest quantity falling in July, August and September.

The city of Charleston, being identified with the above island and military post, the Army statistics are quoted in regard to the situation and health of this locality. "Sullivan's Island is on the north side of the bay which forms Charleston harbor. It is a sandy island, and is but slightly elevated above the level of the sea; several storms having been known to carry the waves over it so as almost to submerge it. There appear to be advocates for the salubrity of Sullivan's Island, while others show that several diseases have originated here as well as in the city of Charleston—as cholera, infantum dysentery, intermittent fever, remittent fever, and yellow fever." In speaking of the fever of 1852 the surgeon remarks:—"The conclusion is irresistible, that yellow fever was not introduced from Castle Pinckney, neither from Charleston, but that it originated on Sullivan's Island."

DEW POINT.—With high range of the thermometer in the low country of the south, humidity and a high dew-point are always associated. This is the case in Charleston harbor. The meteorological register of Fort Moultrie, for three years, shows this.

MEAN OF THE DEW-POINT.

Months.	1849.	1852.	1853.
May,	70.19	70.51	66.50
June,	75.36	72.94	71.40
July,	76.27	78.70	77.33
August,	77.69	75.77	75.32
September	71.70	71.74	71.85

"Humidity and a high dew-point play an important part in the causation of febrile diseases. It is not the sole cause, but there is no question that a high dew-point is a powerful agent; and we may conclude, in the language of another, 'that a high dew-point has a tendency to produce injurious effects on the system; that it is often found to exist in unhealthy localities, or during pestilential times; and that it must assist much in the development of autumnal and periodic fevers, are facts which no one acquainted with the subject will question. We may go further, and affirm that yellow fever never prevails in a place, endemically or epidemically, unless there is a high dew-point. Indeed, heat, humidity, and a high dew-point, are always present in summer, as is evident to the most common observer, in the rapidity with which butcher's meat takes on the putrefactive process; and the humidity of the climate is shown by the rapid oxidation of all articles of clothing, the rusting of keys in one's pocket, the mildew on linen clothing, and the injury done to cloth generally, the mould on leather, &c." Which fact in regard to yellow fever seems to be well proven at all the military posts from Fort Brown to Key West on the Gulf coast, and from the southern point of Florida to Norfolk on the Atlantic coast.

HEAVY RAINS.—These were a cause—predisposing and exciting—of the malignant fever of 1852, both in Charleston and Sullivan's Island, producing, in combination with other causes, yellow and severe bilious fevers. The summer of this year was very wet, the quantity of rain in each month being as follows :—

Months.	Mean Temp.	SULLIVAN'S ISLAND. Rain.	CHARLESTON. Rain.
May, .	73.80 Fahr.	4.17 inches.	4.22 inches.
June, .	76.61 "	8.86 "	5.18 "
July, .	81.40 "	5.43 "	6.93 "
August, .	79.79 "	4.15 "	4.21 "
September,	75.76 "	11.70 "	12.27 "
Total quantity, . .		34.31 inches.	32.81 inches.

The whole quantity for the year :—Sullivan's Island, 51.26 inches ; Charleston, 49,72 inches. Mean annual temperature, 66° Fahrenheit.

"*Fevers* are the proper endemics of Carolina, and occur oftener than any, probably than all, other diseases. These are the effects of its warm, moist climate, of its low grounds, and stagnant waters. In their mildest season, they assume the type of intermittents ; in their next grade, they are bilious remittents ; and, under particular circumstances, in their highest grade, constitute yellow fever."—*Dr. Ramsay.*

Charleston, S. C., and its vicinity.—"We have known almost all kinds of fever to originate on Sullivan's Island—intermittent, remittent, congestive and yellow fever ; but the most usual form of fever is the common bilious remittent. Not a summer passes without it, more or less ; sometimes it is mild, at others severe.

"Cholera infantum is indigenous, as might be expected from the proximity of the island to the city of Charleston, in the low country, and in a hot climate, with a humid atmosphere and high dew-point.

"Chronic diarrhœa and dysentery, in adults or children, are serious complaints in summer, either in the city or on the island, and such patients should have a change of climate without delay.

"Sullivan's Island is an improper residence for persons affected with chronic bronchitis or phthisis pulmonalis. In summer it is too hot, and the winds are too bleak and damp ; in winter, the cold and strong winds render it a very unadvisable resort. Chronic rheumatism and neuralgic pains are not often benefited by a residence on the island, but the contrary. The climate, both winter and summer, is too severe for persons afflicted with these complaints.

"From the situation and physical characteristics of the country around Charleston, including Sullivan's Island, the summer climate must be enervating, and most persons would improve by annually spending July, August and September in a more elevated region."

"We have considered the principal apparent causes of yellow fever in 1852, which appear to have been the same as those of previous epidemics; and they are: 1. Intemperance; 2. Fatigue and exposure; 3. Imperfect ventilation; 4. High solar heat; 5. Humidity and a high dew-point; 6. Defective drainage; 7. Bad water; 8. Heavy rains."—*Medical Statistics U. S. Army.*

St. Augustine, Fla., situated facing the Atlantic Ocean in north latitude 29° 48′, west longitude 81° 35′, has a mean annual temperature of 70° Fahrenheit. The coldest winter month (January) had a mean of 57°, and the warmest summer month (July) 81° Fahr. Average annual fall of rain, 48 inches.

"*Fort Marion* is in the city of St. Augustine, which is situated on the bay of the same name, being distant about two miles from the Atlantic Ocean. The site of the city is slightly elevated, being about twelve feet above the level of the ponds and marshes in the vicinity. The adjacent country is level and generally sandy, some parts being sufficiently rich in calcareous and vegetable matter to produce most of the vegetables cultivated at the North. Oranges flourish here most luxuriantly; but, in the early part of 1855, all the groves in the northern half of the peninsula were wholly destroyed by frost—an occurrence previously unknown.

"St. Augustine has long been celebrated as a winter residence for pulmonary invalids; but the city itself has claims upon the traveller's attention, not the least being the fact that it is the oldest town in the United States. The fort is also one of the oldest in the country. It was finished, as appears by its now nearly illegible inscription, in 1756, in the reign of Ferdinand the Sixth. The walls consist of a concretion of sea-shells obtained from quarries in Anastasia Island, and as the material, under a bombardment, crumbles away without suffering fractures, the fort, duly manned, would be almost impregnable.

"This post has been at all times justly esteemed for its salubrity. Compared with the average mortality of southern posts in general, this station is found to exhibit a much lower ratio. It is seldom that diseases of a malignant character appear at St. Augustine. Toward the close of the present year (1839,) yellow fever, which ravaged the principal cities of our Southern States, made its appearance at this station. This is only the second time that this epidemic has prevailed in this city within the period of twenty years; while at Charleston, we are told by Prof. Dickson, that in twenty-four years' practice, but three have passed without his knowing the occurrence of the yellow fever. As regards the essential cause of yellow fever, we still remain in the dark. It is manifest, however, that to develop the cause, and keep up its action, requires a high range of atmospheric temperature; and as this condition seldom obtains on the coast of Florida, it would seem to afford an apparent explanation of its infrequent occurrence in this region. At Key West, or Thompson's Island, however, which is the hottest and most southern station in Florida, as in the islands generally of the West Indies, yellow fever has prevailed with much malignity."

New Orleans, La.—This important military post and commercial depot is situated on the left bank of the Mississippi River, distant 105 miles by the channel from its mouth, and 80 miles in a southeast course. It is fifty miles from the Gulf of Mexico, south; 14 miles from Lake Borgne, east; and 6 miles from Lake Ponchartrain, north: in north latitude 29° 57′; longitude 90° west from Greenwich, and 13° from Washington. Mean annual temperature 70° Fahrenheit. The coldest winter month (January,) 55°; the warmest summer month (July,) 83°. Annual average fall of rain, 50 inches.

"There are no hills in the vicinity of the city, being built on an inclined plane, descending gently from the river to the lakes. When the river is full, the streets are three or four feet below its surface. Inundations are prevented by a dyke, or levee. The well-water of the city is not used either for washing or for culinary purposes, as it contains the muriates of lime, magnesia, and soda, and the bi-carbonate of lime, and also iron; rain and river water are consequently used by all.

"The southwest and southeast winds prevail during the five months from April to August, and the northeast in September. It is to be remarked that the east-northeast and southeast winds come from the Gulf of Mexico over an immense tract of low swamps, and that the prevalence of north and east winds in July, August and September, is always attended with the epidemic of yellow fever. In fact, these three months are the only ones that can be considered as proper seasons of disease—that is, the cause of epidemic yellow fever is produced during those months. Its ravages may, and do, extend into October; but when there has been no epidemic during August and September, strangers are not as liable to disease in October. The yellow fever of this climate, then, may be traced to the following combined causes:—1. Low stage of water in the river, leaving its banks, with the deposits brought from the upper country, exposed to the action of the sun ; 2. Decomposition of vegetable matter in the swamps in the rear of the city ; and, 3. The prevalence of east and northeast winds. These winds come not only loaded with miasmata from the swamps which they traverse, but are cold, and tend to produce chills, rendering the system more liable to be impressed with other causes incident to the climate, such as sudden alternations from cold showers to a burning sun. In confirmation of this opinion, it is remarked, that a contrary state of things—to wit, high stage of water in the river, and the prevalence of southwest and west winds,—are not attended with epidemic fever."

Yellow Fever.

Extract from the Medical Statistics U. States Army. Compiled by Surgeon S. P. MOORE, stationed at FORT BROWN, Texas, in 1853.

"It may be asked if the yellow fever of the United States is the same disease as the *fiebre amarilla*, or, as it is more frequently called, *vomito prieto*, of the Mexicans? There can be no doubt of it. It is the same disease as the yellow fever of Charleston, S. C., of New Orleans, and Pensacola, I having seen the disease in these cities ; in the late epidemic that prevailed here (Fort Brown) and in Matamoras (on the opposite side of the Rio Grande,) the symptoms were too evident to admit of a doubt.

"It has been suggested that the recent wide-spread epidemic, which has devastated the southwestern States, began in Rio Janeiro in 1850, and has been creeping northward each successive year ; in other words, that it is a new disease. It is probable that in almost every country in which this disease has committed its ravages, it has received a new name. From its depredations in the West Indies, it has been called the St. Domingo, Barbadoes and Jamaica fevers ; on the Guinea coast, and adjacent ports, the Bulam fever ; in British India, it is distinguished by the name of the jungle fever, the Hoogly fever ; and still further east, by that of the Mal de Siam ; and in the south of Spain, the Andalusian pestilence. In the present day its more common name is *Yellow fever*, and, when the attack upon new-comers is slight, *acclimating*.

"From its appearing in different parts of the world, and under different circumstances, it is not surprising that it should often be accompanied with a diversity of symptoms. It is supposed by some writers that the cause of yellow fever and bilious fevers are the same—that is, it proceeds from marsh miasmata. It is evident, however, that these diseases are quite distinct, and arise from different causes.

"The question of contagion is a very important one, and has occupied the attention of physicians and philanthropists for a long period, without definitely settling it; there is no hesitation in giving a decided opinion that it is not. *The disease is of domestic origin.* The arguments for contagion are opposed by facts; these are well known and need not be stated.

"Setting aside the vexed question of *quarantine*, hospitals should be established in healthy situations; all sources of noxious effluvia should be removed; and by correcting such effluvia, when known to exist, by appropriate fumigations, and by excluding persons not exempt from the disease from the infected district. These and such like efforts should be made upon the first appearance of any epidemic." This writer further remarks: "I think yellow fever a peculiar and distinct disease, and the precise pathological conditions essential to it are at present unknown;" and further, "Apart from the epidemic influence on man, nothing was observed remarkable in the animal or vegetable kingdoms."

Another army surgeon, stationed at *Fort Moultrie*, Charleston harbor, remarks: "Little will be said concerning the nature of yellow fever. When the disease first occurred to me, it was regarded as *sui generis*—as different from all other southern fevers; but it must be confessed that this opinion has been considerably modified; and, at the present time, it is believed that intermittent, remittent, continued congestive, and yellow fevers are nearly related, if not modifications of the same fever—all being southern bilious fevers—the nervous system in some, and the blood in others, being pre-eminently affected. Whether the difference in these varieties of southern fever, in different seasons and in the same season, depends on a simple difference of intensity in the predisposing and exciting causes with the same *materia morbi*; whether different causes exist at the same, developing the different forms of fever; or whether there is a blending and conversion of types, as is manifested by Dr. Dickson—all remains to be determined."

FORT YUMA, Cal., situated in north latitude 32° 32′, west longitude 114° 36′, is elevated 350 feet above the waters of the Colorado River. The mean annual temperature, according to the Army Records, is 74° Fahrenheit. The coldest month (January) had a mean of 56°, and the warmest month (July) 92° Fahr. Average annual fall of rain, 4 inches, being the least of any post in the United States.

"Fort Yuma is situated on a high rocky hill on the west bank of the Colorado, opposite the mouth of the Rio Gila, and eighty miles from the head of the Gulf of California. The Valley of the Colorado averages seven miles in width, and is bounded on either side by rocky barren mountains and sand-hills, which separate it from the immense deserts by which it is surrounded. This locality is noted for its excessive temperature and absence of rain; the thermometer occasionally rising to 116°. Although such is the official record of meteorological observations at this post, it appears that the actual temperature is even more excessive than above stated. The principal number of cases of diarrhœa, dysentery and scorbutis reported in the abstract for 1852 occurred at this post, the men being for the time destitute of vegetables, and deprived of the ordinary necessaries of life."

San Diego, Cal., situated in north latitude 32° 42′, and west longitude 117° 14′, has a mean annual temperature of 62° Fahr. The coldest month (January) had a mean of 52°, and the warmest month (August) 74°. Annual average fall of rain, 10 inches.

"The military post at San Diego is situated near the head of a valley, perhaps three quarters to one and a half mile in width, six miles distant from the old Presidio, and eight miles from the sea-shore. The height of the hills and table-land on either side of the valley is about 250 feet. Some 15 or 20 miles to the east of the post is a range of mountains running north and south, broken in places, with some pretty valleys intervening, which mountains extend over a distance in width some 40 miles, and bound the desert on the west side of the Rio Colorado.

"The diseases which have occurred at this post have not been influenced *particularly* by the climate. In some particular places they suffer from intermittent and bilious fevers, of which many die; but in this immediate vicinity a case of intermittent or remittent fever is seldom ever seen, unless contracted elsewhere."

Annual Measurement of Rain

At the different Military Stations in the United States.

The entire amount of water falling in Rain and Snow is in all cases intended to be included in the summaries given in the original record, and of which the results for separate years are here consolidated to determine the mean for a series of years.

NOTE.—For Latitude and Longitude of the different Stations, see *Meteorological Table of Temperatures*, pages 132, 141, 150 and 159.

Military Stations.	Rain, &c., in Inches and Hundredths.				
	Spring.	Sum'er.	Aut'mn.	Winter.	Year.
Fort Kent, Maine, . . .	5.46	11.65	9.64	9.71	36.46
Hancock Barracks, Maine, .	7.62	11.92	9.95	7.48	36.97
Fort Sullivan, Eastport, " .	8.88	10.05	9.85	10.61	39.39
Fort Preble, Portland, " .	12.11	10.28	11.93	10.93	45.25
Ft. Constit'n, P'rtsm'th, N. H.	9.03	9.21	8.95	8.38	35.57
Ft. Indep'dence, Boston Har.,	8.60	8.42	9.27	9.01	35.30
Watertown Arsenal, Mass., .	10.75	10.66	10.83	9.83	42.07
Fort Adams, Newport, R. I., .	13.89	11.44	13.66	13.47	52.46
Ft. Trumbull, N. London, Con.	10.99	10.65	13.16	10.98	45.69
Ft. Columbus, N. Y. Harbor,	11.55	11.33	10.30	9.63	43.23
Fort Hamilton, N. Y., . .	11.69	11.64	9.93	10.39	43.65
West Point, N. Y. . . .	12.57	12.43	10.74	10.79	46.53
Watervliet Arsenal, N. Y., .	8.66	10.34	9.17	6.38	34.55
Plattsburgh Barracks, N. Y.,	8.36	10.03	10.05	4.95	33.39
Madison Barracks, N. Y., .	9.94	10.28	12.51	8.10	39.78
Fort Ontario, Oswego, N. Y.	6.18	7.63	9.77	7.30	30.88
Fort Niagara, N. Y. . .	6.87	9.81	8.68	6.41	31.77
Buffalo Barracks, N. Y. .	8.59	9.23	13.54	7.53	38.80
Alleghany Ar., Pittsburgh, Pa.	9.38	9.87	8.23	7.48	34.96
Carlisle Barracks, Pa., . .	9.05	9.67	7.68	7.61	34.01
Ft. Mifflin, near Philadelphia,	12.97	12.62	10.42	9.26	45.27
Ft. McHenry, Baltimore, Md.	11.13	11.04	10.52	9.31	42.00
Washington City, D. C., .	10.45	10.43	10.15	10.07	41.20
Fort Washington, Md., . .	12.57	12.84	10.22	9.39	45.02
Ft. Monroe, near Norfolk, Va.	9.77	15.08	10.16	10.17	45.18
Fort Johnston, N. C., . .	6.83	15.52	16.32	7.34	46.01
Ft. Moultrie, Charleston Har.,	9.89	17.45	10.06	7.52	44.92
Augusta Arsenal, Ga., . .	6.78	3.66	4.51	8.05	23.00
Oglethorpe Barracks, Ga., .	13.45	23.50	7.21	9.17	53.33
Ft. Marion, St. Augustine, Fla.,	5.90	10.54	9.56	5.80	31.80
Fort Pierce, Fla., ...	11.13	26.25	16.84	8.76	62.98
Key West, Fla., ...	8.34	16.59	15.35	7.37	47.65
Fort Myers, Fla., ...	11.02	32.15	11.96	8.06	63.19
Fort Brooke, Fla., ...	8.56	28.24	10.63	8.04	55.57

Annual Measurement of Rain—*Continued.*

Military Stations.	Rain, &c., in Inches and Hundredths.				
	Spring.	Sum'er.	Aut'mn.	Winter.	Year.
Fort Meade, Fla., . . .	8.76	20.68	6.91	3.87	40.22
Cedar Keys, Fla., . . .	4.10	22.35	11.94	10.11	48.50
Ft. Barrancas, Pensacola, Fla.,	12.86	18.69	13.71	11.72	56.98
Ft. Mitchell, n'r Montg'ry, Ala.,	17.60	14.65	4.61	9.29	46.15
Mount Vernon Arsenal, Ala.,	13.42	18.84	13.15	18.09	63.50
Fort Wood, La., . . .	16.13	17.30	15.60	11.60	60.63
Fort Pike, La., . . .	16.70	23.61	18.96	12.65	71.92
New Orleans Barracks, . .	11.29	17.28	9.62	12.71	50.90
Baton Rouge, La., . . .	15.08	19.14	12.48	15.40	62.10
Fort Jesup, La., . . .	13.68	10.94	9.74	11.49	45.85
Fort Smith, Ark., . . .	12.48	13.03	9.93	6.66	42.10
Fort Towson, In. Ter., . .	15.55	14.36	12.23	8.94	51.08
Fort Washita, " . .	13.19	11.27	10.78	6.42	41.66
Fort Gibson, " . .	11.38	9.68	9.25	6.15	36.46
Fort Arbuckle, " . .	8.15	8.98	8.90	4.54	30.57
Fort Scott, Kansas, . .	12.57	16.37	8.39	4.79	42.12
Fort Leavenworth, Kansas, .	7.97	12.24	7.33	2.75	30.29
Jefferson Barracks, Mo., .	10.56	12.88	8.02	6.37	37.83
St. Louis Arsenal, Mo., . .	12.86	14.09	8.71	6.29	41.95
Detroit Barracks, Mich., .	8.51	9.29	7.41	4.86	30.07
Fort Gratiot, " .	8.02	9.99	8.86	5.75	32.62
Fort Mackinac, " .	4.67	8.88	7.01	3.31	23.87
Fort Brady, " .	5.44	9.97	10.76	5.18	31.35
Fort Howard, " .	9.00	14.45	7.84	3.36	34.65
Fort Winnebago, " .	5.58	11.46	7.63	2.82	27.49
Fort Crawford, " .	7.63	11.87	7.90	4.00	31.40
Fort Atkinson, Iowa, . .	12.22	20.43	4.82	2.27	39.74
Fort Des Moines, " . .	8.86	10.93	4.90	3.87	26.56
Fort Dodge, " . .	7.92	8.15	8.19	3.06	27.32
Fort Snelling, Minn., . .	6.61	10.92	5.98	1.92	25.43
Fort Ripley, " . .	6.31	12.62	8.42	2.13	29.48
Fort Kearny, Neb., . .	10.80	12.05	3.82	1.31	27.98
Fort Laramie, Dakota, . .	13.68	7.15	12.00	2.13	35.00
Fort Belknap, Texas, . .	7.09	6.31	6.85	1.75	22.00
Fort Worth, " . .	14.50	8.80	9.49	8.07	40.86
Fort Chadbourne, " . .	8.52	10.46	8.99	3.91	31.88
Fort Graham, " . .	11.98	6.02	9.77	11.91	40.58
Fort Croghan, " . .	11.61	7.80	8.24	8.91	36.56
San Antonio, " . .	8.63	10.22	7.57	7.35	33.77
Fort Brown, " . .	3.97	9.26	15.08	5.34	33.65
Ringgold Bar'ks, " . .	4.49	7.10	6.31	3.05	20.95
Fort McIntosh, " . .	4.07	7.33	5.06	2.20	18.66

Annual Measurement of Rain—*Continued.*

Military Stations.	Rain, &c., in Inches and Hundredths.				
	Spring	Sum'er.	Aut'mn	Winter.	Year.
Fort Duncan, Texas,	3.55	9.91	6.32	2.42	22.20
Fort Inge, "	6.06	11.06	6.99	3.88	27.99
Fort Clarke, "	4.60	8.53	6.36	2.31	21.80
Fort Bliss and El Paso,	0.70	3.56	5.25	1.70	11.21
Fort Fillmore,	0.75	4.44	3.30	0.74	9.23
Albuquerque, N. M.,	1.10	5.45	2.07	0.80	9.42
Santa Fe, "	2.83	8.90	6.02	2.08	19.83
Fort Union & Las Vegas, N.M.	2.47	9.62	5.12	2.03	19.24
Fort Massachusetts, "	3.50	5.38	8.83	2.83	20.54
Fort Defiance, "	2.91	6.41	4.57	2.65	16.64
San Diego, Cal.,	2.74	0.55	1.24	5.90	10.43
Fort Yuma, "	0.27	1.31	0.86	0.80	3.24
Monterey, "	4.43	0.21	1.65	5.91	12.20
Fort Miller, "	9.57	0.02	3.59	11.34	24.51
San Francisco, Cal.,	8.81	0.03	3.37	11.38	23.59
Benecia Barracks, Cal.,	6.40	0.01	2.65	7.56	16.62
Sacramento, "	9.02	0.00	3.74	8.56	21.32
Fort Reading, "	11.30	0.39	4.89	12.44	29.02
Fort Jones, "	5.38	0.89	5.30	5.20	16.77
Fort Orford, Oregon,	19.12	3.00	19.60	26.80	68.52
Dalles of Columbia, Oregon,	2.63	0.42	4.16	7.11	14.32
Astoria, "	...	...	...	...	60.00
Fort Vancouver, W. T.,	9.28	6.23	10.30	19.69	45.50
Fort Steilacoom, "	11.19	3.85	15.20	21.51	51.75

RECAPITULATION.

Annual fall of	Rain, &c.,	in the	New England States,	41	inches.
"	"	"	State of New York,	36	"
"	"	"	Middle States,	40½	"
"	"	"	State of Ohio,	40	"
"	"	"	Southern States,	51	"
"	"	"	S.W. States & In. Ter.	39½	"
"	"	"	West'rn States & Ter.	30	"
"	"	"	Texas & New Mexico,	24½	"
"	"	"	State of California,	18½	"
"	"	"	Oregon & Wash. Ter.	50	"

Average annual fall of Rain, &c., in the United States, 36 inches.

From the above Army Record, running through a number of years, it appears that the greatest fall of rain in the U. States occurs in the Southern States bordering on the Atlantic and Gulf of Mexico; the average annual fall being 51 inches. The next greatest fall of rain and snow occurs in the Northwestern States and Territories bordering on the Pacific Ocean; the average annual fall being 50 inches, although immediately in the vicinity of the sea-coast and near Puget Sound (say for 100 miles inland), the average annual fall of rain, &c., will amount to 56 inches and upwards—showing a singular coincidence in regard to these two extremes of territory, one receiving its climatic influence from the Gulf Stream as it ascends northward, and the other from a similar current of water and air approaching the northwest coast from the Pacific Ocean.

Consolidated Table,

Exhibiting the annual amount of Sickness and Mortality in the U. States Army,* brought down to January, 1860.

Regions.	Ratio per 1000 Men.	
	Treated.	Died.
1. Coast of New England,	1,755	8.8
2. Harbor of New York,	3,181	18.4
3. West Point, N. Y.,	4,619	4.0
4. North Interior—East,	1,808	10.9
5. The Great Lakes,	2,183	13.1
6. North Interior—West,	2,265	12.0
7. Middle Atlantic coast,	2,336	11.1
8. Middle Interior—East,	3,180	14.9
9. Newport Barracks, Ken.,	2,692	29.3
10. Jefferson Barracks and St. Louis Arsenal,	3,603	43.7
11. Middle Interior—West,	2,622	22.4
12. South Atlantic Coast,	2,658	27.3
13. South Interior—East,	2,989	40.5
14. South Interior—West,	3,354	22.0
15. Atlantic Coast of Florida,	3,515	24.0
16. Interior and Gulf Coast of Florida,	4,902	30.2
17. Texas, Southern Frontier,	3,580	49.6
18. Texas, Western Frontier,	3,063	19.6
19. New Mexico,	2,590	18.5
20. California, Southern,	2,105	18.0
21. California, Northern,	2,784	25.6
22. Oregon and Washington,	2,302	9.8
23. Utah Territory,	1,845	8.2
Average, per 1,000,		21.0

NOTE.—The mortality per 1,000 British troops in Canada is 20; Jamaica, West Indies, 143.

* Asiatic cholera and gun-shot wounds excluded.

PART IX.

CLIMATE OF THE NORTHERN, MIDDLE AND WESTERN STATES.

Climate, Topography, and Productions.

New England States.—This northeast section of the United States lies between 41° and 47° north latitude, extending from 67° to 73° 30′ west longitude. It has for the most part a favored climate as regards health and longevity, although along the sea-coast, when northeast winds prevail, cases of consumption are very prevalent.

The coldest part is the north of Maine, where the mean annual temperature is 37° Fahrenheit. The coldest month being January, and the warmest July. The average mean temperature of the seasons in this region are as follows:—Spring 35°, Summer 62°, Autumn 40°, Winter 12°, Fahr.; the mercury occasionally getting to —36° below zero, and rising to 90° above. The country is elevated, and mostly covered with a dense forest, bordering on the St. John's River and its tributaries.

On the eastern border of Maine, at Eastport, the mean annual temperature is 43° Fahr.; the mercury sometimes falling to —6° below zero, and rising in summer to 85° above. This post lies facing the Bay of Fundy, the temperature being modified by the waters of the Atlantic Ocean. The season of vegetation in this region of country is at least three months later than in South Carolina and Georgia.

The northwestern bounds of New England, running along the parallel of 45° north latitude, extending to Lake Champlain, embraces a fine agricultural section of country, being separated from the Eastern Townships of Lower Canada. Vermont is justly celebrated for a healthy climate and fruitful soil. The interior of the country, on the upper parts of the Connecticut River, embracing the northern part of New Hampshire, is much colder, although no correct data can be obtained of its precise influence. The mean annual temperature of the northern part of Vermont is 44° Fahr. The seasons are as follows :—Spring 42°, Summer 67°, Autumn 47°, Winter 20° ; the mercury occasionally falling to —20° below zero on the shores of Lake Champlain.

THE GREEN MOUNTAINS of Vermont exercise a great influence on the climate of this region, rising from 3,000 to 4,000 feet and upwards above Lake Champlain. They are, however, for the most part cultivable to their summits. The following table gives the altitude of the several peaks and passes ;—

GREEN MOUNTAINS.

PEAKS.	Elevation above the Sea. Feet.	PASSES.	Elevation above the Sea. Feet.
Chin, Mansfield Mt.,	4,348	Lincoln,	2,415
Nose, " "	4,044	Granville,	2,340
S'th Peak, " "	3,882	Peru,	2,115
Camel's Hump,	4,083	Sherburne,	1,882
Jay Peak,	4,018	Walden,	1,615
Shrewsbury Peak,	4,086	Mt. Holley (Railroad,)	1,415
Killington Peak,	3,924	Roxbury "	912
Ascutney, Windsor,	3,320	Williamstown,	908

MEAN TEMPERATURE AT BURLINGTON, VT.

North latitude, 44° 29′. West longitude 73° 11′.

Months.	Degrees Fahr.	*Months.*	Rain in Inches.
January,	19.93	January,	1.58
February,	20.46	February,	1.52
March,	30.83	March,	1.96
April,	42.12	April	1.62
May,	55.10	May,	2.90
June,	64.80	June,	3.59
July,	69.00	July,	4.12
August,	67.73	August,	2.51
September,	59.32	September,	2.95
October,	47.40	October,	4.23
November,	39.39	November,	2.43
December,	23.69	December,	2.41
Mean Annual Temp.,	44.74	Total Inches,	31.82

The southern portion of the New England States, including a part of Massachusetts, Rhode Island and Connecticut, are all alike favored with a healthy and delightful climate for the most part of the year, the Atlantic Ocean and Long Island Sound washing its entire southern coast. The mean annual temperature of this whole stretch of country, lying parallel to the 41st degree north latitude, may be given as is found to exist at *Fort Adams*, near Newport, R. I. The mean annual temperature at this post is 50° Fahr. The seasons as follows:—Spring 46°, Summer 69° 46′, Autumn 53° 56′, Winter 32°; the coldest month being January, and the warmest month July; the mercury occasionally falling to 3° and rising to 90° Fahr.

The varied products of this part of New England are mostly consumed at home; the population being dense and actively employed in commerce and manufactures. Within the bounds of Maine, New Hampshire, and Massachusetts, along the seaboard, are annually built a large number of vessels of different kinds, giving employment to great numbers of landsmen and sailors. The fisheries along this coast, also, are important and profitable—shad, herring, mackerel, halibut, lobsters and salmon being annually taken in large quantities. The resorts and habits of the finny tribe show conclusively that they are governed by the influence of cool and healthy waters, perhaps as much so as men and the inferior animals are by a pure and healthy climate.

The average annual quantity of rain and snow that falls in the six New England States is 41 inches, being nearly equally divided between the different seasons. The smallest quantity, 36 inches, falls in the northern part of Maine, and the largest quantity, 52 inches, falls at Newport, R. I.

Northeast winds and storms prevail in all the northern Atlantic States, producing a cool and damp atmosphere, being most frequently attended by drenching rains, which usually continue for several days in succession. This is considered the most unhealthy wind that occurs, often producing colds, influenza, catarrh affections, and pulmonary complaints. For *Health Statistics*, see different Military Posts, Part X.

The topography of New England presents an interesting feature to the admirer of nature, while it materially affects the temperature where mountain ranges and peaks predominate. The *Green Mountains* of Vermont and the *White Mountains* of New Hampshire have grand and distinctive features; they both tend to cool the atmosphere of the surrounding country, affording most healthy and delightful summer resorts for invalids and the seekers of pleasure.

The soil and climate of New England, although not well adapted to agriculture in general, are particularly genial to the growth of timber. "However rocky and barren the soil may be, if it is not too precipitous, it is always covered by a dense growth of timber; and every little crevice in the rocks affords sufficient hold for some gnarled member of the forest to fix its roots and obtain a subsistence.

"It appears to be a peculiarity of the primary soils of New England that the pine, the elm, the maple, the beach, and the spruce, grow together in social equality. In many places it would have been difficult to find out which of these varieties predominate. In height, however, the pines towered above all the others; and in all those parts of the forest which had been somewhat recently cleared by the fires, the birch was by far the most common. The birch is a rapid grower, but it soon attains maturity or limit of growth, so that in the long run it cannot compete with those which ultimately rise to a greater height, and, overtopping it, shut it out from the sun's rays. Thus, in the older portions of the forest, few birch trees are seen. It is the great variety of trees in the New England forests which affords such a gorgeous spectacle when autumn tinges the leaves with so many brilliant hues." In the pine forests of the State of Maine immense forests of giant pines are still to be found, giving profitable employment to the lumberman.

"The apple tree, the pear tree, the cherry tree, and some other kinds of the fruit-bearing species thrive in favored localities, as well as the cranberry, the whortleberry, and other small fruits."

"The sugar maple, which is peculiar to the climate of the Northern States and Canada, produces, annually, large amounts of maple sugar. The sugar is obtained from the trees in March and April by making incisions in the trunk. The sap, being collected in wooden troughs, is boiled down to a certain consistency, after which it crystallizes on cooling. It is commonly used in a rough and unpurified state, and though retaining the peculiar flavor of the maple, is far from being disagreeable. Large quantities, however, are purified and sold at a remunerating price; also, converted into syrup of a fine flavor." "During the collecting season parties go into the woods, and camp out for several weeks, when the process of bleeding the trees and boiling down the sap are jointly carried on. The maple is often seen growing on very barren soils; its trunk is seldom more than a foot and a half in diameter." The timber is in demand for many kinds of purposes, being extensively used in the manufacture of furniture.

Meteorological Observations in the State of New York,

GIVING THE MEAN ANNUAL TEMPERATURE AND VARIATIONS.

Stations, &c.	Latitude.	Altitude. Feet.	Yearly Mean. ° Fahr.	High'st. ° Fahr.	Low'st. ° Fahr.	Range. ° Fahr.
Albany, . .	42° 31′	130	48.60	97	—23	120
Auburn, . .	42° 55′	650	46.62	96	—14	110
Buffalo, . .	42° 53′	620	47.14	92	—12	104
Canandaigua, .	42° 50′	600	45.73	94	—11	105
Cazenovia, .	42° 55′	—	43.65	97	—28	125
Cherry Valley, .	42° 48′	1,335	44.27	98	—30	128
Clinton, . .	43° 00′	500	45.96	96	—24	120
Delhi, . .	42° 16′	1,384	46.66	93	—17	110
Fishkill, . .	41° 34′	42	49.74	96	— 4	100
Flatbush, L. I., .	40° 37′	40	51.62	96	— 4	100
Fort Edward, .	43° 13′	—	45.08	90	—18	108
Goshen, . .	41° 20′	425	48.56	98	—20	128
Hamilton, . .	42° 49′	1,127	45.00	96	—34	130
Homer, . .	42° 38′	1,096	44.67	95	—28	123
Hudson, . .	42° 15′	150	48.00	99	—24	123
Ithaca, . .	42° 27′	417	48.38	98	—18	116
Jamaica, L. I., .	40° 41′	—	50.00	100	— 7	107
Johnstown, .	43° 00′	—	45.00	96	—30	126
Kinderhook, .	42° 22′	125	47.00	100	—30	130
Kingston, . .	41° 55′	188	49.37	100	—20	120
Lansingburgh, .	42° 47′	30	47.62	100	—28	128
Lewiston, . .	43° 09′	280	47.88	97	— 6	103
Lowville, . .	43° 47′	800	44.00	100	—40	140
Malone, . .	44° 50′	700	43.54	94	—24	118
Newburgh, .	41° 39′	150	49.67	100	—15	115
New York City,	40° 42′	50	51.00	96	—10	106
Ogdensburgh, .	44° 43′	280	44.00	92	—20	112
Oswego, . .	43° 28′	234	46.42	86	—18	104
Oyster Bay, L. I.	40° 50′	—	50.80	95	3	92
Penn Yan, .	42° 42′	750	46.50	95	—15	110
Plattsburgh, .	44° 42′	180	44.17	98	—20	118
Potsdam, . .	44° 40′	394	43.61	96	—34	130
Poughkeepsie, .	41° 41′	50	50.00	100	—22	122
Redhook, . .	42° 02′	60	48.36	98	—28	126
Rochester, .	43° 08′	520	47.00	98	— 9	107
Salem, . .	43° 15′	600	46.53	98	—38	136
Saratoga Springs	43° 06′	960	46.87	90	—38	128
Schenectady, .	42° 48′	—	46.82	91	—16	107
Syracuse, . .	43° 01′	400	47.30	94	— 3	97
Troy, . .	42° 43′	50	47.80	98	—26	124
Utica, . .	43° 06′	173	46.00	97	—27	124
West Point, .	41° 23′	167	50.50	100	—10	110

Climate of the State of New York.

The State of New York, lying to the westward of the New England States, is situated between 40° 30′ and 45° north latitude. "It extends over a hundred miles along the sea-coast, and stretches to the Great Lakes, presenting every variety of surface, from Alpine peaks to sandy plains, exposed to the soft breezes of the Atlantic and the chilling but bracing winds of the north, presenting all the modifications of climate which these varied circumstances can produce. The blossoming of plants, and other harbingers of spring, occur from two to three weeks earlier on Long Island, and the vicinity of the city of New York, than in the northern and western parts of the State, while in the latter, the first frost and snow, indicating the approach of winter, are seen nearly a month sooner. The progress of vegetation in midsummer, as indicated by the harvests, is found to vary but little, thus indicating the more rapid progress of vegetation in the colder sections of the State, and an approach to the short and hot summers of polar climates."

The northern section of the State, lying between Lake Champlain and the River St. Lawrence, has a mean annual temperature ranging from 43° to 44° Fahr.; the coldest month being January, and the warmest month July. The mean of the seasons is as follows:—Spring 42°, Summer 67°, Autumn 46°, Winter 20°. The Adirondack Mountains, lying mostly in the Counties of Clinton, Essex, and Franklin, have an altitude exceeding the Green Mountains of Vermont; Mt. Marcy, the highest peak, being elevated 5,467 feet above the sea. This high and mountainous section of the State, extending many miles westward, is still an unbroken forest, being extremely cold considering its latitude. The mercury frequently falls to 40° below zero and rises to 90°, showing a range of 130 degrees.

The middle section of the State, from Albany westward, is favored in every particular, having a rich soil and healthy climate. Here are to be found a succession of flourishing cities and towns surpassing almost any other section of the Union. The mean annual temperature at the Dudley Observatory, Albany, is 48° 60′ Fahr.; the mercury sometimes falling to 20° below zero, and rising to 98° above. The mean temperature of the seasons is as follows:—Spring 47°, Summer 70°, Autumn 50°, Winter 26° Fahr. The temperature falls as you approach Lake Erie. At Buffalo the seasons range as follows:—Spring 43°, Summer 67°, Autumn 48°, Winter 27°; the mean annual temperature being 46° 45′; the mercury sometimes falling to zero and rising to 86° above. Indian corn, wheat and other cereals flourish in this whole section of country, while the grasses yield in great abundance the most nutritious food for cattle—butter and cheese being produced in great quantities. Fruit of different kinds also flourish, particularly apples, they being largely exported, as well as converted into cider.

The Catskill Mountains, and the Valleys of the Hudson and Mohawk, are the most remarkable topographical features of this part of the State.

A late writer, in speaking of the soil and climate of Western New York, which is the garden of the State, remarks :—"Oak and hickory are the principal trees in the forest, where the soil is the most suitable for the growth of wheat. The butternut and the walnut are only sparingly distributed in the forests. In other parts of this region, where the subsoil is of a compact sand, the maple and the beech divide the land betwixt them. Maple and beech land is not so good for wheat, as there is usually more accumulation of vegetable matter, which renders it too soft, so that the plants are more liable to be thrown by the spring frosts, and the crop on such land is more subject to rust and mildew. The beech and maple land, however, is well adapted for spring and summer crops, such as Indian corn, barley and potatoes. Dr. Lindley, I believe, was the first to suggest that the distribution of forest trees over particular soils was regulated more by the physical condition than by the chemical composition of soils. In the general truth of this opinion I quite concur, and it is amply borne out in the facts which I have just stated regarding the oak and hickory, and the beech and maple soils, inasmuch as the adaptation of the first to winter wheat, and the last to spring crops, shows that it is the physical condition that determines the fitness of the soil for cultivated crops ; for we have only to bear in mind that winter wheat, barley, oats and Indian corn are identical in chemical composition."

The Southern section of the State, below the "Highlands," is a most favored spot. Here, and on Long Island, are annually produced the cereals, grasses, early vegetables, and fruit in abundance, most of which is taken to the New York market. The mean annual temperature of the City of New York is 51° ; the seasons ranging as follows :—Spring 48° 70', Summer 70° 10', Autumn 54° 50', Winter 31° 40' ; the mercury sometimes falling below zero, and rising to 96° ; showing a range of temperature less than in most of the other parts of the State.

The average annual fall of rain and snow in the City of New York is 44 inches ; Newburgh, 36 inches ; Albany, 41 inches ; Plattsburgh, 38 inches ; Utica, 40 inches ; Syracuse, 33 inches ; Rochester, 31 inches ; Buffalo, 28 inches ; Lewiston, 23 inches.

The average quantity throughout the State, 36 inches. In the City of New York about 12 inches falls in Spring, 12 in Summer, 10 in Autumn, and 10 in Winter.

The prevailing winds passing over this State are very variable, although westerly and northwesterly predominate. "One of the most striking results of the observations upon the winds is the correspondence between their direction and that of the valleys in which the stations are located. At most of those on the Hudson, northerly and southerly winds were recorded in the greatest number; in the Mohawk Valley, easterly and westerly or northwesterly winds; and at every other place, the prevailing direction of the neighboring hills and valleys was found to influence that of the surface current. Northeasterly winds prevail in the southern part of the State along the seacoast."

TABLE,

Showing the progress of the Seasons in the State of New York, as indicated by the opening of the Hudson River.

Navigation of the Hudson River.				Navigation of the Hudson River.			
Year.	Began.	Ended.	Days op'n	Year.	Began.	Ended.	Days op'n
1826,	Feb. 26.	Dec. 24.	300	1846,	Mar. 22.	Dec. 15.	268
1827,	Mar. 20.	" 25.	280	1847,	" 19.	" 24.	280
1828,	Feb. 8.	" 23.	320	1848,	" 9.	" 27.	293
1829,	April 1.	Jan. 11.	296	1849,	Feb. 25.	" 25.	303
1830,	Mar. 15.	Dec. 23.	283	1850,	Mar. 10.	" 17.	268
1831,	" 15.	" 5.	269	1851,	Feb. 25.	" 11.	288
1832,	" 25.	" 23.	273	1852,	Mar. 28.	Did not close this Y'r.	
1833,	" 21.	" 13.	267	1853,	" 22.	Jan. 4.	289
1834,	Feb. 21.	" 15.	297	1854,	" 17.	Dec. 7.	265
1835,	Mar. 25.	Nov. 30.	250	1855,	" 27.	" 21.	270
1836,	April 4.	Dec. 7.	247	1856,	April 7.	" 10.	240
1837,	Mar. 28.	" 13.	260	1857,	Mar. 18.	" 27.	285
1838,	" 19.	Nov. 25.	251	1858,	" 19.	" 18.	275
1839,	" 21.	Dec. 18.	272	1859,	" 23.	" 14.	267
1840,	Feb. 21.	" 5.	297	1860,	" 5.		
1841,	Mar. 24.	" 19.	270	1861,	" 6.		
1842,	Feb. 4.	Nov. 29.	298	1862,	" 3.		
1843,	April 13.	Dec. 9.	240	1863,	April 2.		
1844,	Mar. 14.	" 11.	272	1864,	Mar. 12.		
1845,	Feb. 24.	" 4.	282	1865,	" 14.		

NOTE.—The earliest date on which the Hudson has been open at Albany, was Feb. 4, in 1842, and the latest April 13, in 1843. Average time of opening about the middle of March; Average time of closing about the middle of December.

THE CLIMATE OF BUFFALO, N. Y.,

Lying at the foot of Lake Erie, in N. Latitude, 42° 53′; W. Longitude, 78° 58′; Altitude 600 feet.

From certain natural causes, no doubt produced by the waters of Lake Erie, the Winters are less severe, the Summers less hot, the temperature, night and day, at all seasons, more equable, and the transitions from heat to cold less rapid at Buffalo than at any other locality within the temperate zone of the United States, as will be seen by the following table :—

TEMPERATURE OF ALBANY, ROCHESTER, BUFFALO, CLEVELAND, DETROIT, CHICAGO, AND ST. PAUL, DURING THE FOUR SEASONS OF THE YEAR.

Cities.	Spring.	Summer.	Autumn.	Winter.	Yearly Mean.	Range.
Albany, N. Y.,	. 46°	71°	50°	24°	48°	47°
Rochester,* N. Y.,	. 47°	67°	49°	27°	47½°	42°
Buffalo, N. Y.,	. 46°	66°	48°	28°	47°	42°
Cleveland, Ohio,	. 45°	71°	51°	26°	48½°	45°
Detroit, Mich.,	. 46°	67°	49°	26°	47°	41°
Chicago, Ill., .	. 45°	68°	49°	26°	47°	42°
St. Paul, Minn.,	. 45°	70°	46°	16°	45°	54°

By a careful examination of the above table, it will be seen that during the *Summer months*, the temperature of Buffalo is from 2° to 10° cooler than that of any other point, east, south, or west, of the foot of Lake Erie, while the refreshing and invigorating lake breeze is always felt both night and day.

The *Winter Months* compare favorably with Albany and all the lake ports. The thermometer rarely indicates zero, and the mean for January, 1858, was 20° above ; the usual range during the year being from 0° to 90° Fahrenheit ; yet, during extreme hot Summers and cold Winters, the extreme range has been as high as 104°.

"An equally important fact, is the gradual transition from cold to heat, and from heat to cold, in the Spring and Autumn months. In most localities south the temperature suddenly changes at these seasons, showing a change from 25° to 50°. This is debilitating to the constitution and gives rise to diseases, almost as fatal as contagion, which are unknown here.

"The equableness of the temperature of Buffalo is owing to the prevailing direction of the winds, and the fact that the lakes are never completely frozen over. We learn from Mr. Ives, the Librarian of the Young Men's Association, that of 2,100 observations of the course of the winds in 1858–9, from eight points of the compass, 780, or 37 per cent. of the winds were from the southwest, or Lake Erie. Of 860 observations made in the months of November, December, January, and February of the same years, 57 per cent. were from the southwest, west, and

* The temperature of Rochester and the surrounding country is favorably modified by the waters of Lake Ontario ; hence its adaptation to the growth of cereals and fruit of almost every description.

northwest, or from Lakes Erie and Ontario. The remainder, 43 per cent., were about equally divided between the other five points of the compass. In other words, most of the Winter winds are south and southwest, and most of the Summer winds are west and northwest. The Meteorological Register for the winter months of 1859 shows the following:—West wind, 12 days; northwest, 5 days; north, 9 days; northeast, 11 days; east, 7 days; southeast, 4 days; south 10 days; southwest, 28 days; showing that the northerly winds in winter compare with the southerly winds as 6 to 14. There is an average of a little more than one day to a month of strictly north wind.

"Our attention has been called to another fact, illustrating the superiority of the climate of this vicinity. It is well known among professional culturists, that the flavor of fruit, and its perfection, is, perhaps, the most delicate and satisfactory test for health; and it is also known that the fruit produced in localities protected on the north, northeast, and northwest, by the never-freezing waters of Lake Ontario, which temper the severe northern winds, and shielded by the Alleghany ridge from devastating storms, enjoys a world-wide reputation for richness of flavor. Fruit grown south of the lakes is of an inferior quality and cannot be eaten with impunity.

"There are other interesting facts," says the above writer, "in this connection to which it would be pleasant to allude. Our city is pre-eminently the residence of healthy, vigorous, working men. *Labor is actually worth twenty per cent. more here than in manufacturing cities exposed to the south and southwest hot, dry winds of summer, and which are not cooled by any large body of water.*"

Climate of the Middle States.

The Middle States, lying southwest of New York, are comprised of New Jersey, Pennsylvania, Delaware and Maryland. Pennsylvania, extending the most northwardly, is bounded by 42° N. Latitude; and Maryland, the most southwardly, is bounded by 38° N. Latitude on its eastern limits. This section of country extends from the Atlantic Ocean to Lake Erie, on the western confines of Pennsylvania, 80° 40′ W. Long. On the north it has a mean annual temperature of from 47° to 48° Fahrenheit. The mean of the seasons is about as follows:—Spring 45°, Summer 67°, Autumn 48°, Winter 27° Fahr. The Alleghany range of mountains extend through Pennsylvania and Maryland, lowering the temperature in some elevated places very materially. This portion of the States is mostly covered with a heavy growth of timber of different kinds, such as pine, spruce, hemlock, oak, maple, beech, chestnut, walnut, &c.

The middle section, in the vicinity of Harrisburg, Pa., has a mean annual temperature of 50° Fahrenheit. The seasons are as follows:—Spring 49°, Summer 70°, Autumn 50°, Winter 29°. January is usually the coldest month, and July the warmest. Indian corn, wheat, rye, and oats, together with nutritious grasses, vegetables, and fruit of different kinds, flourish in this favored section of country—Pennsylvania being one of the richest agricultural and mineral States in the Union, while New Jersey and Delaware are both justly celebrated for producing delicious peaches, strawberries, cranberries, &c., together with most kinds of vegetables which find a ready sale in the Philadelphia and New York markets.

The southern section of the Middle States has a mean temperature of 58° Fahr. The seasons are as follows:—Spring 58°, Summer 76°, Autumn 59°, Winter 37°. The southern portion of Maryland, lying on the Chesapeake Bay, is a level, sandy section of country, producing Indian corn, wheat, tobacco and sweet potatoes. The country here changes materially from that portion lying above Mason and Dixon's line, or the north boundary of Maryland bordering on Pennsylvania.

The average annual fall of rain in the Middle States is 40 inches, the largest quantity falling near the sea-board or in the vicinity of Chesapeake and Delaware Bays. At Philadelphia, 45 inches is the usual annual fall, while at Pittsburgh but 35 inches usually fall. In both instances the quantity is nearly divided between the four seasons.

Intermittent and other fevers prevail in the southeast part of this region, in the vicinity of Chesapeake Bay, while the climate found on the sea-shore, from New Jersey to Virginia, is celebrated for its health-restoring and invigorating qualities.

Climate of the Western States.

This healthy and fertile section of the Union possesses great advantages as regards climate and soil, having all the elements that tend to increase wealth, knowledge, and refinement. It embraces the States of Ohio, Indiana, Illinois, Iowa, and a portion of the Territories west of the Missouri River, where is found about the same mean annual temperature. Lying in the middle of the Temperate Zone and near the centre of the Mississippi Valley, it is susceptible of sustaining a vast population. Lakes Erie and Michigan lie on the north, the Ohio River on its south border, and the Mississippi and Missouri

Rivers on the west; affording direct communication by lake, river, and canal, with every part of the United States and Canada.

The mean annual temperature ranges from 46° on the northern confines of Illinois and Iowa, to 54° Fahr. in southern Illinois, running through 6½ degrees of latitude. Here is the greatest yield of Indian corn and wheat of any other section of the Union.

The mean annual temperature of Chicago, Ill., is 47° Fahr. The seasons are as follows:—Spring 45°, Summer 68°, Autumn 49°, Winter 26° Fahr.; the temperature being modified by the waters of Lake Michigan. This favored city lying near the head of the lake, is one of the greatest grain and lumber markets in the world, having a water communication with the Atlantic Ocean by means of the Lakes and St. Lawrence River; and with the Gulf of Mexico, by means of canal and river navigation. The system of railroads also centring here affords altogether unrivalled facilities for transhipping the products of this whole immense region of fertile country.

The southern portions of Ohio, Indiana and Illinois are all alike favored by a good climate and rich soil; the mean annual temperature ranging from 52° to 55° Fahrenheit. The mean annual temperature of Cincinnati is 55° Fahr. The seasons are as follows:—Spring 54°, Summer 73°, Autumn 53°, Winter 33° Fahr. January is the coldest, and July the hottest month; the mercury occasionally falling to zero and rising to 96° Fahr. Average annual fall of rain, 42 inches. Here Indian corn, the grape of different kinds, and most of the cereals flourish in great perfection. No portion of the Union is richer or more favored than the Valley of the Ohio, running nearly east and west for near one thousand miles.

The grasses of Ohio, Indiana, Illinois, and Iowa, supply food for great quantities of cattle—the prairie lands, in particular, affording good pasturage. The eastern markets are to a great extent supplied with beef cattle from this section.

Ohio.—In order to show the great amount of agricultural products annually raised in the Valley of the Ohio and Mississippi, compared to the Canadas, we extract the following interesting summary:—

"The northern half of the State of Ohio and the eastern borders are best suited for the growth of wheat; the southern for Indian corn, tobacco, and for grass. There is comparatively little alluvial land along the Ohio, as it has cut a deep channel out of the table-land, and, as already observed, the whole country on both sides is broken into hillocks. A surface so irregular, being less suited for cultivating on a large scale, naturally became occupied with small proprietors, who usually plant crops, such as tobacco and vines, which require more hand labor than those that are more generally raised. The farms are larger on the more level and fertile description of lands."

"There is little unreclaimable land in Ohio, though a large proportion is still in wood. This State is about 200 miles in length, and nearly as many in breadth; the annual temperature varying from 48°, on the shores of Lake Erie, to 54° Fahrenheit, extending south towards the Ohio River. It covers an area of 39,964 square miles, or 25,576,960 acres, of which 9,851,493 were reclaimed in 1850.

"To show the particular direction that agricultural production takes north and south of the Lakes the statistics of Ohio may be compared with those of the Canadas. In 1851 there were 7,300,839 acres of reclaimed land in the Canadas out of 155,188,425 acres. The population of the Canadas was then 1,842,265; of Ohio, 1,980,427. The amount of their chief products were:—

Products.	Ohio. Bushels, &c.	Canada. Bushels, &c.	Products.	Ohio. Lbs.	Canada. Lbs.
Wheat, ...	14,487,351	16,155,946	Butter, ...	34,449,379	25,613,467
Other Cereals, etc.	15,981,191	28,052,301	Cheese, . . .	20,819,542	2,737,790
Indian Corn, . .	59,078,695	2,029,544	Tobacco, . . .	10,455,449	1,253,128
Sheep, ...	3,942,929	1,597,849	Maple Sugar, .	4,588,209	9,772,199
			Wool, ...	10,196,371	4,130,740

The State of Ohio, it appears from the above table, raises a greater amount of agricultural produce than the whole of the Canadas, and in all probability it will continue to do so for many years to come. The statistics indicate the prominent place that Indian corn* occupies in the productions of Ohio, and the small quantity grown in the Canadas; conclusively going to show that the climate of the former section of country is better adapted by nature to the above important production; so in regard to tobacco.

* In Canada East, where the thermometer ranges below 40° Fahr., mean annual temperature, the crop of corn and most of the cereals cease to be cultivated.

TABLE OF METEOROLOGICAL OBSERVATIONS AT COLUMBUS, OHIO.

Lat., 39° 57′ N.; Long., 83° 3′ W. Altitude, 740 feet above the Atlantic Ocean.
By J. B. RICHARD.

Months. 1857.	Thermometer. Mean, ° Fahr.	Rain, Inches.	Snow, Inches.
January,	21.0	1.36	10⅞
February,	43.2	2.60	½
March,	39.3	1.36	4
April,	43.9	2.51	3¾
May,	59.7	5.50	
June,	70.4	4.75	
July,	74.2	3.24	
August,	73.7	3.58	
September,	69.7	2.12	
October,	54.5	4.84	
November,	41.4	6.35	3¼
December,	41.0	3.54	⅞
Annual mean,	52.7	41.75	23½

NOTE.—The total quantity of moisture, including 23½ inches melted snow, was 41.75 inches, being 1.81 inches above the average quantity.

From the Ohio to the Lakes, extending westward to the Rocky Mountains, the advantage of free labor is fully exemplified by the show of an increase of industry and of all the elements of education, which tend to make a nation wealthy and happy. Here freedom has her favorite abode, made sacred by bonds which no legislation can disturb without sacrificing the best interests of the community. The rapid growth in population and wealth of this portion of the Union is without precedent. Already does it assume gigantic proportions, which is, no doubt, soon destined to exercise a governing influence in the political affairs of the Great Republic of modern times. Already has the centre of population of the Union crossed the Alleghany chain, and fixed the initial point near Columbus, Ohio, near north latitude 40°, the climate here being of the most favored character, ranging in the vicinity of 51° Fahrenheit, yearly mean.

When the east and the west, extending from ocean to ocean, are united by a continuous railroad and line of settlements, running along the most favored temperature, then will the sway of power be found near the centre of the Great Mississippi Valley, capable of sustaining tens of millions of freemen, and at the same time of affording a surplus of food for less favored sections of the world.

Climate of the Northwestern States and Territories.

This section of the United States, lying partly in the Basin of the Great Lakes and partly in the upper part of the Valley of the Mississippi, embraces the States of Michigan, Wisconsin, Minnesota, and the Territories of Dakota and Montana, extending westward to the Rocky Mountains. It is bounded on the north by the Great Lakes of America and the 49th parallel of latitude, dividing it from the British possessions on the north.

The extremes of mean annual temperature are from 38° to 48° Fahrenheit. The coldest post on record is Fort Ripley, Minn., in N. Lat. 46° 19′, the mean annual temperature being 39° 30′. The seasons are as follows :—Spring 39° 33′, Summer 65°, Autumn 43°, Winter 10° Fahr. The coldest month is January, and the warmest month July; the extremes being from 36 degrees below to 90 degrees above zero. Average annual fall of rain, 30 inches.

The military posts on the line of the northern frontier are Detroit, Fort Gratiot, Mackinac, Fort Brady (Sault St. Marie), and Fort Wilkins (Copper Harbor), all being influenced more or less by the climate peculiar to the Great Lakes. The mean annual temperature of the three latter stations is about 40° Fahr., or that of Quebec, Can., and Houlton, Maine. Fort Wilkins, in N. Lat. 47° 30′, lies on Keweenaw Point, jutting out into Lake Superior, being mostly surrounded by water. The mean of the seasons are as follows :—Spring 38° 47′, Summer 60° 80′, Autumn 43°, Winter 22° ; mean yearly temperature 41° Fahr. This post furnishes good data for the temperature prevailing along the south shore of Lake Superior, one of the largest and purest bodies of fresh water on the face of the globe.

The country north and west of Fort Ripley, situated on the Upper Mississippi, is mostly uninhabited except by Indians. This section abounds in streams and lakes of pure water, abounding in fish of different kinds. The Red River of the North, which drains a portion of Minnesota and Dakota, is bounded on both sides by a fine section of country, here producing wheat and other kinds of grain and vegetables in abundance, the climate being remarkably healthy and invigorating.

The southern and middle portions of the States of Michigan, Wisconsin and Minnesota are very similar in their temperature and agricultural productions, producing large crops of wheat and other cereals, as well as grasses of different kinds; this whole region being well adapted to the raising of cattle, sheep, and hogs. From Racine and Milwaukee west to Janesville and Madison, and thence to the Upper Mississippi, the lands are very productive, and profitably tilled by American and European settlers. The spring and summer months are warmer in the same latitude as you approach the Mississippi River. The same will hold good in reference to St. Paul, Minn., in a higher latitude; thus rendering vegetation earlier on the Mississippi than in the vicinity of the Great Lakes. The mean annual temperature of OMAHA CITY, Nebraska, situated on the west side of the Missouri River, in N. latitude 41½°, is 49.28° Fahr.; the Spring 49.28°, Summer 74°, Autumn 51°, Winter 22°; the Summer months being 7 degrees warmer than Milwaukee, and 4 degrees warmer than the City of New York, having the Summer temperature of Baltimore, Md.; the winter temperature being about the same as Burlington, Vermont.

Crossing the Plains.

In his last letter from DENVER, Colorado, Mr. Bowles, of the Springfield Republican, gives an interesting account of the condition of the soil, climate, etc., along the route of the *Union Pacific Railroad:*

"The Platte River is a broad, shallow but swift stream, furnishing abundant good water for drinking and for limited irrigation, but offering no possibilities of navigation—not even for ferriage. When it is too swift and strong for fording, it must be let alone, and a route on either shore kept to, or the falling waters waited for. The soil of the valley and of the plains, which it crosses, is not by any means mere sand, but rather a tough, cold, sandy loam, with an admixture of clay. It is too cold and dry for corn and vegetables. Wheat and barley may be raised on its best acres, with the help sometimes of a simple irrigation; but the pasture is its manifest destiny and use. There is a steady, imperceptible rise from the Missouri to the Rocky Mountains; half way, we get above the dew falling point; and here at Denver, at the base of the mountains, we are 5,000 feet above the level of the sea. The days are warm, however; the sun pours down over its shadeless level with a hot, burning power; but a cool wind tempers its bitterness, and at night the air is absolutely cold. This is the universal rule of all our western country, from the Mississippi valley, and distinguishes the summers of its whole extent from those of the East."

A TABLE,

Showing the Monthly Mean Temperature of the open air in the shade at MILWAUKEE, Wisconsin, from 1850 to 1864, as observed by CHARLES WINKLER, M. D., and I. A. LAPHAM, LL. D. (Lat. 43° 03′ N.; Long. 87° 56′ W. Elevation above the sea, 600 feet.)

	1850. I. A. L.	1851. I. A. L.	1852. I. A. L.	1854. I. A. L.	1855. C. W.	1856. C. W.	1857. C. W.	1858. C. W.	1859. W.&L.	1860. I. A. L	1861. I. A. L.	1862. I. A. L	1863. I. A. L.	1864. I. A. L.	Mean of the whole.
January	27.65	27.25	20.72	.....	22.58	10.21	7.75	30.90	23.89	23.99	21.22	20.30	29.76	18.77	22.76
February	28.84	31.26	28.03	.....	14.50	15.02	27.72	16.83	26.96	27.44	27.78	18.35	26.88	25.92	25.30
March	32.02	39.38	31.98	36.17	27.97	24.10	27.03	35.25	37.45	40.38	31.96	32.40	31.42	30.65	33.67
April	40.03	43.64	38.17	43.65	46.70	43.68	33.32	41.51	39.50	43.76	44.82	41.60	44.95	40.12	43.98
May	50.10	52.15	54.65	54.24	53.67	51.05	49.19	49.55	55.01	56.24	50.68	53.65	55.55	55.15	53.95
June	65.59	63.20	65.66	65.50	59.71	66.76	61.01	67.23	60.35	64.17	64.53	60.84	63.70	65.22	64.27
July	72.59	69.02	70.32	74.13	67.69	68.88	68.89	70.92	71.59	69.00	68.05	71.09	68.55	71.27	70.21
August	69.58	67.26	68.13	71.81	65.51	63.91	67.48	68.85	70.80	66.82	69.21	70.63	69.03	69.75	68.11
September	59.70	65.52	58.82	65.65	61.45	57.75	61.04	61.66	59.29	58.76	63.71	62.29	59.42	60.97	61.58
October	49.35	49.51	53.12	54.13	44.61	47.60	46.81	49.30	48.16	50.00	50.11	51.00	43.45	45.67	48.65
November	41.95	34.45	33.42	37.79	36.61	34.12	29.57	33.60	40.28	36.15	36.79	35.35	36.81	32.08	35.94
December	24.55	21.98	26.52	27.41	21.87	15.83	30.71	27.86	19.67	24.80	31.11	30.74	29.16	20.13	25.77
Winter	25.82	27.69	23.58	.....	21.50	15.70	17.10	26.15	26.24	23.70	24.60	23.25	29.13	24.62	24.61
Spring	40.72	45.06	41.60	44.69	42,78	39.61	36.51	42.10	43.99	46.79	42.49	42.55	43.93	41.97	43.87
Summer	69.25	66.49	68.04	70.48	64.27	66.52	65.79	69.00	67.58	66.66	67.26	67.52	67.09	68.75	67.53
Autumn	50.33	49.83	48.45	52.52	47.56	46.49	45.81	48.19	49.24	48.30	50.21	49.55	46.56	46.24	48.72
YEAR	46.83	47.05	45.70	.....	43.56	41.57	42.55	46.12	46.09	46.80	46.66	45.69	46.55	44.64	46.18

The above Table, which has been compiled with great care from manuscript record in my possession, will be found valuable by enabling us to compare the climate of Milwaukee with that of other places. It will be seen that the general mean temperature, deduced from all the observations, is 46.18°; that the coldest year was 1856 (41.57°), differing 5.26° from the warmest year, 1850, (46.83°). The coldest month is January (22.76°), though in six cases February was the coldest. January, in 1857, was the coldest month known since the settlement of the place, the mean temperature being only $7\frac{3}{4}$°. July is the warmest month (70.21°). June, in 1854, was the warmest month. The Winter and Spring of 1845 were the most moderate in eighteen years; the Winter of 1856 and the Spring of 1857 were the coldest. The hottest Summer, 1854, was followed by the mildest Autumn. It will be noticed that no *extreme* of temperature, either of heat or cold, has occurred since 1857. If we take the mean annual temperature of five years together, it will be found that the terms ending with 1853 (46.71°), and with 1863 (46.36), were very near the general mean, while the term ending with 1857 was the coldest (42.56°); thus indicating that there may be some truth in the popular theory of cold and warm terms of years.

I. A. LAPHAM.

MINNESOTA—*Its Situation, Climate and Productions.*—The State of Minnesota extends from 43½° to 49° North latitude, and from 89° 29′ to 97° 5′ of West longitude. The State derives its name from its principal river, the Minnesota, which, in the Dakota language, signifies "*sky-tinted waters.*"

This favored State, as regards climate and productions, "occupies the central point of the North American continent, midway between the Frigid and Torrid Zones, midway between Hudson's Bay and the Gulf of Mexico, and midway between the Atlantic and Pacific Oceans.

"According to the latest estimate, the State embraces an area of 84,000 square miles (53,760,000 acres), an extent much greater than the territory comprised in all the New England States, and nearly equal to the combined areas of Ohio and Pennsylvania.

"The general surface of the country is undulating, similar to the rolling prairies of the adjoining States of Iowa and Wisconsin, with greater diversity, beauty and picturesqueness imparted to the scenery by rippling lakes, sparkling waterfalls, high bluffs, wooded ravines, and deeply cut channels, through which rapid currents wend their tortuous way, visiting almost every homestead.

"To this general evenness of the surface, the high lands known as the *Hauteurs des Terres,* form the only exception. These are a chain of drift hills in the northern part of the State, commonly with flat tops, rising from 80 to 100 feet above the level of the surrounding country. Among these hills lie embedded the lakes that give rise to the three great rivers of the continent. The Mississippi, pursuing a southward direction, over ledges of limestone, through fertile prairies and rich savannas, gathering its tributaries from a country of great fertility and nearly equal in extent to one third the area of Europe, pours its waters into the Gulf of Mexico. Eastwardly, through lakes, rivers and foaming cataracts, flow the waters of the St. Lawrence system, finding their way to the Atlantic. Northward runs the Red River, by a circuitous route, to Lake Winnipeg, where it mingles with waters brought from the Rocky Mountains by the Saskatchewan, and rolls onward to Hudson's Bay.

"The summit of the narrow ridge which divides the sources of the Mississippi and Red River, and highest point of land in the State, in latitude 47° and 95° west longitude, is 1,680 feet above the level of the Gulf of Mexico, and 2,896 miles from it by the river's course.

"From this eminence diverge three distinct slopes, which

give to Minnesota the form of a vast pyramid, down whose sides the disparted waters descend to their ocean outlets. In a southeasterly direction extends the great Mississippi slope, resting its broad base upon the Southern Gulf; eastwardly stretches the great Superior slope, walled in by the rocky coast of Labrador; and northward reaches the slope of the Red River, which, uniting with the Saskatchewan Valley, gives this vast interior basin of the continent the form of an irregular triangle, whose centre is in Minnesota.

"That portion of the *Mississippi Valley* included within the limits of Minnesota, has an estimated area of 49,000 square miles, being more than four-fifths of the whole State. The Mississippi River, from its source to the mouth of the Minnesota, a distance of 630 miles by its course, falls 960 feet, whilst the general level of the country sinks 830 feet. From this point the land, which is 850 feet above the level of the sea, gradually rises to an average height, near the Iowa line, of 1,000 feet; while the river, sinking gently at the rate of four inches to the mile, gives that progressive and picturesque elevation to the banks which characterize the Upper Mississippi.

"Towards the Minnesota Valley the surface of the country exhibits an exterior depression, which is continued throughout the Red River Valley and the great transverse basin of the Saskatchewan, and exercises an important influence upon the climate.

"The Mississippi, the principal River, originating in Lake Itasca, in the northern part of the State, flows southeasterly through Minnesota about 900 miles, of which 134 miles wash its eastern boundary. The St. Croix, its principal tributary on the eastern side, rises in Wisconsin, and forms about 130 miles of the eastern boundary of Minnesota.

"The Minnesota, the principal branch on the western side, rises in the Coteau des Prairies in Dakota Territory, and extending into Big Stone Lake, on our western boundary, flows with a vast sweep through the heart of the State, and empties, 470 miles from its source, into the Mississippi, five miles above Saint Paul.

"The *Superior Slope* has an area of 15,000 square miles. It is traversed by ranges of hills parallel with the Superior shore, which stretches westward to the heights of land that separate the Superior from the Mississippi basin. The hills are sandy, with a small growth of wood; the intervening valleys have a good soil and are well wooded, but are often swampy and imperfectly drained. The rivers of this section are numerous, generally short, and often fall in beautiful cascades over ledges of primary rock.

"The *Red River Slope*, whose southern point extends to Lac Traverse, separated from Big Stone Lake by a distance of only three miles, extends northward, maintaining a uniform altitude of nearly 1,000 feet. The Red River has its source in the heights of land near the head-waters of the Mississippi, where it flows southwardly, then making a sudden detour, where its waters become navigable, it runs nearly due north, washing the western boundary of Minnesota for 380 miles.

"The American Valley of the Red River is about 250 miles from north to south, and contains an area of 17,000 square miles. Capt. Pope, in his official report to Congress, says: 'In its whole extent it presents an unbroken level of rich prairie, intersected at right angles by all the heavily timbered tributaries of the Red River, from the east and west, the Red River itself running nearly north through its centre, and heavily timbered on both sides with elm, oak, maple, ash, &c. This valley, from its vast extent, perfect uniformity of surface, richness of soil, and abundant supply of wood and water, is among the finest wheat-growing countries in the world.'

"The central table land, around which the grand primary slopes converge, is a semicircular curve, surrounded by that immense system of reservoirs which for ages have poured their waters through different outlets into the sea. This level is described as an interminable labyrinth of lakes and streams, separated by low savannas and narrow sandy ridges, covered with pine; the alluvial bottoms with dense forests of hard wood.

"*Climate.*—Prominent among the questions proposed by the emigrant seeking a new home in a new country, are those concerning the climate, its temperature, adaptation to the culture of the grand staples of food, and its healthfulness.

"The climate of Minnesota has often been the subject of unjust disparagement. 'It is too far north;' 'the winters are intolerable;' 'corn will not ripen;' 'fruit will not grow.' These and other similar remarks have found expression by those who should have known better. To the old settler of Minnesota, the seasons follow each other in pleasing succession. As the sun approaches his northern altitude, winter relaxes his grasp, streams and lakes are unbound, flowers spring up as if by the touch of some magic wand, and gradually Spring is merged into the bright, beautiful June, with its long, warm days, and short, but cool and refreshing nights. The harvest months follow in rapid succession, till the golden Indian summer of early November foretells the approach of cold and snow; and again winter with its short days of clear, bright sky, and bracing air, and its long nights of cloudless beauty, completes the circle.

"It will be remembered, that though Minnesota has no mountain peaks, its general elevation gives it the characteristics of a mountainous district; that while it is equidistant from the oceans that wash the eastern and western shores of the continent and is therefore comparatively unaffected by oceanic influences, it has a great water system of lakes and rivers within its own borders. These, combining with other influences, give the State a climate in many respects dissimilar to the other northern States.

"One of the most striking of the peculiarities of this climate is the great variation between the extreme cold of winter, when mercury congeals, and the intense heat of midsummer, when it stands, for many consecutive days, at 95° above zero, in the shade.

"But these extremes afford no index to the real character of the climate of Minnesota. Fortunately we have ample means by which to determine its actual temperature, and also its temperature compared with other and more widely known localities.

"From records kept for a series of years, at different places, the Commissioner of Statistics, in his report for 1860, furnishes the data for the following summary:—Central Minnesota has a mean temperature in the spring (45°) equal to Northern Illinois, Southern Michigan, and Massachusetts. Its Summer mean temperature (70°) coincides with that of Central Wisconsin, Pennsylvania, and Southern New York. Its Autumn temperature (49°) is the equivalent of Central Wisconsin, Northern New York, and New Hampshire. Its Winter mean temperature (16°) equals that of Northern Wisconsin, the southern limit of Canada East, Central Vermont, and New Hampshire.

"Its yearly mean temperature (44.6) coincides with that of Central Wisconsin, Michigan, Northern New York, New Hampshire, and Maine, and has a range from the Summer heat of Southern Ohio and Southern Pennsylvania. Thus, in the breadth of four degrees, the Summers of Pennsylvania and Southern New York are followed by the Winter of Canada and Northern Maine.

"It may be remarked that the hilly district in the northern part of the State, comprising about one-fourth of its surface, has less than 65° Summer heat, or the temperature of Canada and New England; whilst the remaining three-fourths, having a southern slope, warmed by the southerly breeze that sweeps up the Valley of the Mississippi, has a general average of 70° Summer heat, or the climate of Pennsylvania and Ohio.

"From rain tables prepared from observations recorded for a series of years at sixteen different places in Canada and the States, it appears that the mean yearly fall of rain for all the places is 35.5 inches; whilst the mean yearly fall at Fort Snelling is 25.4 inches, and the mean Summer fall for all the places is 11.2 inches, whilst the mean Summer fall at Fort Snelling is 10.9 inches.

"Thus it will be seen, that while Minnesota had a yearly fall of rain ten inches less than the mean of all the places, its Summer rain is but a fraction of an inch less than the mean Summer rain of all the places. It may be added, that one half of the Spring rain falls in the month of May, and a fraction more than one-half of the rains of Autumn falls in September, giving more than two-thirds of the whole yearly amount of rain to the season of vegetable growth, and leaving but the small fraction to the remaining seven months of the year.

"Judging from the climate of New England, where the air is loaded with vapor from the ocean, and the ground is for months covered with deep snows; or judging from the more southern of the Western States, where rain and sleet are followed by severe cold, it has been concluded that winter in Minnesota is a season of terrible storm, deep snow and severe cold. The average fall of snow is about six inches per month. This snow falls in small quantities, at different times, and is rarely blown into drifts so as to impede travelling. The first snow-fall of November usually lays on the ground till March, affording protection to the winter grain. Occasionally at midday a slight thaw occurs in places with a southern declivity. Two or three times in the course of eight or ten winters, the ground has been uncovered for a few days. Long driving snow storms are unknown, and rain seldom falls during the winter months.

"With an average temperature of 16°, the dry atmosphere of Winter in Minnesota is less cold to the sense than the warmer, yet damp, climate of States several degrees further south. With the new year commences the extreme cold of our Minnesota Winter, when, for a few days, the mercury ranges from ten to thirty degrees below zero, falling sometimes even below that. Yet the severity of these days is much softened by the brilliancy of the sun and the stillness of the air. Thus, while other States in lower latitudes are being drenched by the cold rain storm, or buried beneath huge drifts of wintry snow, Minnesota enjoys a dry atmosphere, and an almost unbroken succession of bright cloudless days and serene star-lit nights; and when the moon turns her full-orbed face towards the earth, the night scene of Minnesota is one of peerless grandeur.

"*Adaptation of Climate to Agriculture.*—Scientific men have determined that the successful cultivation of Indian corn requires a temperature of 67 deg. for July, and of 65 deg. for the Summer. Minnesota has a Summer temperature of 70 deg. and a temperature for July of 73 deg. The cultivation of wheat is said to require a mean temperature of from 62 to 65 deg. for two of the Summer months. Thus it will be seen that the climate of Minnesota is well adapted to the successful cultivation of all the cereals.

"The fact established by climatalogists, that 'the cultivated plants yield the greatest products near the northernmost limit at which they will grow,' finds abundant illustration in the productions of Minnesota. It is a well known fact that cereals raised in the southern latitude are far inferior in quality to the same kind produced in the cooler climate of the north. Corn, which grows to the height of thirty feet in the West Indies, yields but a few kernels on a spongy cob. In the Southern States the stalks grow fifteen feet high, and yield fifteen bushels per acre; in highly cultivated sections of the north from eighty to one hundred bushels are taken from stalks seven or eight feet in height.

"The warm early Springs of milder latitudes develop the juices, and push forward the leaf and stalk at the expense of the seed; whilst the cool, late Spring weather of the north checks this rank luxuriance of leaf and stem, and reserves the chief development to the ripening period. Minnesota, with its peculiar climate, combining the warm Summers of the southern of the more Middle States, with the cool, backward Springs of New England, exceeds the latter section in the quantity of its products, because its Summers are warmer; and the former region in the quality of its products, because its Springs are cool, and hold back the growth of the plant. Thus this State enjoys the conditions of temperature during the growing season adapted to the production of superior grains, grasses, and esculent roots."

PART X.

CLIMATE OF THE SOUTHERN STATES.

Northern Section, or Border States.

THIS portion of country, lying south of the Potomac and Ohio Rivers, known as the "Border States," constitutes a most favored region. It lies between 35° and 40° north latitude, and embraces the States of Virginia, West Virginia, Kentucky, Tennessee, and Missouri; being bounded on the west by Kansas and the Indian Territory.

The climate varies from 50° to 60° mean annual temperature, in running through five degrees of latitude, being included in the Temperate Zone. The coldest month is January, and the warmest month July.

The temperature of the seasons is as follows:—

Stations.	N. Lat.	Spring.	Summer.	Autumn.	Winter.	Year.
Norfolk, Va., .	37°	57° Fahr.	77°	61°	42°	60°
Lewisburg, W. Va.	38°	54° "	74°	55°	35°	54°
Louisville, Ky., .	38° 08′	55° "	75°	55°	37°	54°
Memphis, Tenn., .	35° 08′	61° "	78°	59°	42°	60°
St. Louis, Mo., .	38° 40′	54° "	76°	55°	32°	54°

Mean annual fall of rain, from 36 to 46 inches.

In the eastern section of this region, facing the Atlantic Ocean for about one hundred miles, and extending westward across the Alleghany range of mountains to the Ohio River, the climate and soil are varied, chiefly owing to altitude. On the sea-board the country is sandy and level, being in most places clothed with a growth of yellow pine when not cleared for cultivation. The numerous streams flowing into Chesapeake Bay drains the most of Virginia east of the Alleghany ridge, affording many navigable rivers. Indian corn, wheat and tobacco are the chief agricultural products, the latter being very extensively cultivated, and exported in large quantities.

Western Virginia, including the Alleghany range and its western slope to the Ohio River, embraces a healthy, rich, and romantic section of country well adapted to free labor. The forest trees are varied and of a large growth, the mountain sides and even the summits being mostly heavily timbered. The valleys have a rich soil, with a mild and invigorating climate. The medicinal fountains here found in great variety, together with the health-restoring atmosphere, give this section a famed celebrity which it most justly deserves.

Climate of West Virginia.

"A study of the causes affecting the climate of West Virginia, forming a part of the Alleghany range of mountains, will be found interesting. In its latitude, lying as it does mainly between 37° and 40° North, it is neither suggestive of hyperborean blasts in winter, or a torrid temperature in summer, of pent-up valleys, blockaded with drifted snow and solid ice for weary months, or sweltering plains, parching and baking under a brazen sky. Its mountains, unlike those of Europe, or the Rocky Mountains in the west, do not very materially affect the conditions of climate, except to reduce the temperature in proportion to altitude. There are local differences, to be sure, the result of peculiar position, but the interior valleys of the Alleghanies have nearly the same temperature as the broad slopes on either side, and these opposite slopes scarcely differ in their climatic peculiarities. Unlike the mountains of Europe, however, the Alleghanies in this latitude have less rain than the plains below."

ALTITUDE.—"The average altitude of the highest summits is 2,500 feet in this section of the Alleghany range, increasing southward. The upper valley of the Kanawha, instead of being an arid desert like the Colorado and other elevated plateaus, is luxuriant in verdure, differing comparatively little in humidity and temperature from the Atlantic coast and the Ohio Valley in the same latitudes; indeed, the elevation of the Kanawha is but 2,500 feet in Southern Virginia near its source, descending more than one hundred miles before it bursts its Alleghanian barrier in Monroe County, West Virginia, where it ranges between 1,800 and 1,300 feet—thence rapidly falling to little more than 600 feet at the foot of the falls near the mouth of the Gauley, whence it flows gently, with the slight descent of a few inches to the mile, to the Ohio River. The following table exhibits the elevation of the Alleghanies and their slopes in this section of that great mountainous range:—

SUMMIT ELEVATION.

Summit in latitude 37½°, ...	2,650 feet.
Summit at crossing of Baltimore & Ohio R.R., .	2,620 "

Western plateau at White Sulphur Springs, .	2,000 feet.
Source of Cheat and Greenbrier River, . .	2,400 "
Blue Ridge, near Harper's Ferry, . . .	1,800 "

ELEVATION OF THE VALLEY OF VIRGINIA.

Near the Potomac River, ...	800 feet.
At Covington, Alleghany County, . . .	902 "
At Staunton, Augusta County, ...	1,222 "

"The first of these divisions, the summit and table-lands of the Alleghanies, comprises a narrow strip little more than the average width of a county, and extends from the Alleghanian backbone to the chain of mountains which are really a continuation of the Cumberland range, and known as Cotton Hill, Gauley, Laurel Hill, &c. The valley between these two ranges lies at a level of 1,350 to 2,000 feet above the sea; the Greenbrier Valley, for instance, for a length of 150 miles, having an average elevation of 1,500 feet. Much of the cultivated land of Greenbrier County, which is one of the summit counties, lies at a height of 1,800 to 2,000 feet, and yet ripens corn and sorghum without difficulty, and enjoys a winter climate of great mildness.

"The second division includes the Valley of Virginia, or the Shenandoah Valley, averaging, perhaps, fifty miles in width, and extending through the old State of Virginia in a southwestern and northwestern direction. Only the mouth of the valley is embraced in West Virginia. Its average elevation in this section is, perhaps, 1,000 feet.

"The section west of the mountains, which may be said practically to represent the elevation of the State, containing at least 16,000 square miles, or two-thirds of its entire area, including and almost bounding on the east the great coal basin, lies between the altitudes of 600 and 1,500 feet. The uplands, a few miles from the Ohio, with an elevation differing considerably at different points, may be averaged at 800 feet."

TEMPERATURE.—The mean temperature of West Virginia, for the year, as may be seen by an examination of the isothemal lines, is lower than any other locality in the same latitude east of Missouri River. It lies between the lines of 50° and 55° Fahr., which embrace the southern and central portions of Ohio, Indiana and Illinois, with contiguous portions of Missouri; on the Atlantic, deflecting northward to include the coast line between New York and Baltimore. The isothermal, indicating a mean temperature of 55°, passes through Baltimore and Washington, circles round the southern boundary of West Virginia, intersects the northern border of Kentucky, and strikes St. Louis, leaving Philadelphia and Cincinnati a very little north of the line. The line of 52° would come near the centre of West Virginia. This would make the average temperature slightly less than that of those two cities.

Rainfall.—The distribution of rain in West Virginia is admirably calculated, in quantity and seasonableness, to insure success to husbandry, and give facility to all its successive operations. The Spring opens early, and with its opening come gentle and frequent showers. The Summer, with less humidity than any surrounding State, is not subject to long-continued droughts. The grasses spring green and fresh upon the summits of the loftiest mountains during the Summer."

"The amount of rain precipitated in West Virginia is from 32 to 36 inches only, as indicated by partial records kept in different parts of the State, and especially in the vicinity of Lewisburg and the White Sulphur Springs, where the same mean annual rainfall decreases to some extent southward from quantity was indicated on both sides of the Alleghanies. The Pittsburg, and its minimum quantity is found in Summer.

Salubrity.—"It would scarcely need the corroboration of sanitary facts to prove the healthfulness of this region. The altitude, the irregularity of surface, the absence of marshy plains, so peculiarly characteristic of West Virginia, would give in connection with its medium temperature, assurances of health and longivity to her population."

The Alleghanies, or Appalachian Chain.—"This mountain range extends nine hundred miles, nearly parallel with the seacoast, consisting of ridges fifty to one hundred miles apart, and parallel with each other, watered and wooded to their summits, with extensive and fertile valleys between.

"The Blue Ridge, Alleghany, and Cumberland, with many other subdivisions, as North Mountain, Laurel Hill, Peaks of Otter, and Greenbrier, are but parts of the great Alleghany system. That portion of this embraced in West Virginia abounds in many a plateau, with an elevation just sufficient to insure a pure and bracing atmosphere, and all conditions essential to vigorous and healthy growth, both in animal and vegetable life."—J. R. Dodge.

Kentucky and Tennessee, lying contiguous, near the centre of the great Mississippi Valley, may be said to be the "Garden of the Union." Indian corn, tobacco and hemp here flourish in perfection, producing a large surplus for exportation. The natural produce of wheat is said to be much smaller on the grazing lands of Kentucky than in Northern Ohio, or the Northwestern States. "The same lands," says a late writer, "which on an average yield 75 bushels of Indian corn to the acre, would not yield more than eighteen bushels of wheat. In Southern Ohio, Kentucky and Tennessee those conditions of climate prevail which are favorable to producing the maximum yield of Indian corn, but are not equally well suited for large crops of wheat."

The grasses of Kentucky are justly celebrated for their nutritious qualities; this State producing fine sheep, cattle, horses and mules. The forests, in many parts, are extensive and heavily timbered with a great variety of forest trees. The grape-vine also flourishes, and is extensively cultivated along the Valley of the Ohio.

MISSOURI, lying west of the Mississippi, is also favorably situated as regards climate and soil. Here are produced similar products to those of Kentucky and Tennessee, while the rich and productive soil, in connection with her mineral wealth, renders her capable of sustaining a dense population on almost every portion of her large territory.

CULTIVATION OF TOBACCO.—The successful raising of tobacco may be said to be the distinctive agricultural feature of the Border States, while the culture of cotton gives a distinctive character to the more Southern States of the Union.

The adaptation of the soil and climate for the cultivation of tobacco can be shown by no better test than the census returns of 1860, giving the products of the States north and south of the Ohio River, including Pennsylvsnia, Maryland and Virginia, lying to the eastward.

States.	Tobacco, lbs.	States.	Tobacco, lbs.
Pennsylvania, . .	3,181,586	Maryland, . .	38,410,965
Ohio, ...	25,092,581	Virginia, . . .	123,968,312
Indiana, ...	7,993,378	Kentucky, . .	108,126,804
Illinois, ...	6,885,262	Tennessee, . .	43,448,097
Total, . . .	43,152,807	Total, . . .	313,954,178

NORTHERN LIMIT OF COTTON GROWING.—The following extract from a letter in the *Chicago Tribune* discloses the cause of the failure of the cotton-planting experiment in Southern Illinois, the crop being ruined by an early frost in the latter part of August, 1863.

"The 36th parallel of latitude passing through the States of North Carolina, Tennessee, and Arkansas, may be said to form the northern limit for cotton growing. Not one of the northern counties in these States raises cotton to any considerable extent. Between the 35th and 36th degrees cotton is cultivated, often successfully, but it is subject to be blighted by frost.

The most northern county in North Carolina where cotton is cultivated extensively and successfully is Edgecombe, south of Roanoke River, which is situated in the lowlands, and has its climate tempered by the sea breezes. The up-land counties in the same latitude cultivate cotton on a small scale, but often have their crops cut off by a frost. But in the western part of the State cotton is not successfully planted, except in the counties bordering on or near to the line of South Carolina. In Tennessee, in the same latitudes and elevation, the results are the same. The cotton statistics of 1860 show that even in the lowlands, between the Tennessee and the Mississippi Rivers, no cotton is produced in counties on the Kentucky border—in Olion, Weakly, Henry, &c.; while the southern counties of Tennessee, between the same rivers, rival the best cotton districts of Mississippi in productiveness. In Arkansas the same law of climate prevails. The southern counties constitute the very heart and centre of the cotton region of the United States, while the northern counties produce very little, and some of them none at all, although the soil is highly fertile."

Climate of the Cotton Growing States.

The cotton-growing region extends over a wide range of country, including most of the States of North Carolina, South Carolina, Georgia, Florida, Alabama, Mississippi, Louisiana, Arkansas and Texas, also the southern portion of Tennessee, running through ten degrees of latitude, from 26° to 36° north. This whole region possesses a very similar climate and soil, being favorably influenced, no doubt, by the warm waters of the Gulf Stream, and may be considered *sub-tropical* in its character and productions, having a mean annual temperature varying from 60° to 76° Fahrenheit. The southern peninsula of Florida, along the Gulf coast, being the hottest part of the United States, where the climate assumes a tropical character.

That portion of country lying on the Atlantic slope is drained by numerous streams, taking their rise in the Alleghanies or Appalachian chain of mountains, flowing eastward into the ocean. The principal rivers are Cape Fear, Great Pedee, Santee, Edisto, Savannah, Altamaha, and St. John's. On the margin of most of these rivers rice is produced in large quantities, while in the same neighborhood, contiguous to the coast, is produced the sea-island cotton in abundance. This section of country, which produces Indian corn, cotton, and sweet potatoes, seems destitute of nutritious grasses; hence the cattle, sheep and hogs are of an inferior quality.

Cotton Culture.—"It would seem," says a late writer on agriculture and climate, "that South Carolina and other cotton growing States bordering on the Atlantic, cannot materially increase the number of bales of cotton, for land capable of growing cotton is limited. The sea-island variety is not cultivated at a greater distance than 30 miles from the sea, and the greater part of it within ten miles. For upwards of 100 miles from the coast, the country consists of pine barrens and swamps. The pine barrens occupy 6,000,000 acres of land, which is of the poorest description and cannot be cultivated by slave labor. The inland swamps occupy at least 1,200,000 acres, which are wholly unproductive. It is well known that the upland cotton is raised on a belt intervening between the pine barrens and the hilly country, at the base of the Blue Ridge Mountains. These up-land cotton soils were naturally moderately fertile, but somewhat easily exhausted and difficult to ameliorate. Their peculiar character will be afterwards described. The exhaustion of the soils in the uplands, where the greater part of the cotton crop is raised, forces the owners of slaves to go farther westward, so that there is not a great increase in the number of slaves, and the amount of produce is also rather declining.

"The low islands along the coast of Carolina and Georgia, with a margin of about thirty miles from the sea, furnish the finest quality of cotton raised in any part of the world. In this comparatively limited area, the climate produces a length, strength and firmness of fibre which cannot be obtained by art in other localities. The northern limit of the sea-island cotton is about the 33rd degree of latitude, and recent trials are favorable to the extension of its culture along the Florida coast.

"The cotton is planted in ridges 4½ feet in width and a foot and a half between each plant in the row; but if the soil is rich, as much sometimes as three feet. The cotton seed is planted from the 20th of March to the 20th of April; and as the plants rise, the soil is thrown up to their roots by the plough and the hoe. The seeds of the cotton-plant, like those of peas and beans, ripen soonest on the branches next the ground; indeed, while the lowest branches of the cotton plant have ripe seeds the upper are bearing flowers. As the seeds ripen the husks expand, and the cotton fibre appears attached to the seeds in the form of a round ball as large as an orange. As soon as the earliest husks are open, which is usually about the last of July, picking commences. This operation is long continued, for a succession of pods ripen until the end of November. As the cotton is gathered, it is dried and stored up till winter, when the separation of the fibre from the seed is affected.

"The soil being so poor upon which the sea-island cotton is raised, the most of it is manured with a compost of cow-pen manure and vegetable stuff from the swamps. Guano has also been applied to a considerable extent in raising cotton. I saw one field which must have been greatly benefited by a quantity of this stuff, as the plants to which it had been applied were nearly double the size of these that were undressed. Guano, however, is more esteemed as a manure for cotton on poor soils than on rich, for on the latter it is apt to send up too much wood. In climates in which frosts do not occur, all the varieties of cotton that are cultivated in the United States are perennial.

"The cotton crop on the sea islands is very precarious. Two or three weeks of showery weather frequently occur during the picking season, and the seed is often shed out and the produce diminished. The common opinion seemed to be, that the average produce of clean cotton of the sea-island variety is not more than 150 lbs. to the acre. On the sea-islands and along the coast, the produce is kept up by manuring the soil with salt marsh mud, which seems to be the best application for obtaining quantity and quality. On the whole, the quantity of sea-island cotton is not increasing, owing to the natural poverty of the soil and the limited region over which it can be cultivated. Latterly, some planters have been improving the salt-tide swamps along the mouths of the rivers, and raising large crops of cotton. Rich land, however, is not so favorable to the production of the finest fibre, for if the plant is stimulated beyond a certain degree, the wool becomes inferior."

The lowland and fine upland cotton soils of the Carolinas, Georgia, Alabama and Mississippi have the great defect of being unsuited to the growth of good grasses for pasturage. This circumstance renders the lands easily exhausted, and of little value unless under crop. There are no good perennial grasses native to these soils, and none have yet been found that grow well upon them.

The Bermuda grass, a native of the Valley of the Ganges, is the most valuable one for the Southern States, though it does not mature its seeds. Owing to this, it is difficult to obtain a sward, which can only be done by breaking up the land where it has got possession, and by planting small pieces of sod, and these spreading, soon fill the ground. It must have the full blaze of the sun for its growth, and it perishes in the shade of other plants. On rich land it is cut three times during Summer, and has been known to yield ten tons of hay to the acre in one season. This valuable grass is only successfully cultivated in the Southern States, from Virginia to the Gulf of Mexico.—*Russell's Agriculture and Climate of U. States.*

RICE CULTURE.—The same writer remarks :—" In Georgia and the Carolinas there are a good many marshy grounds in the pine region, which have arisen from beds of clay sending the water to the surface in springs. They are composed of black vegetable matter, too deficient in earthy materials to be possessed of fertility of any great permanence. It was on such soils, however, that the first settlers raised rice ; but, being easily exhausted, recourse had constantly to be made to new land. Though a considerable quantity of rice is here and there raised over the upper country on such soils, and even on dry cotton lands, for domestic use, none of it is reckoned sufficiently good for exportation. The discovery that the tide-water swamps are peculiarly well adapted for the culture of rice is comparatively recent. At first, the barren sandy soils were more valued than they are now, because indigo was raised upon them, and was one of the great staples of the country. This article can now be brought to the European market at a cheaper rate from India, and its culture has, therefore, been abandoned in the United States.

" It is on the tide-water swamps of the Savannah and the numerous other rivers in Georgia and the Carolinas, that the fine rice known in Europe as the Carolina rice is cultivated. The production of rice for exportation is, in a great measure, confined to these swamps ; and it is further limited to the fresh-water-tide swamps ; for where the tides are salt, or even brackish, they are unsuitable for irrigation. Rice is cultivated about four miles below and twelve miles above Savannah.

" The average produce of rough or unhusked rice on the Savannah swamps is estimated at from 45 to 55 bushels to the acre. Though the fields have been long under cropping, the produce is still large, but no doubt smaller than when the land was first cleared ; still from 70 to 80 bushels are sometimes got on old cultivated fields. Crops of rice are usually taken in succession as long as the land is clean ; but when it becomes foul through weeds, or the 'volunteer rice,' it is laid under dry cultivation for a year. This is attended with great benefit ; for although no manure is applied, and two crops—one of oats and another of potatoes—are taken, yet the land is so much renovated that the succeeding crop of rice is often increased by a half, and sometimes even doubled.

" The 'volunteer rice,' which is interesting in a physiological point of view, causes a great deal of trouble to the planters.

The rice-seeds that are shed when the crop is cut, and lie over the winter, produce an inferior quality of grain, for under these conditions they appear so far to revert to their natural state. Though the husk of the 'volunteer rice' is of the same light-yellow color as that of the finest quality, the kernal is red, and a few grains of this kind in a sample detract from its market value. There are several varieties of 'volunteer rice,' which are usually the most vigorous plants in the field; and as some of them ripen before the main crop, they fall out and increase with great rapidity.

"The rice plant adapts itself in a most wonderful manner to the most opposite conditions in respect to moisture. There is no cultivated plant that bears any resemblance to it. The same variety which grows on the upland cotton soils and on the dry pine barrens, grows in the tide-swamps, where the land is laid under water for weeks at a time; and even in the lower part of the Delta of the Mississippi, where the fields are under water from the time of sowing to that of reaping.

"The rice-grounds are comparatively healthy to white men in Winter, but not so in Summer and Autumn, when the crops are growing and ripening. It has been often remarked that the swamps, in their original state, along the southern rivers of the United States, were by no means so deleterious to the whites as they are now, when brought under cultivation. Though this seems to apply, in a certain degree, to all the rich alluvial soils in the river bottoms, yet it is particularly applicable to the rice-grounds that are irrigated by the tides. Indeed, the undrained swamps remain comparatively healthy so long as they are covered with the natural vegetation. It is said to be attended with extreme danger to a white man to remain, during the hot season, for one night on the rice-grounds of Carolinas. The planters, with their families, invariably leave the rice-grounds during the hot season, and remain in a more healthy part of the country until the crops are harvested. Though the negroes are not liable to those diseases which are so fatal to the white inhabitants in Summer, yet they do not increase in the rice districts; the damp ground and the nature of the labor render them liable to pulmonary diseases and other complaints."

The climate of the Summer months in the vicinity of Charleston and Savannah is intensely hot, and very moist in the low grounds. On an average about twenty-two inches of rain fall, and the temperature obtains a mean of 80° Fahr. Such conditions give great force to vegetation. The early Autumn is also hot, though dryer than Summer, and this seems to promote those miasmatic emanations which are so injurious to the white population.

Rice is also cultivated in Florida, along the Gulf coast, and in Louisiana, near the mouth of the Mississippi River, where, also, the sugar-cane is successfully cultivated.

Climate of Florida, its Productions, &c.

The peninsula of Florida, bounded on the east by the Atlantic Ocean and on the west by the waters of the Gulf of Mexico, the most southern land belonging to the United States, approaches within a degree and a half of the Torrid Zone, of whose climate it largely partakes, and a number of whose productions it yields, being wafted by the warm winds and moisture proceeding from the Gulf of Mexico. According to Dr. Perrine's tables the mean temperature at Key West and Havana, in 1838, was :—

Months.	Key West. ° Fahr.	Havana. ° Fahr.	Months.	Key West. ° Fahr.	Havana. ° Fahr.
January, . . .	69	70	July, ...	82	80
February, . .	70	75	August, . . .	81	80
March, . . .	73	77	September, . .	77	79
April, ...	75	78	October, . . .	74	75
May, ...	79	80	November, . .	72	72
June, ...	81	81	December, . .	70	72

On the night of January 28–29 (1858), the coldest known for many years, the thermometer fell to 44°. At St. Augustine the thermometer has sunk on various occasions to 33°, 30° and 24°; and at Pilatka, in Lat. 29° 38′, to 28° and to 27° ; at Tampa, Lat. 27° 48′, to 28° and 26° ; at Fort King, in the interior, half a degree south of St. Augustine, the climate is more severe than on the coast, and ice an inch thick is sometimes seen in its vicinity. The Summers, however, are hotter than on the coast. While the minimum range at St. Augustine was 39°, and the maximum 92°, at Fort King the minimum was 26°, and the maximum 105°. The Gulf coast, too, has a more severe winter climate than the Atlantic.

From the relative number of deaths occurring annually, according to the census of 1860, it appears that Florida in the south and New York in the north are the two most healthy sections of the United States bordering on the Atlantic coast.

"The climate of South Florida may be set down as the most desirable winter climate in the United States, presenting to the invalid of the north a desirable retreat from the rude blasts peculiar to that region." In addition to St. Augustine and its vicinity, "the Miami, on Key Biscayne Bay, and Key West, may be considered as most desirable points for establishing the necessary facilities for the encouragement of visitors of that class, numbers of whom annually go on to Cuba and other West India Islands, but who would be induced to stop in Florida were there proper accommodations. No place possesses greater advantages for fishing, boating, &c., than those places mentioned. At Miami, on Key Biscayne Bay, has been stationed, during the past eighteen months, a company of United States troops, and it has been a matter of surprise to the Surgeon, that he has had no case of sickness among the soldiers during that time. The inhabitants, some of whom have resided there for many years, are all grateful witnesses of the remarkable healthful ness of that vicinity; and although the Summers are warm, the air, during the entire day, is fanned by the easterly winds prevailing in that season, and rendering it comfortable for the laborers to pursue their vocations at all times. This, in connection with other and peculiar advantages, make it emphatically the home of the man of slender means and enterprising disposition.

"The various tropical fruits are all adapted to this southern portion of Florida, and many of them can be made profitable for export, such as the lime, guava, citron, lemon and cocoa-nut. Cotton, rice, and the sugar-cane can also be cultivated. Preserves made from the three first-named fruits are always in demand, and may be prepared for market extensively. The pine-apple is successfully cultivated at Indian River and other places, but as it requires a peculiar soil, it is confined to certain localities. Many points of this region are adapted to the plantain, banana, orange, &c., where future efforts, governed by experience and discretion, will doubtless cause them to become established products."

Climate of the Southern States.

SUB-TROPICAL CLIMATE.—It seems to be popular of late, with a certain class of writers, to extol the climate of the South: now slavery being abolished, it is argued that Northern people can settle there with impunity and enjoy health. "The climate of Georgia, in the southern part," says a late writer, "is never excessively warm, the range of the thermometer in the month of July does not exceed 81°, and for December and January 52° Fahr. The sky is generally clear, and the atmosphere, if not bracing, certainly not enervating." The following Table is reliable.

TEMPERATURE OF THE SEASONS,

Along the Gulf and Atlantic Coasts, together with the Mean Annual Temperature.

Stations.	N. Latitude.	Spring. ° Fahr.	Su'mer. ° Fahr.	Aut'mn. ° Fahr.	Winter. ° Fahr.	Year. ° F'hr
Key West, Fla., .	24° 32′	76	82	77	69	76
Fort Myers, . .	26° 38′	75	82	77	65	75
Fort Brocke, . .	28° 00′	72	80	73	62	72
St. Augustine, . .	29° 48′	69	80	71	58	69
Pensacola, . .	30° 18′	69	81	70	56	69
Mobile, Ala., . .	31° 12′	67	79	66	52	66
New Orleans, La., .	29° 57′	70	82	71	56	70
Savannah, Ga., .	32° 05′	67	80	68	54	67
Charleston, S. C., .	32° 45′	66	80	68	52	66
Wilmington, N. C., .	34° 00′	64	80	67	50	65
TEMPERATE CLIMATE.						
Fort Monroe, Va., .	37° 00′	57	77	62	40	59
Washington, D. C., .	38° 53′	56	76	56	36	56
Baltimore, Md., .	39° 17′	53	74	56	34	54
Philadelphia, Pa., .	39° 56′	50	74	56	33	53
New York, N. Y., .	40° 42′	48	71	54	32	51
Boston, Mass., . .	42° 20′	46	68	52	28	48½
Portland, Me., . .	43° 40′	43	65	48	25	45
Eastport, " . .	44° 54′	40	61	47	24	43

The annual quantity of rain that falls between the 24th and 35th parallels of latitude, along the Atlantic coast, averages from 50 to 60 inches,* while between the 35th and 45th parallels the quantity of rain and snow averages from 35 to 45 inches. The same excess of moisture occurs along the northwest Pacific coast, embracing Oregon and Washington Territory, while not to exceed 5 and 10 inches of rain falls annually in parts of New Mexico and Arizona.

* The heat and moisture of this region, in many localities, engenders fevers and other diseases.

Monthly Temperatures.

TABLE,

Showing the different Monthly Temperatures, for the first *Six* Months of the Year, from Florida to Maine, on the Atlantic seaboard.

Stations, &c.	N. Lat.	Mean Temperature. Jan. ° F'hr.	Feb. ° F'hr.	Mar. ° F'hr.	April. ° F'hr.	May. ° Fahr.	June. ° F'hr.	Yearly Mean. ° Fahr
Key West, Fla.,	24° 32′	70	72	74	77	79	82	76.00
Fort Dalles, "	25° 55′	66	67	70	75	78	81	74.75
Fort Pierce, "	27° 30′	63	65	70	73	77	80	73.20
St. Augustine, Fla.,	29° 48′	57	60	64	69	74	79	69.61
Pensacola,* "	30° 18′	54	58	60	67	74	80	68.74
Savannah, Ga.,	32° 05′	54	55	59	67	75	80	67.44
Augusta, "	33° 28′	47	51	56	65	72	79	64.00
Charleston, S. C.,	32° 45′	51	56	60	66	73	78	66.61
Fort Johnston, N. C.	34° 00′	49	51	56	61	73	78	65.68
Norfolk, Va.,	37° 00′	42	46	49	56	66	74	59.00
Washington, D. C.,	37° 53′	34	37	46	55	66	74	56.00
Baltimore, Md.,	39° 17′	33	35	43	54	63	71	55.00
Philadelphia,	39° 57′	33	34	41	50	61	70	53.00
New York,	40° 42′	32	34	40	49	58	68	51.00
New London, Conn.,	41° 21′	32	33	39	47	56	65	50.07
Newport, R. I.,	41° 30′	33	34	38	46	55	64	50.27
Boston, Mass.,	42° 21′	23	28	33	42	56	64	48.90
Portsmouth, N. H.,	43° 04′	23	27	33	42	52	60	44.87
Portland, Me.,	43° 39′	20	25	33	42	51	58	44.21
Eastport, "	44° 54′	21	26	32	41	48	54	43.00
Montreal,† Can.,	45° 30′	15	17	29	43	58	68	44.60
Quebec, "	46° 49′	10	13	26	38	52	62	41.00
	Latitude.			Degrees of Fahrenheit.				
Variation,	22° 17′	60	59	48	39	27	20	35.00

From the above Table, running through 22 degrees of latitude, embracing the whole of the Atlantic coast of the United States and north to Quebec, Can., it appears that the temperature of Key West in February is about the same as Charleston in May, and Baltimore in June; the temperature of Pensacola in January is the same as Norfolk in April, and Eastport in June; the temperature of St. Augustine in January is about the same as New York in May, and Portland in June; and Charleston in January has the same temperature as Portland in May, being a difference of four months.

* Lowest in January, +21° Fahr. Highest in July, +98° Fahr.

† Lowest in January, —31° Fahr. Highest in July, +98° Fahr.

TABLE,

Showing the different Monthly Temperatures, for the first *Six* Months of the Year, from Texas to Minnesota, through the Valley of the Mississippi and Great Lakes.

Stations, &c.	N. Lat.	Mean Temperature. Jan. ° Fahr.	Feb. ° F'hr.	Mar. ° F'hr.	April. ° F'hr.	May. ° Fahr.	June. ° F'hr.	Yearly Mean ° Fahr.
Fort Brown, Texas, .	25° 54′	61	65	66	75	80	83	73.75
Corpus Christi, " .	27° 47′	56	57	66	70	78	82	71.00
San Antonio, " .	29° 25′	54	56	60	66	75	79	68.00
New Orleans,* La., .	29° 57′	55	58	64	70	76	81	70.00
Baton Rouge, " .	30° 26′	53	55	61	69	76	80	68.00
Fort Towson, In. Ter.	34° 00′	43	46	53	64	70	77	61.69
Fort Smith, Ark., .	35° 23′	40	44	52	62	70	75	60.00
Fort Scott, Kansas, .	37° 45′	34	36	48	53	69	73	56.00
St. Louis, Mo., . .	38° 40′	33	36	46	52	66	72	54.50
Fort Leavenworth, .	39° 21′	32	35	46	51	65	70	53.00
Chicago, Ill., . .	41° 52′	24	25	33	46	56	64	47.00
Milwaukee, Wis., .	43° 03′	23	25	33	45	54	64	46.20
Detroit, Mich., . .	42° 20′	27	32	39	45	56	66	47.35
Fort Gratiot, . .	42° 55′	25	26	33	44	54	63	46.30
Green Bay, Wis., .	44° 30′	20	25	33	40	50	62	44.50
Fort Snelling,† Minn.	44° 33′	14	22	37	50	57	69	44.54
Fort Ripley, "	46° 19′	8	12	30	41	52	62	39.30
Mackinac, Mich.,	45° 51′	19	25	30	36	45	58	40.65
Sault St. Marie, Mich.	46° 30′	16	22	29	38	49	58	40.37
Fort Wilkins, "	47° 30′	23	21	29	38	48	57	41.00
	Latitude.	Degrees of Fahrenheit.						
Variation, . .	21° 76′	53	53	36	37	32	26	34.45

From the above Table, running through about 22 degrees of latitude, extending from the Rio Grande to the Upper Mississippi, it appears that the temperature of Fort Brown in January is about the same as Fort Ripley in June, being a difference of five months; the temperature of New Orleans in February being about the same as Chicago in May, and Sault St. Mary in June, a difference of four months.

* Lowest in January, +24° Fahr. Highest in July, +98° Fahr.

† Lowest in January, —30° Fahr. Highest in July, +96° Fahr.

Climate of the South-western States.

The Gulf States of Alabama, Mississippi, Louisiana and Texas, including Arkansas, possess for the most part a similar climate and soil, yielding like productions, if we except the sugar-cane, which is most common to Louisiana. This section of country extends westward from the Gulf of Mexico to the confines of New Mexico and the Rocky Mountain range, being known as the Southwestern States. The mean annual temperture varies from 60° to 74° Fahrenheit, being the hottest in Texas near the Rio Grande.

The mean temperature of the seasons is as follows :—Spring, 70°, Summer 82°, Autumn 71°, Winter 56°. Yearly mean, 70° Fahr., agreeing with the Summer temperature of the City of New York and St. Paul, Minn.

The mean fall of rain for the seasons is as follows :—Spring, 11 inches; Summer, 17 inches; Autumn, 10 inches; Winter, 13 inches. Mean annual fall of Rain, 51 inches; subject, however, to great variation from year to year.*

The Delta of the Mississippi, below Baton Rouge, is about two hundred miles in length, averaging seventy-five in breadth. Its estimated area is fifteen thousand square miles, and it is only a very small portion of this extent that is capable of being cultivated ; for the interior is a vast swamp covered with trees, whose tops only are sometimes visible during the flood season. The cultivated land of the Delta is mostly confined to the banks of the river and those of its bayous; indeed, rarely extending beyond a mile or two from the channels. And these have only been reclaimed by the formation of embankments, or "levees," to prevent inundation.

The eastern part of the State of Arkansas, bordering on the Mississippi, and the valleys of the large rivers which empty into it, are low and swampy, with a heavy growth of timber, and is frequently overflowed. In the central part it is undulating and broken, and in the northwestern part the Ozark Mountains,

* New Orleans was overflowed on the 28th Dec., 1863, to an extent never witnessed since 1849. For three days it rained terrifically and incessantly. From St. Charles street to the swamp the thoroughfares were converted into miniature rivers. Hundreds of buildings were completely flooded, and many bridges swept away. On Sunday morning the city was submerged to a depth varying from one to two feet.

rising sometimes to the height of 1,500 feet, extend across the State. The soil is of every variety, from the most productive to the most sterile. Prairies are abundant and of large extent. Cotton, sugar and Indian corn are the staple productions; but much of the country is well calculated for raising cattle.

The northern and central portions of the State of Mississippi become elevated and diversified after leaving the river bottoms; much of the soil, being a deep rich mould, producing, abundantly, cotton, Indian corn, sweet potatoes and grapes. The natural growth of timber consists of cypress, hickory, black-walnut, maple, cotton-wood, magnolia and sassafras. The country generally being healthy and productive. The southern part of the State, for about 100 miles from the Gulf shore, is mostly a sandy pine barren, interspersed with cypress swamps, open prairies and inundated marshes, and a few hills of a moderate elevation. This region is generally healthy, and, by cultivation, produces sugar, cotton, indigo, and Indian corn. The same can be said of much of Louisiana; but the southwestern part of the State consists of a sea-marsh, on the margin of the Gulf, but further inland, of extensive and fertile prairies, which contain many flourishing settlements. The northern part has an undulating surface, and a heavy natural growth of white, red and yellow oak, cotton-wood, hickory, black-walnut, poplar, cypress, magnolia and sassafras. Cotton, sugar, rice, corn and indigo are the chief productions. Yellow and other malignant fevers often prevail along the Gulf coast and along the Delta of the Mississippi, as far north as Vicksburg, Miss.

The general aspect of Texas, extending on the south to the Rio Grande, is that of a vast inclined plane, gradually sloping from the mountains, eastward to the Gulf, and traversed by numerous rivers, all having a southeast direction. It may be naturally divided into three regions: The first, which is generally level, extends along the coast with a breadth varying from 40 to 100 miles, being narrowest at the southwest. The soil of this section is principally a rich alluvium, with scarcely a stone, and singularly free from stagnant swamps. Broad woodlands fringe the banks of the rivers on the northern portion, between which are extensive and rich pasture lands. Cotton and sugar are produced in this region. The second division is the undularing prairie region which extends for 150 to 200 miles further inland; its wide grassy tracts alternating with others that are thickly timbered. Here vast herds of cattle and sheep are raised on many of the large estates. Cotton is also grown in large quantities, as well as Indian corn, &c. The third, or mountainous region, situated principally on the west and southwest, forming part of the Sierra Madre of Mexico, is but little explored, being mostly inhabited by roaming tribes of Indians. The mountain sides are mostly clothed with forests; and there are few, if any districts of country, of the same extent as Texas, with so little unproductive land.

Texas is divided into a wet and dry season; the former lasts from December to March, and the latter from March to December. Snow is seldom seen except on the mountains in the west. The country is in most parts covered with a luxuriant native grass; and in general amply supplied with timber, among which are the live-oak, white, black, and post-oak, hickory, walnut, sycamore, cypress, cotton-wood, &c. The northwestern portion of the State is justly celebrated for its grazing and wheat lands, where is found pure water and a healthy climate.

The "Northers" is a fierce chilly wind peculiar to the Gulf of Mexico, visiting Texas and the coast southward during cold weather. These storms, which often assume the character of tornadoes, make the climate of this part of the country very variable, the atmosphere, during their prevalence, being rendered dry and cold.

TEMPERATURE.—Fort Belknap, Texas, in latitude 33° 8′, has a mean annual temperature of 64° Fahr.; Fort Graham, in latitude 31° 56′, 66° Fahr.; San Antonio, latitude 29° 25′, 69 Fahr.; Corpus Christi, in latitude 27° 47′, 70 Fahr.; Fort Brown (on the Rio Grande), in latitude 25° 54′, 74° Fahr.; thus showing a variation of mean annual temperature from 64° to 74° Fahr., in the eastern part of Texas, while in the interior the land gradually rises and the temperature decreases, affording generally a healthy and invigorating climate.

"The Northers."

"The fierce north winds that prevail from November to March in the Gulf of Mexico and surrounding coasts, usually at intervals of about a week, are termed 'Northers.' These winds are well known to navigators, and as being particularly violent on the Mexican coast, from the Rio Grande to Vera Cruz, often causing shipwreck and fearful loss of life. They even cross the Isthmus of Central America, and Mexico at Tehuantepec, and blow with great violence into the Pacific Ocean. The 'Northers' being the land winds from the Territories of the United States, lower the temperature as they rush south over the warm waters of the Gulf of Mexico. Though they have a tendency to cool the air considerably in Cuba and other West India Islands, yet they are greatly tempered by the large surface of warm water over which they blow before reaching the shores of Cuba, which is, mostly exposed, on the other hand, the 'Northers' are in all their vigor along the States bordering on the Gulf of Mexico, because these land winds sweep across them without meeting any influence in their course to mitigate their coldness and severity. In Southern Texas, especially, which is alternately subject to the warm south winds from the Gulf, and the cold 'Northers,' the fluctuations of temperature are of the most extraordinary character. Than Texas, there is, perhaps, no country on the globe where the inhabitants are exposed to such sudden changes from heat to cold, and to whom the lines of Milton are more applicable. They

> " 'Feel by turns the bitter change
> Of fierce extremes, extremes by change more fierce,
> From beds of raving fire to starve in ice.'

"The temperature, along with the direction of the winds in Southern Texas, in 1855, will give some idea of the nature of the climate in these regions. The observations were obtained at the Smithsonian Collection at Washington.

Days.	7 A. M.	2 P. M.	9 P. M.
19th January, .	52° S. E.	72° S. E.	58° S. E.
20th " .	44° Calm.	81° S. W.	60° S. W.
21st " .	32° North.	44° N. W.	32° N. W.
22d " .	18° N. W.	56° South.	38° South.

"From these figures we perceive that the temperature fell from 81° Fahr. at 2 P. M. of the 20th with the south wind, to 18° Fahr. on the morning of the 22d with the 'Northers,' making a difference of 61 degrees in the course of 41 hours."

"Before leaving the island of Cuba, accounts reached me of a severe storm having occurred at Philadelphia in January, 1855, a fact that drew my attention to the connection between the 'Northers' of the Gulf of Mexico and the winter storms of the United States and Canada. The 'Northers' of the Gulf of Mexico, I now find, *correspond to the cold westerly winds of the United States, and the two are propagated simultaneously from west to east* over the Gulf, and the whole territories of the United States and Canada.

"As has been already shown, the northerly winds cause a great depression of temperature in all the Southern States, in winter. The frosts, by which they are accompanied, kill down the stalks of the sugar-cane and the cotton plant, and, occasionally, orange groves, rendering the fields as wintry in their aspect as those of Britain during the cold season."—"*North America, its Agriculture and Climate*," *by* ROBERT RUSSELL, *Edinburgh*, 1857.

Arizona and New Mexico.

This mountainous portion of the United States forms, in a topographical point of view, a most interesting feature for investigation, it being entirely composed of elevated plains and valleys, encompassed by lofty mountains, forming part of an immense chain known as the Sierra Madre, or Mountains of North America. Fort Fillmore, situated near El Paso, Mexico, in N. Lat. 32° 13′, stands elevated 3,937 feet above the ocean, while Santa Fe, further to the north, is elevated 6,846 feet, and Fort Massachusetts, in the northern part of New Mexico, is elevated 8,365 feet above the ocean level, being surrounded by passes and elevated peaks rising to 10,000 feet and upwards, showing the great altitude of this whole section of country, of which the Rocky Mountains are the culminating point.

In a country thus diversified, running through eight degrees of latitude, from 32° to 40°, but little reliance can be placed upon the temperature of any locality until first ascertaining its topographical peculiarities. The mean annual temperature of Fort Fillmore is 64° Fahr., Santa Fe 52°, and Fort Massachusetts 41°, while Great Salt Lake City, in N. Lat. 40° 46′, attains a mean annual temperature of 53° Fahr. The agricultural productions of this region are as varied as its climate, producing, in the southern parts, Indian corn, tobacco and the grape, while wheat and the more hardy cereals and grasses flourish in colder localities, more elevated or farther north. Horses, cattle and sheep are raised in many parts of the country, the hilly portions being peculiarly adapted to the increase of the latter kind of stock.

The soil for the most part is poor and unfit for cultivation, owing partly to the want of water, the valleys and side hills alone being cultivated to any extent. The former are rendered productive in many places by the process of irrigation, which is extensively pursued in New Mexico and other Territories. In the Territory of Arizona the temperature for the most part is very intense during the greater part of the year, yet still there are fertile valleys where vegetation flourishes.

The mineral productions are rich and varied—gold, silver and copper being found in many localities, while further investigation will, no doubt, reveal immense deposits of the precious metals.

Indian tribes, of a warlike character, roam over the greater portion of this immense region, or have their fixed habitations as they become partially civilized.

PART XI.

CLIMATE OF THE PACIFIC STATES.

California.

The climate of California, running through ten degrees of latitude, is variable, but not unhealthy, and most of the diseases that prevail are not produced by its influence. It is much warmer, of course, than in the same latitude on the Atlantic; and at the south the heat is sometimes intense. Near the Colorado, falling into the Gulf of California, the thermometer often rises to 120° Fahr.; and in the Valley of the Sacramento, to 110°, in the shade. Along the Pacific coast it is not so warm. During the dry season, from the first of March to the first of November, the mornings are clear, and the heat generally intolerable; but at noon the sky becomes overcast, the strong and unpleasant, but bracing, northwesterly gales set in, and condense the vapors which have risen during the morning, and the thermometer falls very rapidly. The nights are almost always cool. During the rainy months the plains and low grounds are usually enveloped in fogs and mists, and every little *arroya* is swollen far beyond its ordinary limits, while the large streams roll down a vast flood to the ocean.

M. de Mopas, one of the most learned and scientific travellers who have visited this country, insists that all that part of California lying between the coast and the Sierra Nevada is of admirable fertility, and perfectly proper for colonization. Capt. Wilkes also informs us that the fertility of the soil is so great that eighty bushels of wheat is the average yield per acre. "The hills and uplands afford the finest pasturage; but they are not calculated to produce anything except gramineous plants. The elevated plains are covered with immense fields of wild oats and wild mustard, of a most thrifty growth, which often climb up the sides of the mountains to a considerable height. The soil of the low grounds is a rich, dark loam, that becomes dry like powder in the Summer season; but the Winter and Spring rains soon convert them into blooming gardens.

"Irrigation will be needed all over the country, in order to render agricultural enterprises eminently successful; but where this is practicable abundant crops will be obtained. The *tute* marshes could readily be converted into rice fields, and the interval lands will produce most of the cereal grains with but a tolerable culture. Hemp and blue flax are well suited to the country. In Southern California the vine (*ritis vinifera*) thrives wonderfully, and great quantities of wine and brandy are made. The volcanic soil is well adapted for vineyards, and the attention of the inhabitants will probably be still more diverted to the cultivation of the grape, whenever the excitement in regard to the gold deposits has subsided.

"California cannot be termed well-wooded, although the high-lying sections between the Pacific and the Sierra Nevada are dotted quite frequently with forests of excellent timber, and the flanks of the mountains and the deep *canons* opening into the valleys beneath, are fringed here and there with strips of woodland. The courses of the streams, also, are usually lined with belts of stately trees, or thickets of shrubby undergrowth. The most valuable of timber trees are live oak, pine, ash, cedar, cypress, sycamore, willow and cotton-wood. Of the fruit trees apples, pears, plums, peaches, oranges, limes, figs and olives thrive with great luxuriance where they receive proper care and attention. All the vegetables found in the same latitudes in other parts of the world flourish here equally well, often attaining a very large size.

"The whole country is rich in flowering plants and creepers. Beautiful mosses exhibit their long trails from the tops of the highest trees, and the mistletoe shelters itself beneath the shade of the noble oak, climbs up its rugged trunk, and nestles amid its tufted canopy. Among the grasses on the flats and the wild oats of the hilly slopes and mountain sides, are mingled the most valuable bulbous roots, and the brighest and sweetest flowers. There are tulips and hyacinths, the lily and the narcissus, golden poppies and delicately tinted daisies, crimson and scarlet pinks, the fragrant graphalium, and the medical canchalagua; and their beauty, too, is enhanced, in a great degree, by the fine contrast presented by snow-crowned peaks of the Sierra Nevada, that glisten like burnished silver on the very border of the dark line of vegetation, and, more than all, by the beautiful ultra-marine tints which, in a clear day, dye the whole landscape, from the ocean surf to the loftiest mountain height."—*Wilkes' Expedition.*

Among the principal wild animals of California are the fierce grizzly bear, the antlered elk, the black-tailed deer, the savage panther and puma, the California lion, the shy antelope, and the noisy coyote or prairie wolf. Hares, squirrels and marmots are abundant. Among the feathered tribes are the eagle, hawk, vulture, crow, pheasant, partridge, goose, duck, pelican, curlew, crane, turkey, pigeon and plover, besides singing birds of different kinds. The streams abound in fine-flavored fish; and the delicate and luscious salmon are quite plenty in the more northern waters.

Fort Yuma—Colorado Desert (Southern California.)

"On the right bank of the Colorado, and in a bend opposite the mouth of the Gila River, rises up a low irregular hill, from 70 to 80 feet in height. On the water side there is a perpendicular cliff. The other sides are less steep, but equally rugged. This hill is of Plutonic origin, and presents a bleak, dreary appearance. The surface is covered with sharp, volcanic rocks, cutting like glass under the tread, and is destitute of every form of vegetation, except the *euphorbia*, a rank poison, and used by the Indians as an antidote against the bite of the rattle-snake. Such is the site of the military post of FORT YUMA, situated in N. Lat. 32° 43′; W. Long. 114° 43′.

"The climate of this region is in accordance with everything else relating to it. Encamped there during the three winter months, we found the weather generally mild, although the changes in temperature were very great. The thermometer during part of this time was as high as 90° Fahr., and then as low as 30°. The days were sometimes uncomfortably warm and the nights intensely cold. Living and sleeping in tents all the time, we seldom had occasion to have a camp-fire except at early dawn. Owing to the clearness of the skies, the radiation is extremely rapid, and ice quickly forms.

"Having returned the following August to Fort Yuma, the thermometer in the shade at the post was found to be 116° Fahr., and over 120° in the shade along the river. The heat, commencing to be excessive in May, becomes almost unendurable in the months of June, July and August. Even in winter the sun is so hot, and the direct as well as reflected light upon the same plains so dazzling, that, excepting a couple of hours after daybreak and an hour before sunset, it is only possible to see objects through the best instrumental telescopes in the most distorted shapes—a thin white pole appearing as a tall column of the whitest fleece.

"In this belt of country rain seldom falls. In the distance dark clouds may be seen hanging over the California and Sonora Mountains, but they seldom visit the intermediate localities. During the whole of one year they had but two inches of rain. After our arrival a few drops from some passing cloud fell in the two winter months, December and January, and in the following February .07 of an inch. The coast rains take place during the winter; and the rainy season, in Sonora, the Mexican State south of the boundary line, in the months of July, August and September. Spring, in the intermediate section, puts forth its thick green foliage in February, without any rains to refresh and cool the parched ground.

"The atmosphere is often so clear that you are able to see at long distances. In the morning a beautiful sight is afforded by the *mirage*. It has the effect, apparently, of raising the mountains and bringing them more plainly to view; and many are the fantastic and peculiar shapes that are represented.

"Instead of storms of rain during the Winter and Spring, they have those of dust and sand. These are caused by high and strong winds, sweeping over the desert plains, coming principally from the northwest, raising and carrying before them, like mist, clouds of pulverized sand and dust. You can watch them in progress as they approach for hours beforehand, and when they reach you the dust penetrates into every crevice, the finest silk not being impervious to it. They last, generally, a day; sometimes three. The winds blow up quickly and violently, and it is useless to attempt to work with nice instruments. These dust storms were our great drawbacks, as it was impossible to see many feet distant, and only at the risk of being blinded. The gusts of wind which produce this unpleasant effect in Winter are in Summer like the *Simooms of Sahara*—they sweep over and scorch the land, burning like the hot blasts of a furnace."—*Emory's Report, U. States and Mexican Boundary*—1857.

The town of Los Angeles, situated in north latitude 34° 5′, west longitude 118° 16′, is the centre of an extensive and rich grazing, agricultural, and grape-growing country. The quantity of grapes and fruit generally shipped to San Francisco, during the proper season, is already enormous, supporting two large coast steamers. Over 100,000 gallons of wine and 5,000 gallons of brandy were produced in 1854, and the culture of the grape bids fair to outstrip all others. Cotton, sugar-cane, tobacco, flax and the cereals yield productive crops; and the olive, as well as most other kinds of fruit, grow in abundance, rendering this section the *garden* of California.

The country farther inland, at the foot of the hills and mountains, is as productive as any other in California. The vast plains are literally covered with cattle, and many of the rancheros count their yearly increase by thousands. These cattle are driven to the mining districts and San Francisco; but during the not unusual droughts of Summer great suffering is experienced, and great numbers of them perish.

The rainy season commences in this region early in November, and continues until the middle of March. The quantity of rain that falls does not average over 15 inches. During that season southeast gales prevail, and sometimes during the Summer months southerly weather will bring up heavy rain.

Agriculture in California.

The amount of land in California adapted to the purposes of agriculture is estimated at 41,622,400 acres, exclusive of the swamp and overflowed lands, estimated at 5,000,000, which, when reclaimed, will produce every variety of crop. On the Sacramento the experiment is being successfully made to cultivate rice with Chinese labor. The amount of grazing land is estimated at 30,000,000 acres. The amount of land under cultivation in 1856 was 578,963 acres; and of that enclosed for the purposes of agriculture about 120,000. The amount in wheat was 176,869 acres, and the product 3,979,032 bushels; in barley 154,674 acres, and the product 4,639,678 bushels; in oats 37,602 acres, and the product 1,263,359 bushels.

The President of the State Agricultural Society, in his address of 1856, says:—"It is now a well ascertained fact, established by several years' experience, that California stands without a rival in respect to her capacity for producing wheat and other cereals. She produces it in larger quantities to the acre, of better quality, with more certainty, and with less labor than any other country in the known world."

The cultivation of the grape and its manufacture into wines and brandies is rapidly assuming a degree of importance, and increasing to such an extent that these products must soon become one of the most reliable and lucrative branches of the resources of the State. The experience of the last few years has proved conclusively that this country produces this fruit in the greatest variety and abundance; and in a few years will, no doubt, surpass the most extensive wine-producing countries of the world.

The two great staples, cotton and flax, will also soon render the country independent of other places for her manufactures; whilst the production of silk bids fair to go hand in hand with both. The true wealth of the country has but commenced its development, and in a short period, no doubt, she will successfully compete with the Atlantic States and Europe for the markets of the Pacific.

Fruit Culture.

"In determining the capacity of California to produce human food," says a late writer, "fruit production must not be overlooked. The annual statistics of the number of trees and vines and their product are carefully taken by the State; and, in so doing, it presents an example worthy of all imitation by every other State and by the General Government. The list shows the remarkable adaptation of California to all fruits, both of the warm and temperate regions. Among the fruit trees are the apricot, quince, nectarine, cherry, plum, pear, peach, apple, cherimoya, persimmon, prune, pineapple, pomegranate, olive, orange, lemon, citron, aloe, and gooseberry; and among the vines are the grape, strawberry and raspberry. There are also the walnut, almond and pecan. The value of the fruit, the tons of grapes, and the gallons of wine and brandy, are also taken in its annual statistics."

From the returns of the counties in the Surveyor General's Report, the soil and climate adapted to fruit-growing appear to be much more extensive than are suited to profitable grain production. The middle of the Sacramento Valley is, perhaps, too hot and dry for fruits, but the slopes at the base of its mountains are most excellent. Smaller valleys, with their higher elevations, are no less suitable. The great length of the State, north and south, in connection with its mountain elevations, adapts the State to the growth of the large and varied list of fruits just stated.

The Grape.—The number of grape vines in California in 1861 was 10,592,688, of which Los Angeles County had 2,570,000 and Sonoma 1,701,661. All European varieties of the grape grow well in this State, as also those of the Atlantic States. This fact is significant of the remarkable adaptation of its climate and soil to the culture of the grape, and indicates that California will become the greatest wine country of the world. Mr. Hittel, in summing up its superiority, says:—"California vineyards produce ordinarily twice as much as the vineyards of any other grape district, if general report be true. The grape crop never fails, as it does in every other country. Vineyards in every other country require more labor, for here the vine is not trained to a stake, but stands alone."

To set forth more particularly the peculiar advantages of California, as well as to place before the vine-grower in the Atlantic States the causes of them, it is proper to dwell more at length on the soil and climate of California as they influence the success of grape-growing.

1. *The Soil.*—"The vine," says Mr. Hittel, "likes a sandy or

gravelly (not very moist) soil, and never thrives in wet, loamy, or stiff clay soil." Rich land does not seem to be well adapted to the vine there. He remarks :—"The soil of the vineyards at Los Angeles and Anaheim is a deep, light, warm sand. To the inexperienced eye it looks as though it were too poor to produce any valuable vegetable growth. In Sonoma and Napa Valleys the vineyards are planted in a red, gravelly clay near the foot of the mountains, or in a light, sandy loam in the centre of the valley. Of late the vine-growers of these valleys have done without irrigation. In Santa Clara Valley most of the vines have been placed in a rich, black loam, but the vineyards are unhealthy. The Sacramento vines are planted in sandy loam; those of the Sierra Nevada in sandy loam or in gravelly clay."

These soils are very general in California, and Mr. Hittel, speaking of the extent of the grape region there, says :—"The grape region extends from the southern boundary a distance of five hundred and ninety-five miles north, with an average breadth from east to west of about one hundred miles."

2. *Climate.*—The influence of climate, in its altitudes, heats and rains, on grape-growing, has not received that systematic consideration which is due to its importance. The Western States have at times, in the Summer months, a moist, sweltry atmosphere, during which the grape rot is most fatal. The general elevation of these States is from five hundred to seven hundred feet above the sea level. Whether a greater elevation, from one thousand to two thousand feet above it, would not be free from the rot, is a question not yet determined. In a dry climate, like that of California, the altitude is immaterial, for the dryness is sufficient in the lowest localities to shield the grape from rot. If these localities have a rich and moist soil, then, as we have seen, the vines are unhealthy in California. Mr. Hittel, alluding to the *oidium*, says :—"This disease, which has done such great damage in France, appeared in 1859, but has done no injury as yet save in a few small young vineyards. I have heard of it only in Santa Clara, Sonoma, and Alameda Counties, where the vines are planted in a wet, black loam or stiff clay."

This disease seems to be one resembling a combination of our blight and mildew.

As California appears to be free from the rot, a comparison of its climate as to dryness with the Atlantic climate, both in the older States and in Europe, may not be useless, either in showing the superiority of California, or in directing attention in these older States to the true cause of the rot. The following tables are taken from Mr. Blodget partly, and also from the meteorology of the Smithsonian publications :—

AMERICAN PACIFIC CLIMATES.

	Inches of Rain.				
CALIFORNIA.	Spring.	Summer.	Autumn.	Winter.	Total.
Sacramento,	3.3	0.1	3.2	6.9	13.5
San Francisco,	4.6	0.7	3.7	8.8	17.8
Los Angeles,	2.5	0.1	1.6	5.5	9.7
NEW MEXICO.					
El Paso,	0.6	6.6	4.9	0.3	12.4
Albuquerque,	0.6	5.6	1.2	1.0	8.4
AMERICAN ATLANTIC CLIMATES.					
Cincinnati, Ohio,	11.9	14.2	10.0	11.3	47.5
Cleveland, "	9.1	11.6	9.8	6.9	27.4
Ann Arbor, Mich.,	7.3	11.2	7.0	3.1	28.6
Pittsburgh, Penn.,	9.5	12.3	7.6	7.4	36.8
St. Louis, Mo.,	12.7	14.6	8.7	7.0	42.5
EUROPEAN CLIMATES.					
Turin, Piedmont,	8.2	9.0	11.5	7.8	36.5
Valley of the Rhone,	10.2	9.5	10.4	4.3	34.4
Vevay, Switzerland,	7.9	10.8	11.1	3.9	33.8
Manheim, Rhine,	6.3	8.0	7.4	5.3	27.0
Bordeaux, West France,	7.3	7.4	10.3	9.0	34.0
Dijon, East France,	7.1	7.5	9.3	7.3	31.2
Chalons, N. East France,	5.4	6.2	6.1	5.6	23.3

These Tables exhibit an average fall of rain during Summer in California of 0.3 of an inch, and in the Atlantic States of 13 inches nearly, and in European vine-growing countries of 7.7 inches. The climate of California would be more favorable if it had more rain in Summer; but in moist situations, or where irrigation may be employed, it presents all that invites to grape production.

San Jose Valley, California.

37° North Latitude.

"The climate in this valley," says a late writer, "is much more uniform in temperature than any I have yet seen in the United States. In the latter part of February, 1864, the thermometer rose to 75° Fahr., and has ranged from 60° to 90° for quite a number of weeks; in fact I have not seen it range as low as 40° at any time within the past year in the middle of the day, and on two occasions only have I seen it as low as 28° (very early in the morning), and it always moderates to about 50° or 60° in the middle of the day in the coldest weather. Ice seldom forms at all, and never lasts till noon. In October, 1863, the range was from 70° to 90°; November, 60° to 80°; December, 56° to 66°; January, 1864, from 50° to 70°. The first week in January, when it was so cold in the Northwestern States, 50° Fahr. was the coldest in the middle of the day. I have not seen a flake of snow for a year except a little on the mountains of the coast range, which here rise from 3,000 to 4,000 feet: there never falls any in the valley. Roses and honeysuckles bloom all the year round in the open air. Rain seldom falls between April and November, thus rendering irrigation necessary through this otherwise favored section of country.

The Seasons in California.

(Copied from an Official Document.)

"There are but two seasons on the Pacific coast, usually denominated the *dry* and *rainy* seasons; the former corresponding to the Atlantic summer, the latter to the winter; but much error exists in regard to them, especially as to the amount of rain falling during the rainy season. The following totals of rain that fell at San Francisco during each wet season, from 1851 to 1857, will show that the yearly amount is not great.

During the wet season of			1851–52,	there fell	18.0	inches.
"	"	"	1852–53,	"	33.2	"
"	"	"	1853–54,	"	23.0	"
"	"	"	1854–55,	"	24.6	"
"	"	"	1855–56,	"	21.3	"
"	"	"	1856–57,*	"	18.7	"

The following table will show how these amounts were distributed each month from 1851 to 1857.

MEAN MONTHLY FALL OF RAIN.

January, . .	4.52	inches.	July, . .	.00	inches.
February, . .	3.37	"	August . .	.00	"
March, . .	3.32	"	September .	.18	"
April, . .	3.07	"	October, . .	.45	"
May, . .	.73	"	November, .	2.08	"
June, . .	.00	"	December, .	4.45	"

Giving an yearly average of . . . . 22.17 "

These figures show clearly what months constitute each of these two characteristic seasons.

"During the seasons we passed about San Francisco, we never heard thunder or saw lightning; and never but once saw snow fall, and then only at an elevation of 400 feet; the line being distinctly marked, and the elevation being well determined by the knowledge of the height of the hills."

The following statement will give a general idea of the tem-

* To end of March.

perature of the sea-board. The interior is much warmer, but, on account of the dryness of the atmosphere, the effect is not so enervating to the system as a lower temperature on the Atlantic.

MEAN TEMPERATURE AT SUNRISE AND NOON, FOR SIX YEARS,

From 1851 to 1856, computed from the *California State Register* for 1857.

MONTHS.	SUNRISE. ° Fahr.	NOON. ° Fahr.	MONTHS.	SUNRISE. ° Fahr.	NOON. ° Fahr.
January,	44.0	57.7	July,	52.6	67.8
February,	46.9	60.5	August,	53.7	68.2
March,	47.6	63.1	September,	54.0	69.9
April,	49.3	65.6	October,	52.7	68.4
May,	49.9	64.5	November,	49.8	61.9
June,	51.4	68.1	December,	45.2	55.7
Mean average,				49.7	54.3

The lowest temperature experienced in San Francisco in the above six years was 25° Fahr., in January, 1854. In 1852, '53, '56, the temperature was always above freezing, falling no lower than 40° in 1853. The highest temperature was 98°, in September, 1852, and that may be considered remarkably high; 90° having been reached but once in any other year.

In a late letter from San Francisco, Cal., the writer says:

"We pay very close attention here to climatic changes in the winter, because our wealth depends upon them. We begin our conversations, not by remarking upon the weather, but upon the *climate*—a wider subject, and requiring a traveled person to appreciate it. As we have a dozen climates in California, we can all talk about them. While Napa is wet as a swamp, Santa Clara is dry as timber; and while the Sacramentans are blistering with heat, we in this city shiver with our overcoats buttoned up to our chins. March has brought rough winds with him, and during this week the meekness of our winter's sun has been succeeded by winds as cold and harsh as those of July and August.

"The prospect for fruit is excellent just now, but the crop will be in great danger for three weeks to come. Two men in Napa County, each having about 100 acres of orchard, have each $50,000 at least depending on the weather between this and the 20th of April. If there be no severe frosts, the produce of their two orchards will sell for at least $100,000; if the frosts be severe, the produce will be worth little or nothing. They keep men on the watch every night, and if there is danger of frost, numerous fires are built between the rows throughout the orchard. In other counties near the bay equally large interests are endangered by Jack Frost."

The *Sierra Nevada* of California forms part of the great mountain chain, which, under different names, at unequal heights, but in a uniform direction, spreads from the Sierra de San Bernardino to Russian America, being second only to the Rocky Mountain Range, affording no other apertures than those through which the Columbia and the Frazer flow into the Pacific. This range is remarkable for its extent, its parallelism with the sea-shore, its volcanic peaks, and the elevation of its isolated mountains, some of which rise above the highest summit of the Rocky Mountains. The greatest part of these peaks, like pyramids, are placed on an immense *plateau*, overgrown with magnificent forests, and stretching as far as the frozen regions of eternal snow, and rising from 12,000 to 17,000 feet above the level of the sea. The Sierra Nevada exercises a visible influence on the climate and productions of Northern California. Distant 150 miles from the coast, this gigantic wall receives the hot winds, loaded with vapors, that blow from the ocean and fall in rain and snow on the western part of the range, leaving the opposite declivity exposed to drought and cold blasts. Consequently, you may find at the same season, in the same latitude, and at the same height, mildness of climate, fertility, and in fact summer, reigning on one side, whilst sterility, cold atmosphere and frost exist with more or less intensity on the opposite slope of these mountains, whose sublime beauty is perhaps unequalled throughout the world. Here, too, are found, on the western slope, the most extensive *gold field* of any known region.

Meteorological Table for Sacramento City, Cal.

N. Lat. 38° 34′, W. Long. 121° 27′. Altitude, 40 feet.

By Thomas M. Logan, M.D.

1859–60.	Maximum.	Minimum.	Mean. Temp.
March, . . .	64° Fahr.	39° Fahr.	53.00° Fahr.
April, . . .	76 "	40 "	57.11 "
May, . . .	80 "	53 "	63.00 "
June, . . .	96 "	61 "	74.85 "
July, . . .	87 "	60 "	69.07 "
August, . . .	85 "	58 "	67.16 "
September, . .	82 "	56 "	65.89 "
October, . . .	83 "	49 "	63.28 "
November, . .	68 "	42 "	54.05 "
December, . .	53 "	34 "	43.52 "
January, . .	56 "	37 "	46.20 "
February, . .	65 "	37 "	49.83 "

Mean annual temperature, 58.92° Fahr.

During the above period 18.74 inches of rain fell, which was a little less than the usual average. 171 days were clear; 195 days cloudy, and 61 days rainy.

Table of Rains at Sacramento, Cal.

Showing the quantity, in Inches, of each month during nine years.

Months.	'51-2 Inch.	'52-3 Inch.	'53-4 Inch.	'54-5 Inch.	'55-6 Inch.	'56-7 Inch.	'57-8 Inch.	'58-9 Inch.	'59-60 Inch.	Mean Inch.
July,	0.00	0.00	0.00	0.00	0.00	0.00	0.01	0.00	0.03	0.00
Aug.,	0.02	0.00	0.00	Spri.	0.00	0.00	Spri.	Spri.	0.00	0.00
Sept.,	1.00	0.00	0.00	Spri.	Spri.	Spri.	0.00	Spri.	0.02	0.11
Oct.,	0.18	0.00	0.00	1.01	0.00	0.19	0.65	3.01	0.00	0.59
Nov.,	2.14	6.00	1.50	0.65	0.75	0.65	2.40	0.14	6.48	2.08
Dec.,	7.07	13.41	1.54	1.15	2.00	2.39	2.63	4.33	1.83	4.44
Jan.,	0.58	3.00	3.25	2.67	4.91	1.37	2.44	0.96	2.31	2.42
Feb.,	0.12	2.00	8.50	3.46	0.69	4.80	2.46	3.90	0.93	2.52
March,	6.40	7.00	3.25	4.20	1.40	0.67	2.87	1.63	5.11	4.03
April,	0.19	3.50	1.50	4.32	2.13	Spri.	1.21	0.98	2.87	2.82
May,	0.30	1.45	0.21	1.15	1.84	Spri.	0.20	1.03	2.49	0.87
June,	0.00	0.00	0.31	0.01	0.03	0.35	0.09	0.00	Spri.	0.07
Totals,	18.00	36.36	20.06	18.62	13.77	10.44	15.00	16.02	22.09	20.00

Average fall of rain for 10 years, 19 inches.

NOTE.—The most important feature of the above Meteorological observations, in a practical point of view, lies in the periodical rain of California, by which it will be seen that the agriculturalist cannot depend with any certainty upon the rains alone, but must be prepared to supply their deficiency, whenever it occurs, by irrigation.

Oregon, and Washington Territory.

"Oregon boasts of a fine climate, not more favorable to the health of the inhabitants than to the growth of agricultural products. The range of the thermometer in the valley of the Willamette, is from 30° to 96° up to the 45th parallel, and above this it is not often much colder. The winter is short, commencing the last of December and continuing until February. During this time, south of the above parallel, snow falls but rarely, never to the depth of more than three or four inches, and soon disappears. Rains are quite frequent, especially from November till March, though not often heavy.

"It is well known that isothermal lines, or lines of equal temperature, traverse the earth with varied eccentricity; and that it is much warmer on the Pacific coast than in the same latitude on the Atlantic. Hence fruit trees blossom early in April at Nisqually (about north latitude 47°), and green peas and strawberries are abundant in May; while south of the Columbia River grass grows all the winter long, and the cattle are not housed, and only confined in pens at night to protect them from wolves and other wild animals.

"Fever and ague, occasioned by the decomposition of the vegetable matter turned up by the plow on the prairies, and some pulmonary complaints, are the principal diseases to which the inhabitants are subject. The first is quite fatal to the Indians, solely on the account of bad treatment, however; and small-pox has made dreadful ravages among them.

"Most conspicuous among the productions of Oregon, and Washington Territory, are the enormous timber trees. These are truly giants. Near Astoria, in the primeval forest, there are fir trees over forty feet in circumference, three hundred feet long, and rising to the height of one hundred and fifty feet without giving off a single branch. Among the evergreens are the Douglass pine, fir, spruce, arbutus, cedar, yew, and arbor vitæ. The principal deciduous trees are red and white oaks, hard and soft maples, the alder, poplar, elm and cherry. The ash, here and there, scatters its winged seeds upon the wind; and in the forests of Southern Oregon, the long string of balls of the sycamore, and the feathery scones of the cotton-wood, wave above a dense undergrowth of willows, hazels, and wild roses, amid which occasionally glisten the silvery trunks of the birches, 'the ladies of the wood.' South of the Columbia River, however, there is, comparatively speaking, but little forest land. But in Washington Territory, north of the 46th parallel, there is an abundance of timber for home consumption, as well as for exportation; and since the discovery of the gold mines of California and the rapid population of that State, the value of the timber has enhanced in a wonderful degree.

"All kinds of grass—timothy, clover and blue grass—grow with the greatest luxuriance in the valleys of the Columbia, Willamette and Umpqua, and other streams in the eastern section. Indeed, this country seems to be peculiarly well adapted to their growth, and it can scarcely be excelled in the Union for good pasturage. There are two crops of rich, juicy grass produced on the river prairies—one in the Spring, and the other after the overflow subsides, in July and August. Yet there is very little hay made, except for exportation: the scythe and the rake, and the toil and sweat of the mower, are rendered almost unnecessary by the kindness of nature. The growth of the grass is so rapid in the early Summer that the subsequent heats convert it readily, into hay, where it stands without the loss of any of its juices. Upon the second crop the stock feed during the Fall and Winter.

"The soil of the prairies and interval lands contains an abundance of silex, and where it is sufficiently dry produces fine crops of wheat—the yield varying from thirty to fifty bushels per acre, often of more than sixty pounds weight. There is no such thing as a complete failure of the wheat crop; but, as the waters of the rivers are quite cold, and possess little or no fertilizing properties, it is liable to be injured by the inundations in all low exposures. Indian corn and oats do not succeed very well, the former suffering much during the cold nights, and the latter producing small heads in comparison with the stalk. For peas, beans, potatoes, cabbages, and most garden vegetables, the soil is superior, producing abundantly and of an enormous size.

"Oregon and Washington Territory are not deficient in fruits. Apples, pears, plums, gooseberries and currants, have a thrifty growth, and yield plentifully; and the indigenous fruits, including strawberries, blackberries, serviceberries, cranberries, crab apples, wild cherries, wild peas and thorn apples, are very prolific.

"The streams flowing into the Pacific produce excellent fish, and great quantities of salmon are annually taken in the rivers discharging their waters into Puget's Sound. Most all the birds commonly found on the Atlantic coasts in about the same latitude are found here; and on the ocean shores there are an abundance of gulls, frigate-birds, villula, and other aquatic fowl."—*Wilkes' Expedition.*

Scenery and Climate of Washington Territory.

"The natural features of Washington Territory are strikingly different throughout from those of a corresponding portion of the Atlantic coast, owing both to its mountainous character and peculiar products. To a traveller approaching the coast by sea, the whole country appears mountainous and densely clothed with dark green forests from the water level to the limits of perpetual snow. Far above this tower, in indescribable majesty and beauty, the brilliant snow-clad peaks of the Cascade Range, in strong relief against the deep blue sky, and seemingly close to the sea, although Mt. St. Helens, the nearest, is one hundred miles inland. At sunset the softening mist, which often hangs over them, becomes tinted with the most delicate hues, until in the moonlight they become like monuments of shining silver.

"On nearing land, this noble scenery is found to be accompanied by a proportionately gigantic vegetation, and, indeed, everything seems planned on a gigantic scale of twice the dimensions to which we have been accustomed. The Columbia, unequalled in grandeur even by the 'Father of Waters,' is bordered by lofty cliffs and mountains, clothed from base to summit with perpetual verdure, and supporting on almost every foot of surface trees of astonishing magnitude. At every bend constantly varying scenes of the wildest beauty burst upon the view, while the calm silence is often unbroken, save by the screaming of the panther or the shrill cry of the eagle soaring far overhead.

"The country bordering on the lower Columbia has been celebrated ever since its discovery for the gigantic growth of its forests. Even species so nearly resembling those of the Atlantic States as to be generally considered identical, attain a much greater size. The milder climate and abundant moisture, causing a longer growing season, may be considered, perhaps, as one cause of this increase. It seems certainly to have an influence upon many smaller plants, and most strikingly so on cultivated vegetables, whose seeds we knew to have been brought from the East. The great height to which trees grow may also be due to the rarity of lightning, as it is well known that thunder-storms, though common on the mountains, are very rare in the valleys.

"Entering by the Straits of Fuca, the scenery is quite different, but no less interesting. The calm blue waters of the sound lie placid as a lake in the basin formed by their steep shores with an ever-varying outline of points and bays, and dotted with islands of every form and size. Prairies are often visible to the water's edge, interspersed with evergreen forests, and extending as an elevated plateau to the base of the rugged and snowy mountains that rise like walls on the east and west.

"With all this magnificence there is not wanting scenery of a milder and more home-like aspect. The smooth prairies, dotted with groves of oaks, which in the distance look like orchards, seem so much like old farms that it is hard to resist the illusion that we are in a land cultivated for hundreds of years, and adorned by the highest art, though the luxuriant and brilliant vegetation far excels any natural growth in the East. Nothing seems wanting but the presence of civilized man, though it must be acknowledged that he oftener mars than improves the lovely face of nature.

"The sea-beach, too, has peculiar attractions for one accustomed to live in its vicinity. Its broad hard sand forms an excellent road, smooth and solid as the floor, on which are often to be found objects of interest and value, free gifts from the domains of Neptune. The constant roar of the surf forms a pleasant relief to the silence of the surrounding forests, and in solemn tones unceasingly it speaks of that Power who created all these things, 'whose path is in the great waters, and whose footsteps are not known.'"—*Stevens' Pacific Rail Road Report.*

The immense section of country lying east of the Cascade Range of mountains, extending to the Rocky Mountains, and between 46°′ and 49° north latitude, being the upper valley of the Columbia River, is peculiar as a dry, healthy climate, being destitute of forest trees except on the mountains, which are crowned with pines of a large size. This whole country is made up of hills and valleys covered with rich grasses, affording an extensive range for countless flocks and herds.

During the spring months the country presents a beautiful green aspect, variegated by flowers, and watered by occasional showers. The summers are very dry and cloudless, the temperature rising high during the day-time, followed by cool nights. The cultivation is most successful where irrigation is applied, while the grasses afford at all times sufficient nourishment for stock of every kind. The autumn months are dry and pleasant, while the winter months are frosty and cold; snow falling occasionally in the valleys, but more frequently in the mountains.

METEOROLOGICAL TABLE,

Relating to Washington Territory.

Stations.	Lat.	Long.	Mean Temperature—Fahr. Spr'g.	Sum.	Aut'n.	Wint.	Year.
Fort Walla-Walla, .	46°30′	118°20′	52.07	77.14	53.96	31.27	53.60
Olympia, . .	47°00′	122°30′	—	—	—	—	51.00
Fort Steilacoom, .	47°10′	122°25′	46.54	62.52	48.84	36.28	49.29
Fort Bellingham, .	48°45′	122°30′	46.59	62.00	—	—	—
Camp Semiahmoo, .	49°00′	122°45′	49.00	64.91	50.00	—	49.00

"The climate of this country," says a medical writer, "as regards temperature, possesses a medium between hyperborean cold and intertropical heat. The seasons may be said to be divided into the rainy and dry. From the middle of October until the first of April is the rainy season. During this time the sky is almost constantly obscured by clouds, and rain greater part of the time falling. During April and May there are frequent showers, after which there are occasional showers; but rain sufficient to wet the ground very seldom falls; this, however, applies more particularly to the coast region, west of the Cascade Mountains. To show the average temperature during the year, and the general characteristics of climate, we have made the following extracts from the Meteorological Register kept at *Fort Steilacoom,* north latitude 47°10′, during the year 1851.

"December was the coldest month. The mean temperature as follows: Sunrise, 37.67; 9 A.M., 41.55; 3 P.M., 44.09; 9 P.M., 41.29. The maximum was 52° at 3 P.M. on the 7th, and the minimum 22° at sunrise on the 22d.

"August was the warmest month. The mean temperature as follows: Sunrise, 55.80; 9 A.M., 66.58; 3 P.M., 77.70; 9 P.M., 64.22. The maximum was 92° at 3 P.M. on the 20th, and the minimum 46° at sunrise on the 31st.

"The maximum temperature during the year was 92° at 3 P.M. on the 18th of July, and the minimum 22° at sunrise on the 22d of December. During the year 40 inches of rain fell. The maximum quantity in one month was 15.30, in January; and the minimum, 0.36 inch in July. The last frost in the spring was on the 8th April, and the first killing frost of autumn on the 11th October. Snow falls to a greater or less extent every winter, but seldom remains on the ground over two or three days. It has fallen once during the last three years to the depth of twelve inches, and remained on the ground four or five days. Ice seldom forms over half an inch thick. The prevailing winds during the rainy season are southerly; and during the dry, northerly. Southerly winds are always indicative of rainy weather, and northerly of dry.

"The country generally being high and dry, the lakes, all of pure fresh water, no marshes or alluvial bottoms being in the vicinity, diseases of a malarious origin are almost entirely unknown. Catarrhs, rheumatism, and diseases incident to exposure to cold, combined with moisture, are quite common during the rainy season."

Pacific Coast—Washington Territory.

Extract from a letter, dated Port Townsend, Washington Ter., July 8, 1862.

"As the long, dry, and hot summer of California began to affect my health, I removed early in 1859 to this place, where I shall probably remain, for the following reasons:

"1. There is no dry season here. There are rains through the year, little or no snow in winter, and west of the Cascade Mountains, precisely the climate of Old England, save only that I have not discovered the same tendency to affect the lungs.

"2. This country is destined to become the most interesting portion of our glorious Union, and soon to take position as the great Northwest State.

"3. No country on earth equals this in majestic forests of thickly studded trees, rising from eighty to three hundred feet in height, of the very best quality for spars and ship timber generally, easy of access, and now beginning to supply the ships of every known part of the world, which from time to time pass in front view of my office. Fir, cedar, pine, maple, hackmatack (hard as iron and very lofty). The density of the forests have left comparatively little scope for agriculture; yet there is a considerable amount of prairie, yielding the largest crops of wheat, oats, potatoes, pumpkins, barley, apples, pears, plums, raspberries, gooseberries, blackberries, strawberries, currants, growing luxuriantly in their natural homes throughout the Territoty. Peaches do not ripen, nor tomatoes, nor Indian corn, nor grapes (the just boast of California), nor melons, save in a few selected spots. The grains best suited to this country (west of the Cascades) are red-top, orchard, timothy, and whatever else is suited to Old England.

"4. The world does not present any comparison with the archipelago of islands appurtenant to the continent, and lying immediately south of the 49th parallel and east of the Island of Vancouver. San Juan has luxuriant prairies, with park-like appearance, several fine harbors, and for sheep cannot be excelled. There is not a wolf or "bete feroce" on the island. Lopez has less prairie, any quantity of stalwart wolves, deer, bears, &c. Orcas and others much the same as Lopez, with here and there a hunter occupied in supplying our British neighbors of Victoria, which place, being by the policy of the British Government a seaport, attracts from up a large trade that otherwise would center at Port Townsend.

"These islands must, within a few years, become densely settled, and present all the charms of rich cultivation. They have large harbors, around which rove in their season the salmon, halibut, rock, cod, and myriads of smaller fish of delicious taste. The dog-fish and shark are also taken in great quantity for oil; the oil of the former medicinally surpasses the famed 'cod-liver oil,' when prepared by steam or water-heated cylinders placed within the larger. For whaling depots the harbors of Puget Sound are all that possibly could be desired. For national navy yards, dry docks, &c., they must become the great United States rendezvous of the Pacific.

Idaho Territory.

The Governor of Idaho gives the following in his recent message:

"The immense structural wealth embodied within our confines, so nicely balanced, of mineral, farming and grazing interests, with mountain forests and timber land and water power of every description, eminently adapt us for a self-supporting community. The fertile bottom lands of the St. Joseph, Cœur D'Alene, Spokane, La-Toh, Palouse, Lapwaii, Koos-

koos-kia, Nas-so, Payette, Wiser, Boise, Malade, and their tributaries, would alone sustain, properly cultivated, a population larger than most of the Atlantic States; while ranges of nutritrious "bunch grass," suitable for herds, cover millions of acres. Add to this placer diggings, of greater or less richness, extending for hundreds of square miles, with well-defined gold and silver-bearing quartz ledges, unrivalled by those of Mexico or Peru; a glorious climate, with Syrian summers and Italian winters, bespeak the permanence of our untold resources, and the prosperity that surely and positively awaits their development."

Rocky Mountains—Climate, Snow, etc.

(*Copied from* Capt. MULLAN'S *Report.*)

"There has been no one subject so little understood or so much misrepresented as the climate of the northern valleys of the Rocky Mountains and the plains extending to their either base. I am frank to admit that the section of our road from the Cœur d'Aléne Mission to the Bitter Root ferry does interpose the obstruction of snow to such an extent that I despair of seeing it travelled in winter, unless a daily mail coach is placed upon the line, when the snow being beaten down twice a day, would, I think, keep the line constantly open. But all the remaining sections are mild, with so little snow that travelling with horses can be kept up all winter. And although the climate in the region first referred to is severe, by going north to the Clark's Fork, we at once enter a milder section, and one that offers every advantage to travel. The temperature of Walla-Walla, in 46°, is similar to that of Washington City, in 38° N. latitude; that of the Clark's Fork, in 48°, to that of St. Joseph's, Missouri, in latitude 41°; that of the Bitter Root valley, in 46°, is similar to that of Philadelphia, in latitude 40°, with about the same amount of snow, and with the exception of a few days of intense cold, about the same average temperature. This condition of facts is not accidental, but arises from the truths of meteorological laws that are as unvarying as they are wonderful and useful. As early as the winter of 1853, which I spent in these mountains, my attention was called to the mild open region lying between the Deer Lodge valley and Fort Laramie, Dakota Ter., where the buffalo roamed in millions through the winter, and which, during that season, constituted the great hunting grounds of the Crows, Blackfeet, and other mountain tribes. Upon investigating the peculiarities of the country, I learned from the Indians, and afterwards confirmed by my own explorations, the fact of the existence of an infinite number of hot springs at the headwaters of the Missouri,

Columbia and Yellowstone Rivers, and that hot geysers, similar to those of California, existed at the head of the Yellowstone; that this line of hot springs was traced to the Big Horn, where a coal-oil spring, similar in all respects to those worked in West Pennsylvania and Ohio, exists, and where I am sanguine in believing that the whole country is underlaid with immense coal fields. Here, then, was a feature sufficient to create great modifications of climate, not local in its effect, but which even extends for several hundred miles from the Red Buttes, on the Platte, to the plains of the Columbia. The meteorological statistics collected during a great number of years have enabled us to trace an isochimenal line across the continent, from St. Joseph's, Missouri, to the Pacific; and the direction taken by this line is wonderful, and worthy the most important attention in all future legislation that looks towards the travel and settlement of this country. This line, which leaves St. Joseph's in latitude 40°, follows the general line of the Platte to Fort Laramie, where, from newly introduced causes, it tends north-westwardly, between the Wind River chain and the Black Hills, crossing the summit of the Rocky Mountains in latitude 47°; showing that in the interval from St. Joseph's it had gained six degrees of latitude. Tracing it still further westward, it goes as high as 48°, and develops itself in a fan-like shape in the plains of the Columbia. From Fort Laramie to the Clark's Fork, I call this an *atmospheric river of heat*, varying in width from one to one hundred miles. On its either side, north and south, are walls of cold air, and which are so clearly perceptible, that you always detect when you are upon its shores.

"It would seem natural that the large volume of air in motion between the Wind River chain and the Black Hills must receive a certain amount of heat as it passes over the line of hot boiling springs here found, which, added to the great heat evolved from the large volumes of water here existing, which is constantly cumulative, must all tend to modify its temperature to the extent that the thermometer detects. The prevalent direction of the winds, the physical face of the country, its altitude, and the large volume of water, all, doubtless, enter to create this modification; but from whatsoever cause it arises, it exists as a fact that must for all time enter as an element worthy of every attention in lines of travel and communication from the eastern plains to the north Pacific. A comparison of the altitude of the South Pass, with the country on its every side, with Mullan's Pass, further to the north, may be useful in this connection. The South Pass has an altitude of seven thousand four hundred and eighty-nine feet above the level of the sea. The Wind River chain, to its north, rises till it attains, at Fremont's Peak, an elevation of 13,570 feet, while to the north, the mountains increase in altitude till they culminate to an elevation of 15,000 feet; while the plains to the east have an elevation of 6,000 feet, and the mountains to the west, forming the east rim of the great basin, have an elevation of 8,234 feet, and the country between it and the South Pass an elevation of 6,234 feet above the level of the sea. The highest point on the road in the summit line at Mullan's Pass has an elevation of 6,000 feet, which is lower by 1,489 feet than the South Pass, and allowing what we find to be here the case, viz.: 280 feet of altitude for each degree of temperature, we see that Mullan's Pass enjoys six degrees of milder temperature, due to this difference of altitude alone. At the South Pass are many high snow peaks, as Fremont's Peak, Three Tetons, Laramie Peak, Long's Peak and others, all of which must tend to modify the temperature; whereas, to the north we have no high snow peaks, but the mountains have a general elevation of from 5,000 to 8,000 feet above the level of the sea, and of most marked uniformity in point of altitude.

"The high range of the Wind River chain stands as a curvilinear wall to deflect and direct the currents of the atmosphere as they sweep across the continent. (By-the-by, whence arises the name of the Wind River chain?) All their slopes are well located to reflect back the direct rays of the heat of the sun to the valleys that lay at their bases. These valleys, already warm by virtue of the hot springs existing among them, receive this accumulative heat, which driven by the new currents of cold air from the plains, rises and moves onward in the form of a *river* towards the valleys of the Rocky Mountains, where it joins the milder current from the Pacific, and diffuses over the whole region a mild, healthy, invigorating and useful climate."

AGRICULTURAL AND GRAZING CAPABILITIES.—The amount of agricultural land between the 46th and 48th parallels of North latitude, may be safely estimated at thousands of square miles, extending from Walla-Walla valley eastward, through the Bitter Root valley to Fort Benton, situated at the head of navigation on the Missouri River. This route possesses a favorable climate, being on the line of proposed Northern Pacific Railroad, extending from St. Paul, Minn., to Columbia River, or Puget's Sound.

"The experience of all persons travelling through this region," says Capt. Mullan, "has been that, from the Columbia to the Missouri Rivers, finer grasses have never anywhere been seen; the number and condition of the stocks that feed upon the wild grass alone shows both their abundance and nutrition.

Wild hay can be, and is, cut from thousands of acres. The grass is mostly a wild bunch grass, growing from twelve to eighteen inches high, and covering the entire country. Horses and horned-stock by thousands, and sheep by hundreds, all bespeak the wealth that is wrapped up in the native grasses of the North Pacific region, and I confidently look forward to seeing the wealth of these beautiful mountain valleys yet consist in the thousands of fleecy flocks to be here sheared; and if the streams of the Rocky Mountains are themselves caught and harnessed to the spindles and looms of wool manufactories to be there erected, that the annual shipments of wool to eastern markets will constitute a trade replete with wealth and magnitude."

Northern Pacific Railroad Route.

(*Copied from* Gov. Isaac I. Stevens' *Report.*)

"In an examination of that country (now forming part of Montana Territory), which I made in 1853, '54, '55, the passes of the Rocky Mountains, 'Hell Gate' and 'Cadot's Pass,' were crossed by my parties in the months of December, January, February and March, in the years 1853–54, and in no one of these passes did they find more than fifteen or twenty inches of snow.

"The mean Winter temperature of Fort Benton (47°49′ N. lat.), in 1853–54, was **25**° above zero.* The average at Montreal, on the line of the Grand Trunk Railroad, for the same year was **13**°, and for a mean of ten years, 17° above zero. At Quebec it was, in 1853–54, **11**° above zero, and for a mean of ten years, 13° above zero.

"At Fort Snelling (44°53′ N. lat.), on the great lines through Minnesota from St. Paul to Pembina, and from St. Paul to Breckinridge, the mean Winter temperature of 1853–54 was **12**°, and the mean of thirty-five winters 16° above zero. Thus, in the winter of 1853–54, an unusually cold winter, Fort Benton was **12**° Fahr. warmer than Montreal, **14**° warmer than Quebec, and **13**° warmer than Fort Snelling. Looking to the Bitter Root valley, we find its average temperature in the winter of 1853–54 to be **25**°, and in 1854–55, 30° above zero. The greatest cold in the winter of 1853–54 was —29° below zero at Cantonment Stevens (46°20′ N. lat.) At Fort Snelling it was —36°, at Montreal —29°, and at Quebec —34° below zero; from these results it appears, that on this route, the greatest cold is not equal to the greatest cold on the railroad routes of Lower Canada. The same fact is unquestionably true of the great artery of Russia from Moscow to St. Petersburg.

* Mean annual temperature, 48° Fahr.

"Take the number of cold days when the average temperature was below zero, and we find the following result: The average temperature was below zero twelve days at Fort Benton, ten days at Cantonment Stevens, eighteen days at Fort Snelling, eighteen days at Montreal, and twenty-three days at Quebec. Thus, you see that there were more cold days on the line of the Grand Trunk Railway, and of the railroads in Minnesota, than on this Northern route. Having compared the average winter temperatures and the number of cold days, let us look at the climate in another point of view. Take the number of warm days when the average temperature was above the freezing point, and I find that at Fort Benton the thermometer was forty-three out of ninety days, and at Cantonment Stevens thirty-two out of ninety days above the freezing point, against only six days out of ninety at Fort Snelling, five days out of ninety at Quebec, and eight days out of ninety at Montreal—all in the winter of 1853–54.

"But it may be objected, that the temperature of Fort Benton and Cantonment Stevens is not the measure of the temperature of the intermediate rocky range through which the route passes, and which must be much lower. Fortunately, the party of Lieut. Grover, which has been already referred to in connection with the depth of snow, made observations of temperature on the route, and it has been found by careful comparison that the party made the passage during the extreme cold weather of that winter, and the temperatures observed, therefore, indicate the extremest cold of the pass, and not the usual cold. The mean temperature in the pass from January 12th to January 23d, 12 days, was —10° below zero. At Cantonment Stevens, the mean temperature was —5°, and at Fort Benton, —7° below zero. The greatest mean cold of any day observed in the pass was —22°, against —24° at Fort Snelling, and a still lower figure at Pembina (49° N. lat.)

"That the winter of 1853–54 was unusually cold in the mountain region of the Northern route, is shown from the fact, in the Bitter Root valley, the thermometer never went down to zero in the winter of 1854–55, whilst it fell as low as —29° below zero in the winter of 1853–54. The average mean temperature of this valley in the winter of 1853–54 was 25°, whereas, in 1854–55, it was 30° Fahr. The same general result, determined by observation, as regards the temperature of the pass, would be arrived at by using the formula, that every 1,000 feet in altitude would depress the temperature three degrees. Now, only six miles of the pass is more than 5,000 feet above the sea, the greatest altitude being but 6,044 feet, and the average height of the pass is but about 4,000 feet."

The distance from St. Paul and the western end of Lake Superior, *via* Fort Benton, to the shores of Puget's Sound is, in round numbers, 1,900 miles ; or a little over 3,000 miles from the City of New York, extending for a great part of the distance on the Isothermal line of 50° mean annual temperature.

Meteorological Abstract for Esquimalt, Vancouver's Island,

For the Year 1860–61. North Latitude, 48°30′.

Months.	Maximum. ° Fahr.	Minimum. ° Fahr.	Mean. Temp. ° Fahr.	M'thly Range. ° Fahr.	No. of fine days.
March, . .	59.00	34.00	45.31	25.00	13
April, . .	61.50	43.50	51.74	18.00	19
May, . .	62.50	46.50	55.50	16.00	20
June, . .	68.00	52.50	59.44	15.50	24
July, . .	68.50	54.50	61.00	15.00	22
August, . .	72.00	55.00	62.10	17.00	27
September, .	65.50	50.00	58.00	15.50	14
October, . .	60.50	45.50	54.10	15.00	9
November, .	61.00	40.50	49.16	20.50	10
December, .	59.00	28.50	42.62	30.50	13
January, . .	51.50	23.50	39.20	28.00	9
February, .	50.50	29.50	43.17	21.00	7

Mean annual temperature, 51° Fahr.

SYNOPSIS OF WINDS.

Southerly Winds, 62 per cent.	Easterly Winds, . 7 per cent.
Northerly " 25 "	Westerly & Varia'e, 6 "

NOTE.—Of the 365 days of the year, no fewer than 183, or 50 per cent., were fine, the remainder being dull, showery, rainy, &c. Snow fell on 12 days in small quantities ; the thermometer seldom falling below freezing.

PART XII.

MEXICO AND CENTRAL AMERICA.

THE climate of Mexico and Central America, and their inhabitants, are of an heterogeneous character; being difficult to delineate. The temperature of the climate, and its influence on the human race, varies according to the altitude of the country—thus you encounter the tropical, sub-tropical and temperate climates in a journey of a few hundred miles from the coast. Here we find a degree of culture, refinement and haughtiness, with indolence, ignorance, and a ferocity of character, which engenders feuds and civil war, unfitting the inhabitants from maintaining a purely republican form of government. The experiment of making Mexico a monarchy or an empire, will, no doubt, soon be solved, and a renewed attempt made to establish a pure republican form of government.

The Mexican family consists of several branches, besides the Castilian race, whose blood is intermingled with the natives.* The pure Mexicans or Aztecs occupy nearly the entire extent of the territory stretching from the 25th parallel to the lake of Nicaragua. The numerous monuments of their early attainments in arts and science, in their pyramids and roads, their idiographic writing, division of time, and religious institutions, assign to this people the highest rank in the intellectual scale of the numerous races which belong to the New World.

The population of Mexico is about 8,000,000, of whom about 4,500,000 are Indians, 2,500,000 Mestizos or mixed races, 1,000,000 whites, and 10,000 negroes. The three original races are the Mexican Indians, European whites, and African negroes, and from their intermixture, in different degrees, no less than

---

* The aboriginal Indians and Africans or negroes, who were formerly in a state of slavery, constitute a great portion of the population. Besides, there are various mixed races—Mestizos, Zambos, Mulattos, Quadroons, &c.

twenty different castes are specified, besides the produce of other unions which have no specific name. In no portion of the earth, except in a tropical region, would this demoralizing mixture of the races occur to so great an extent.

MEXICO, lying between 15°58′ and 32° north latitude, is bounded on the east by the Gulf of Mexico, and on the west by the Pacific Ocean; being about 1,900 miles in length, and 1,100 miles in its greatest breadth. A large portion of Mexico is traversed by a continuation of the *Cordilleras de los Andes*, which runs through its whole length, and renders the surface extremely varied. On the north of Guatemala, the mountains diverge into two chains, one of which follows the coast of the Pacific Ocean, and the other that of the Gulf of Mexico. The vast tract between them, comprising about three-fifths of the whole area, consists of a table-land, called *Anahuac*, being an extensive plateau 6,000 to 8,000 feet above the level of the sea; and, owing to that great elevation, possessing a decided temperate climate, though lying within the tropics. Some very high mountains, however, rise above the table-land, and it is also divided in several places by well-defined ridges; but, in general, the surface is broken by a few transverse valleys, and in some directions it is quite unbroken by either depressions or elevations.

"The physical geography of the country is very extraordinary. Perhaps no region of the globe presents such varieties of surface or climate within the same extent of territory. Along the coast there is a narrow fringe of lowland. Advancing into the interior, the ground rapidly rises, sometimes mountain ranges stretch precipitously like a mighty wall for many leagues, sometimes the ascent is more gradual, and slopes upward at a scarcely perceptible angle. But whatever may be the gradient the ascent is continuous till an elevation of from 5,000 to 8,000 feet above the level of the sea is attained, and a vast plateau of table-land extends for many hundred miles. These broad sweeps of level plains on the tops of the mountain chains occupy almost the entire area of Mexico, and form the platform from which the volcanic and other mountains rise into the region of perpetual snow.

"These table-lands rise to different heights, and enjoy a varied climate, subject, however, to very slight annual changes from season to season. This plain is widest at the latitude of the capital, where it spreads out to 300 or 400 miles in breadth.

Its eastern or Atlantic side is 7,500 feet above the level of the sea, and it gradually decreases to a height of about 4,000 feet as it approaches the shore of the Pacific. Of course, so enormous a space does not literally present an unbroken surface; but this is actually the case for many leagues together, and carriages may roll down from the City of Mexico to El Paso, in Chihuahua, a distance of twelve hundred miles; that, too, without encountering much change in the temperature of the seasons. The mountains which enclose the great plain of Anahuac on the eastern side are called the Sierra Madre. In some places they rise to the height of 17,000 feet above the level of the sea."

The Mountains of Mexico exhibit numerous peaks of great elevation. The loftiest are as follows:

		Feet.
Popocatepetl,*	Mexico,	17,735
Pico d'Orizava,*	Vera Cruz,	17,388
Yxtaccihuatl,*	Mexico,	15,700
Cerro de Ajusco,	"	15,800
Nevado Toluca,	"	15,156
Cofre de Perote,*	Vera Cruz,	13,514
Volcano de Colima,*	Colima,	12,200
Zempoaltepetl,*	Oaxaca,	11,300
Pico de Quincoa,	Michoacan,	11,000
Soconasco,*	Chiapas,	8,000

As regards *climate*, Mexico is divided into three regions, the *tierras calientes*, the *tierras templadas*, and the *tierras frias*, or the hot, the temperate, and the cool regions. The first, the low grounds along the coasts of the two seas; and the mean annual temperature is about 78° Fahr. It is especially suited for the cultivation of sugar, cotton, indigo and banannas. Here all the products of the tropics are to be met with in great luxuriance, and their rank growth in the hot, damp atmosphere begets the terrible *vomito* (yellow fever), which desolates the coast. From the virulence of this disease, Vera Cruz has been aptly named "the city of death." Its harbor is shallow and exposed, being swept from October to April by dangerous winds, called "nortes," when the air is filled with sand, the sky is dark with clouds, and the whole coast line is one unbroken sheet of foam. During the unhealthy season of the vomito, from May to November, the merchants and their families usually retire to Jalapa, where "reigns eternal spring." "Here hundreds of trees, plants, shrubs, cereals and parasites spring

* Volcanoes.

almost spontaneously from the soil, and render the necessary labor of man insignificant."

The second regions are of comparatively limited extent, occupying only the slopes that rise above the coast-lands, with an elevation of from 3,000 to 5,000 feet, and a mean temperature of from 68° to 70°, the extremes of heat and cold being equally unknown. The Mexican oak and most of the fruits and cerealia of Europe flourish in this genial clime, the humidity of which produces great beauty and strength of vegetation. The cold region includes all the vast table-land 5,000 feet and upwards above the level of the sea.

In the City of Mexico, at an elevation of 7,400 feet, the thermometer has sometimes fallen below the freezing point. In the coldest season, the mean temperature of the day varies from 55° to 70°, while in summer the thermometer seldom rises in the shade above 76. The mean temperature of the year may be taken at 60°, being about that of Rome. In the latitude of Mexico City, the snow-line varies from 14,000 to 15,000 feet; but, whenever the elevation is greater than 8,000 feet, the climate is rough and disagreeable. Owing to the rarity, or thinness of the air, vegetation is not so vigorous on the table-land as in the tierras calientes, and the plants of Europe do not succeed so well as in the tierras templadas.

In the tropical and central regions, as far north as 28°, there are only two seasons; that of rain, lasting from July to the middle of September, and the dry season, continuing from October till the end of May. From the 24th to the 30th parallel, rain falls less frequently; but this deficiency is compensated by the abundance of snow in January and February. The southern portion of Mexico, however, is decidedly tropical in its character.

In the eastern Cordillera, granite, though forming the body of the mountains, is seldom met with on the surface. It is overlaid with porphyry, greenstone, amygdaloid, basalt and other igneous rocks. In the western chain, however, granite appears on the surface. The great central plateau, between 15° and 20° N. latitude, is a mass of porphyry, characterized by the constant presence of hornblende and the complete absence of quartz. Here are found large deposits of gold and silver. Iron is found in great abundance in Guadalaxara, Michoacan and Zacatecas, and copper in Michoacan and Guanaxuato. Tin is obtained partly from mines, but principally from the washings of the ravines. Zinc, antimony and arsenic have been discovered. Quicksilver is found in Queretaro. The *soils* of Mexico are noted for their richness and fertility; but over so large a territory, there must necessarily be a great variety, and much that is unfit for cultivation. The principal barren regions occur in the north, beyond the 29th parallel.

VALLEY OF MEXICO.—It is difficult to picture a fairer scene than that presented by the Mexican capital, lying, as it does, near to the waters of Lake Tezcucan, in the heart of this beautiful, healthy and fertile valley. On which ever side you turn, there rise the serrated ridges of the Cordilleras, encompassing the city with a gigantic azure belt. The streets run in long and unbroken straight lines, bisecting one another at right angles; and in the clear atmosphere of the table-land, the varied color of the houses is beautifully toned down by the back-ground of the purple hills. To the south, two volcanoes, which overtop the other peaks of the Sierra, raise their majestic summits covered with eternal snow, which, in the light of the evening sun, put on a pale purple tint, here and there flecked with delicate ruby. Lakes Chalco and Xochimilco lie to the south of the capital, Tezcuco adjoins it, and Lakes San Christoval and Zumpanzo are on the north. These lakes are drained by means of an artificial canal running into the River Yuba, which flows into the Atlantic Ocean.

THE CLIMATE OF MEXICO.—A late writer remarks: "It would require more space than I can claim to describe the ever-changing scenery along the road from the foot-hills of the Mexican Cordillera to the capital. In three days, as many climates will have been passed through, with the attendant variations in foliage. From Vera Cruz to Paso del Macho, 70 or 80 miles inland, the climate, scenery and verdure are all tropical. The landscape seems to seethe and glow under the heat of a torrid sun, whose rays, glittering in a thousand curious forms of dense foliage and strange plants and flowers, render the country a vast hot-bed, bursting continually into new forms of life and beauty. Hastening through this enchanting but unhealthy region, the cars soon bear us to the commencement of the stage-travel, where, in a veritable Concord coach, and drawn by nine horses harnessed as only a Mexican knows how, we enter the foot-hills of the Cordillera, and by night have reached the temperate region where peaches, grapes and the northern cereals flourish side by side with the tropical fruits and cotton, cocoa, coffee, rice and tobacco. Jalapa, situated 4,000 feet above the ocean, may be said to enjoy perpetual spring, being favored with a healthy and invigorating climate. At Orizaba and Cordova we are almost under the shadow of the vast volcanic cone of Orizaba, snow-crowned to its peak, and green with dense foliage up to the snow-line.

"After passing Puebla we have mounted to an elevation of

9,000 feet, and are in a cold, mountainous region, whose natural features of sterile wastes, rocky passes, and solemn pines sighing mournfully in the blast, contrast sadly with the fairy land we were in but yesterday. You leave Puebla before daylight, and can thus enjoy the majestic sight of the sunrise tints upon the summits of the great volcanoes of Popocatapetl and Iztaccihuatl—the first 18,000 and the last 16,000 feet above the sea. Popocatapetl with us was in sight all day—in fact, the road leads almost around its very base, along which the diligence was whirling off the leagues, giving us ample opportunity to feast our eyes on the dizzy peak reaching apparently into the unclouded heavens, and crowned with 2,000 feet of a snow-cap, dazzling to behold.

"The descent from the highlands around the great valley of Mexico into the plains beneath is made at a full trot. From these hights, Cortez and his mailed cavaliers first saw the Aztec capital, said by the Spanish chroniclers to have contained its millions of people. However that may have been, the stranger who now visits the Mexican capital, will find a far more populous and beautiful city than he would have imagined from any previous acquaintance he may have had with places of Spanish origin. The streets are generally wide, well paved and clean, and in all respects superior to the old part of Havana, which was laid out at nearly the same time.

"The City of Mexico is full of interest to the stranger, and offers innumerable historical reminiscences. The localities commemorating the great Indian Empire of the Montezumas, which, under Prescott's word painting have become classical ground, are visited with renewed pleasure after reading those vivid descriptions. A remarkable fact connected with the advent of the Europeans here it the great increase in the population of the capital. In 1861, about the time of the French intervention, the city was commonly estimated at about 180,000. These had been the figures, with some few fluctuations, for 20 years. The population at present is variously estimated at from 250,000 to 280,000, and it is possible that a correct census would reach even beyond the last-named estimate."

This increase of population is mainly attributable to its healthy climate; although situated within the tropics, in north latitude 19°25′, it is elevated 7,500 feet above the ocean, enjoying a mean annual temperature of 60° Fahrenheit, varying but a few degrees from season to season. Here the wealthy and the gay, together with the shopkeepers and artizans of all kinds, the musical and literary celebrities, delight to congregate.

Climate and Topography of Central America.

"In its physical aspect and configuration of surface," says E. G. Squier, "it has very justly been observed that it is an epitome of all other countries and climates of the globe. High mountain ranges, isolated volcanic peaks, elevated table-lands, deep valleys, broad and fertile plains and extensive alluvions are here found grouped together, relieved by large and beautiful lakes and majestic rivers; the whole teeming with animal and vegetable life, and possessing every variety of climate, from torrid heats to the cool and bracing temperature of eternal spring. Situated between 8° and 17° north latitude, were it not for these topographical features, the general temperature would be somewhat higher than that of the West Indies. As it is, the climate of the coast is nearly the same with that of the islands alluded to, and exceedingly uniform. It is modified somewhat by the shape and position of the shore, and by the proximity of the mountains, as well as by the prevailing winds. The heat on the Pacific coast is not, however, so oppressive as on the Atlantic; less, perhaps, because of any considerable difference of temperature than on account of the greater dryness and purity of the atmosphere.

"In the northern part of the State of Guatemala, in what is called 'Los Altos,' the Highlands, the average temperature is lower than in any other part of the country. Snow sometimes falls in the vicinity of Quezaltenango, the capital of this department, as well as on the high plains of Intibucat in Honduras, but soon disappears, as the thermometer seldom remains at the freezing point for any considerable length of time. In the vicinity of the city of Guatemala, the range of the thermometer is from 55° to 80°, averaging about 72° Fahrenheit. Vera Paz, the north-eastern department of Guatemala, and embracing the coast below Yucatan to the Gulf of Dulce, is nearly ten degrees warmer. This coast, from Belize downward to Izabal and Santo Tomas, is hot and unhealthy. The same remark applies, in a less degree, to the northern and eastern coast of Honduras, from Omoa to Cape Gracias à Dios." The climate is hot and moist in the lowlands, where there are dense forests; but is milder and more salubrious on the elevated table-lands. Earthquakes are frequent, especially in the plateaus, and numerous volcanoes exist, more particularly along the Pacific coast or a short distance inland.

"The State of SAN SALVADOR," says Gordon, "lies, on an average, considerably lower than that of Guatemala, but the heat is never oppressive except near the coast. The average temperature of the City of San Salvador may be equal in the dry season to the South of France, the wet season being about eight degrees colder than the dry. Many of the large towns in this state, as Sonsonate and San Miguel, are situated very little above the level of the sea, and have an oppressively hot climate, varying from 80° to 90° in the wet and dry seasons."

COSTA RICA.—"The climate of Costa Rica is very humid, the rain falling for six months of the year. It is cool and healthy on the Pacific declivity, excepting the immediate coast; hot, wet, and unhealthy on the Atlantic; cold and salubrious on the table-lands of the interior, where the thermometer ranges from 65° to 75° Fahrenheit in the course of the year. It must be observed that the rainy season on the Pacific and in the interior is from April to November; but, upon the Atlantic coast this order of things is reversed, and the rainy season is from November to February." Another writer remarks, that the "climate of Costa Rica is exceedingly varied, ranging from 50° to 80° of Fahr., according to the elevation."

HONDURAS.—"The northern and eastern coast of Honduras has a higher temperature than any other portion of the State; it, however, diminishes rapidly as we penetrate inland. The modifying influences of the neighboring mountains is felt even before the increase in altitude becomes perceptible. Her table-lands have, of course, a climate varying with their hight above the sea, and their exposure to the prevailing winds. Consequently, there can be no generalization on the subject of the climate of Honduras, except so far as to say that it has a variety adapted to every caprice, and a temperature suitable for the cultivation of the products of almost every zone."

TABLE OF METEOROLOGICAL OBSERVATIONS,
Made in the City of Guatemala, for the year 1857.

	Thermometer.					Rain
Months.	Maximum.	Minimum.	Yearly Mean.	Rainy Days.	Fogs.	in Inches.
January,	73.5	38.9	57.5	4	4	.20
February.	81.0	43.0	63.0	0	6	.00
March,	81.1	46.0	63.6	5	7	.55
April,	88.7	51.6	68.9	9	10	2.07
May,	102.3	52.5	68.1	17	7	5.28
June,	82.3	54.6	67.1	24	5	13.28
July,	81.3	53.7	66.2	25	7	11.72
August,	80.9	53.6	66.2	20	10	11.12
September,	77.9	54.5	66.0	18	10	5.40
October,	82.4	53.6	65.6	17	9	3.55
November,	80.2	49.5	64.6	11	8	1.11
December,	77.0	46.5	62.6	5	4	.24
For the year,	88.7	48.9	65.0	155	87	54.52

From information collected in Guatemala, M. De Puydt constructed the following Table, illustrative of the seasons as marked in that Republic:

Localities.	Days. Rain.	No Rain.	No Rain during changes.	Variable Weather.
Atlantic Coast, . . .	105	110	30	120
Pacific Coast, . . .	90	125	40	110
Interior, ...	100	130	45	90

Of the plateau of Guatemala, he remarks: "Here, as generally throughout the interior, the mean temperature is 17° of Reaumer (70° Fahrenheit), during the summer. The prevailing winds are from the north; so that the climate, as compared with the coast, where the mean temperature is 22° of Reaumer (81.5° Fahrenheit), is almost cold, or at least so regarded by the inhabitants of the country."

METEOROLOGICAL OBSERVATIONS AT RIVAS, NICARAGUA, 1850–51.

Date of Observation.	Av. Ther. °	Highest. °	Lowest. °	Range. °
September, 1850, . . .	78.12	88	71	17
October, " . . .	77.00	86	70	16
November, " . . .	78.42	86	74	12
December, " . . .	77.11	84	72	12
January, 1851, . . .	76.40	87	69	18
February, " . . .	76.00	84	70	14
March, " . . .	77.00	84	72	12
April, " . . .	78.83	88	72	16
May, " . . .	78.29	91	68	23
June, " . . .	77.12	88	71	17
July, " . . .	76.98	86	71	15
August, " . . .	76.20	86	71	15
Total Mean, . . .	77.42	86.45	71.15	15.30

Here it will be observed that the maximum range was in the month of May, and was 23° Fahrenheit. The mean range for the year, however, was only 15.30°. The heat, it will be perceived, at no time of the year is as great as it is during July and August, in the city of New York.

The Rain which fell during the same period is as follows:

Month	Year	Amount		Month	Year	Amount	
September,	1850,	15.240	inches.	April,	1851, .	.430	inches.
October,	"	17.860	"	May,	" .	9.145	"
November,	"	1.395	"	June,	" .	14.210	"
December,	"	3.210	"	July,	" .	22.640	"
January,	1851,	.380	"	August,	" .	11.810	"
February,	"	.000	"				
March,	"	1.410	"	Total Inches, .		97.730	

Belize.—The British establishment of Belize, situated near the southern extremity of the Peninsula of Yucatan, on the Bay of Honduras, in Lat. 17°39′ north, and Long. 88°12′ west, has a temperature and climate which may be regarded as common to the entire eastern coast of Guatemala and Yucatan, and probably not far different from that of the islands off the same coast in the Bay of Honduras. Observations made here, under the authority of the governor, for the year 1848, gave the following results :

TABLE OF THERMOMETRICAL OBSERVATIONS

Made at Belize (British Honduras).

Months.	Average Maximum.	Average Minimum.	Fall of Rain.
January,	82° Fahr.	66° Fahr.	2.7 inches.
February,	85 "	73 "	4.2 "
March,	83 "	75 "	.0 "
April,	89 "	74 "	.0 "
May,	89 "	75 "	2.5 "
June,	90 "	77 "	4.3 "
July,	90 "	78 "	3.3 "
August,	90 "	78 "	.6 "
September,	91 "	76 "	8.2 "
October,	87 "	75 "	4.8 "
November,	85 "	68 "	9.9 "
December,	86 "	75 "	6.7 "
Total for the Year,			47.2 inches.

The average mean temperature for the year 1848, was 79° Fahrenheit.

" The climate of this part of the American continent (bordering on the Bay of Honduras), is greatly superior to that of most other parts of the same vast portion of the globe, either in higher or lower degrees of latitude. It is equally superior to the climate of the West India Islands generally, for persons, whose health and constitutions have become impaired from the effects of the latter, very frequently acquire a sudden restoration of both after an arrival in Honduras. With the exception of a few months of the year, this country is constantly refreshed by regular sea-breezes, accompanied by an average heat that may be taken at the temperature of 80° Fahrenheit.—*Henderson's Honduras.*

The principal productions of Central America are coffee, indigo, cochineal, Brazil wood, mahogany, dye-woods, sarsaparilla, India rubber, balsam of Peru, hides, tallow, wool, tortoise-shell, gold and silver. Below the elevation of 3,000 feet, indigo, cotton, sugar and cacao are the chief crops; between 3,000 and 5,000 feet, the cochineal plant is abundantly cultivated. Maize is generally raised, but wheat only on the high table-land in the north. In some parts the chicozapote, a fruit yielding a good deal of nourishment, supplies the place of corn; other products are tobacco, dragon's blood, mastic, various balsams and drugs, tamarinds, pepper, cassia, ginger, vanilla, and all the fruits of a tropical region. The country is very productive of the precious metals, abounding in gold, silver, copper, iron, lead and zinc.

Horses, asses, goats, sheep, hogs, having been introduced by the Spaniards, are now very numerous; vast herds of cattle are pastured in the grazing farms of Nicaragua, and large flocks of sheep on the plateau of Quesaltenango, being reared almost wholly for their wool. Cattle and sheep breeding are, with agriculture, the main occupation of the population; but the productions of coarse woolens, cottons, blankets, caps, hats, earthenware, furniture, cabinet-work, employs a good many hands, and the Indians weave mats of different colors, which are used as carpets.

The vegetable productions of the country, owing to difference of altitude, are represented as more varied than almost any other part of the world; and, if in the possession of an industrious and enterprising people, it would not fail to be one of the richest on the globe. Cotton of a superior quality can be raised in Nicaragua, but like all other articles produced in this State, the cultivation is now at a very low ebb, and almost entirely neglected. Sugar also is susceptible of profitable cultivation in many parts of Central America.

Taking the natural divisions of the continent alone into consideration, Central America may be regarded as lying between the Isthmus of Panama and Darien and the Isthmus of Tehuantepec, and consequently in a tropical climate. This narrow, tortuous strip of land, which unites the continents of North and South America, stretches from S. E. to N. W. about 1,200 miles, varying in breadth from 25 to 300 or 400 miles, thus including a portion of Mexico and New Granada.

"As a regular chain, the Andes descend suddenly at the Isthmus of Panama, but as a mass of high land they continue through Central America and Mexico, in an irregular mixture of table-lands and mountains. The mass of high lands which forms the central ridge of the country, and the watershed between the two oceans, is very steep on its western side, and runs near the coast of the Pacific, where Central America is narrow; but to the north, where it becomes wider, the high land recedes to a greater distance from the shore than the Andes do in any other part between Cape Horn and Mexico.

"This country consists of three distinct groups, divided by valleys which run from sea to sea, namely: Costa Rica, the group of Honduras and Nicaragua, and the group of Guatemala.

"The plains of Panama, very little raised above the sea, and in some parts studded with hills, follow the direction of the isthmus for 280 miles, and end at the Bay of Parita. From thence the forest-covered Cordillera of Paraguay, supposed to be 9,000 feet high, extends to a small but elevated table-land of Costa Rica, surrounded by volcanoes, and terminates at the plain of Nicaragua, which, together with its lake, occupies an area of 30,000 square miles, and forms the second break in the great Andean chain. The lake is only 128 feet above the Pacific, from which it is separated by a line of active volcanoes. The River San Juan de Nicaragua flows from its eastern end into the Caribbean Sea, and at its northern extremity it is connected with the smaller lake of Managua or Leon by the river Penaloya. By this water-line it has been projected to unite the two seas. The high land begins again, after an interval of 170 miles, with the Mosquito country and Honduras, which mostly consist of table-lands and high mountains, some of which are volcanoes.

"Guatemala is a table-land intersected by deep valleys, which lies between the plain of Comayagua and the Isthmus of Tehuantepec. It spreads to the east in the peninsula of Yucatan, which terminates at Cape Catoche, and encompasses the Bay of Honduras with terraces of high mountains. The table-land of Guatemala consist of undulating, verdant plains of great extent, of the absolute height of 5,000 feet, fragrant with flowers. In the southern part of the table-land, the cities of Old and New Guatemala are situated twelve miles apart. The portion of the plain, on which the new city stands, is bounded on the west by the three volcanoes of Pacayo, del Fuego and de Agua; these, rising from 7,000 to 10,000 feet above the plain, lie close to the new city on the west, and form a scene of wonderful boldness and beauty. The Volcano de Agua, at the foot of which Old Guatemala stands, is a perfect cone, verdant to its summit, which occasionally pours forth torrents of boiling water and stones. The old city has been twice destroyed by it, and is now nearly deserted on account of earthquakes.

"Though there are large savannahs on the plains of Guatemala, there are also magnificent primeval forests, as the name of the country implies, Guatemala signifying, in the Mexican language, 'a place covered with trees.'

"As the climate is cool in the elevated plains and highlands, the vegetation of the temperate zone is there in perfection. On the lowlands, as in other countries where heat and moisture are in excess, and where nature is for the most part undisturbed, vegetation is vigorous to rankness; forests of gigantic timber seek the foul air above an impenetrable undergrowth, and the mouths of the rivers are dense masses of jungle with mangroves and reeds 100 feet high, yet delightful savannahs vary the scene, and wooded mountains dip into the water."—*Mrs. Somerville.*

Isthmus of Panama.—"The majority of the natives of Panama and its vicinity are a mongrel race, in whose veins, white, Indian and negro blood is mingled in every conceivable proportion. Yet these are every way superior in physical development to the few who boast an unmixed Castilian descent. It is fearfully probable that no race of whites can escape deterioration upon the Isthmus. The indomitable energy which braves every hardship, and overcomes every visible obstacle, yields to the fatal influence of climate, and each generation sinks lower than the one that preceded it.

"The pestilential climate, with which no race of men and no strength of constitution can contend, and against which no measure of precaution and no process of acclimation is a safeguard, is of the most fatal character. At certain seasons no man can expect to escape the terrible 'Panama fever' for more than a few weeks or months at most."

MEMORANDUM OF FALL OF RAIN AT ASPINWALL, NEW GRANADA.

Months.	1860. Inches.	1861. Inches.	1862. Inches.
January,	—	3.91	5.42
February,	—	2.31	1.94
March,	—	2.88	.70
April,	—	3.61	2.51
May,	8.30	19.01	4.27
June,	—	12.28	—
July,	—	13.82	—
August,	8.70	14.99	—
September,	11.37	9.62	15.51
October,	20.83	7.10	13.10
November,	19.88	26.80	—
December,	12.68	18.08	—
Total,		134.41	

The mean annual temperature surrounding Central America, along the sea-coast, varies from 75° to 80° Fahr.; while on the table-lands and elevated localities it varies from 60° to 70°, thus producing all the tropical productions with many of the temperate zone.

Yellow Fever and Rainy Weather.

A letter from Belize, Honduras, of a late date, speaking of the ravages of the Yellow Fever on that coast, says:

"Our old, acclimated population and the colored people have generally escaped, but it has made fearful havoc on those who had been only a short time here. We have had nothing like it for over sixteen years, and, in fact, there is no one who can remember such a sickly season before. The doctors themselves say that its fatality is beyond their skill. I think that not more than two who have had it have recovered."

Temperature of the Air and the Ocean.

Record of a Passenger on the Mail Steamship.

	1865.	Air. °	Ocean. °	Barometer.	N. Lat. ° ′
Panama,	Nov. 10,	87	86	30.22	8 56
	" 11,	78	80	30.25	—
Off Costa Rica,	" 12,	82	82	30.30	9 55
	" 13,	82	82	30.27	12 18
Off Guatemala,	" 14,	82	82	30.32	15 6
	" 15,	82	82	30.24	15 42
Acapulco,	" 16,	84	82	30.30	16 50
	" 17,	82	82	30.24	18 29
	" 18,	81	82	30.32	20 04
Cape St. Lucas,	" 19,	76	77	30.38	22 28
	" 20,	75	76	30.47	25 12
	" 21,	64	66	30.44	28 28
San Diego, Cal.,	" 22,	64	64	30.38	31 41
Los Angeles,	" 23,	58	59	30.41	34 25
San Francisco,	" 24,	54	55	30.60	37 46
Variation,		30	31	00.38	28 90

The above shows a close affinity between the air and the ocean. It is supposed a current from the North reduces the temperature of the ocean at California.

The lofty mountains of Central America, of Mexico, and of Lower California were in sight nearly all the way.—*Journal of Commerce, March* 24, 1866.

Mean Temperature in Tropical America.

Cities, etc.	Population.	N. Lat. ° ′	W. Long. ° ′	Year. ° Fahr.
Mexico (Alt. 7,500 feet),	250,000	19 26	99 00	60.50
Vera Cruz,	20,000	19 12	96 08	77.00
Jalapa (4,000 feet),	10,000	19 30	97 00	70.00
Acapulco,	5,000	16 50	99 49	78.00
Belize,	10,000	17 29	88 10	79.00
Guatemala (4,000 feet),	40,000	14 36	90 30	65.00
Rivas, Nicaragua,	20,000	11 00	—	78.00
Havana, Cuba,	150,000	23 09	82 22	76.00
Matanzas, "	28,000	23 02	81 38	78.30
San Domingo, Hayti,	15,000	18 28	69 50	78.00
Kingston, Jamaica,	35,000	17 58	76 47	78.70
Panama, N. G.,	6,000	8 56	79 31	80.00
Bogota (8,650 feet),	40,000	4 35	74 10	58.00
Caracas, Ven., (2,880 feet),	60,000	10 30	67 00	72.00
George Town, Br. Guiana,	25,000	6 49	58 11	81.00
Para, Brazil,	10,000	1 28 S.	48 30	82.00
Maranham, Brazil,	30,000	2 31 "	44 18	81.00
Pernambuco, "	30,000	8 06 "	34 51	80.00
Bahia, "	100,000	13 00 "	38 31	79.00
Rio Janeiro, "	300,000	22 54 "	43 09	74.00
Quito, Ecuador (9,543 feet)	80,000	0 13 "	78 50	70.00
Guayaquil "	20,000	2 20 "	79 43	82.00
Truxillo, Peru	10,000	8 30 "	79 09	80.00
Lima, "	100,000	12 02 "	77 06	76.00

Climate of Caracas, South America.

CARACAS, or CARACCAS, Venezuela, situated in N. lat 10°30′, W. long. 67°, being twelve miles inland from its port, La Guayra, and elevated nearly 3,000 feet above the level of the Caribbean Sea, is freed, in consequence of its elevation, from the excessive heats of the tropical regions. Though delightfully cool in the mornings and evenings, the heat of noon is very great. Rain is abundant during April, May and June. The climate of Caracas has been called a perpetual spring. "What, indeed," says Humboldt, "can we imagine more delightful than a temperature which, during the day, keeps between 10° and 28° Reaumur; and at night, between 12° and 14°; and which is equally favorable for the cultivation of the plantain, the orange tree, the coffee plant, the apricot, the apple and corn? It is to be regetted, however, that this climate is generally unconstant and variable. The inhabitants complain of having several seasons in the course of the same day, and of the rapid transition from one season to another."

PART XIII.

ANTILLES, OR WEST INDIA ISLANDS.

This extensive and important group of Islands in the Atlantic extends in a semi-circular form between the two continents of America, and constitutes a sort of barrier to the Caribbean Sea and Gulf of Mexico. These islands describe nearly the diagonal of a parallelogram, of which the sides are the meridians of 59° and 86° W. longitude, and the parallels of 10° and 23° N. latitude; but this diagonal is of very variable breadth; and there are several small islands off the neighboring coasts not included in the limits now traced. "These islands," says Malte Brun, "have been vaguely denominated the West Indies, from the term India, originally given to America by Columbus."

They are divided into the Greater and Lesser Antilles; and have been nautically classified by the British under the general denominations of the Windward and Leeward Islands. With reference to the trade-wind, however, the whole group are windward islands. The greater Antilles, consisting of the four large islands of Cuba, Jamaica, St. Domingo or Hayti, and Porto Rico, stretch from near the coast of Florida on the west, towards the lesser Antilles or Caribbee Islands on the east, with which they are connected by the Virgin group. The total number of islands and islets which compose these groups is upwards of eight hundred; but many of them are bare, uninhabitable rocks. With the exception of St. Domingo—which is an independent government—these islands mostly belong to different European powers.

The Greater Antilles appear to be of primitive formation, and their highest summits are granite. Most of the Lesser Antilles have indications of volcanic origin; but in many of them the volcanic rocks are covered with calcareous formations of a thickness varying from 25 to 1,000 feet.

Climate—Health, Winds, etc.

"The north wind blows here from November to February, and sometimes lowers the mercury of the thermometer to 16° of Reaumur, or 69° of Fahrenheit, and the needle of the hygrometer to between 60° or 70°. Its prevalence is marked by epidemic rheumatism and catarrhic affections. The south wind is warm and humid; it blows from July to October, but with less force and continuity than that of the N. and E. It raises the thermometer to 28° of Reaumur, and 95° of Fahrenheit. Its influence is dangerous and malignant. To it is attributable the exhalation from the marshes of Saint Lucia, and the elevation of the waters of the Orinoco, which rise to the height of 39 or 41 feet, and inundate the country 200 leagues E. to W. The hurricane season is reckoned from the 15th of July to the 15th of October; and the hurricane region is included between the parallels of 10° and 28° N., and the meridians of 58° and 86° W. The east wind prevails in March, April, May and June. It resembles the north, to which quarter it generally more or less inclines, but is not so dry or warm, for in traversing the Atlantic it loses, before reaching the Antilles, a part of the heat it acquired in passing over the African deserts. During its continuance, the climate is favorable to Creoles and Europeans. The west wind is the severest of all, and inclines more to the north than to the south. The seasons, however, alter with the winds, and are strictly confinable to two—the wet season, from November to April, passing from S. to E.—the dry from May to October, passing from E. to S. Spring commences in April; and from May till October is the reign of summer, during which the medium height of the thermometer is about 80°. The autumnal rains commence in October, and continue till the middle of December; from 60 to 65 cubic inches are the medium fall. This humidity M. de Jonnes attributes, 1st. to the situation of the islands in the midst of a vast body of water, the daily evaporation from which amounts to more than 33,000,000 tons of water for a degree square; 2d. to the proximity of the different islands of the archipelago which form a chain of 200 leagues, disposed in the form of a right angle, the direction of which is towards the prevailing winds; 3d. to the mineralogical mass of these islands, which exercises a superior influence on the atmosphere to such insulated solitary islands as Saint Helena, Ascension, or the Isle of Paquas; 4th. to the conflict between opposing currents of wind during the winter season; 5th. to the elevation of the mountains, which rise 300 or 400 toises into the region of the clouds, which hang during the rainy season at less than 2,000 feet above the level of the sea, beginning from the 14th parallel; and 6th. to the conic or pyramidal form of the mountains, which sensibly augments their action on the electric clouds. An official document on the health of the British troops stationed in the West Indies, exhibits, as regards British Guayana, Trinidad, Tobago, Grenada, St. Vincent's, Barbadoes, St. Lucia, Dominica, Antigua and St. Kitt's, the following general results: Tobago is the most remarkable for fever; Dominica for diseases of the bowels and of the brain; Barbadoes for those of the lungs; Grenada for those of the liver; while Trinidad is most noted for its dropsies. It may be observed that the mortality of all the islands, except Antigua, Grenada and Tobago, is higher among the troops than among the black population; and this is the more remarkable, as the mortality of the negro slave population was calculated upon persons of all ages, including old men and infants, sickly and healthy; whereas, that of the troops was calculated upon persons in the prime of life only. It appears that in these colonies, as well as in Jamaica, the most sickly as well as the most fatal period of the year extends from August to December, and that the only months comparatively healthy are March, April and May.

"VEGETATION.—The atmosphere of the Antilles resembles that of Africa more than that of Europe. Hence, while European productions degenerate here, those of Africa attain singular luxuriance. The sugar-cane—which now covers these islands—came originally from one of the African islands; the coffee, from Arabia; part of the alimentary plants, from the coast of Guinea; and the finest grapes of the savannahs, and flowers from the same source. The dates are those of Atlas; and from Senegal were transplanted those tamarinds whose thick shade suffocates the American trees with which they are surrounded. The numerous race of negroes, too, originally brought from Africa, has here usurped the place of the aborigines. 'Trees similar to those that we have admired in other tropical countries,' says Malte Brun, 'grow in equal luxuriance on these islands. A canoe made from a single trunk of the wild cotton-tree has been known to contain a hundred persons; and the leaf of a particular kind of palm-tree affords a shade to five or six men. The royal palmetto, or mountain cabbage, grows to the extraordinary height of 200 feet, and its verdant summit is shaken by the lightest breeze. Many of the plantations are enclosed by rows of Campeachy and Brazilian trees;

the corab is alike prized for its thick shade and its excellent fruit; and the fibrous bark of the great cecropia is converted into strong cordage. The trees most valuable on account of their timber, are the tamarind, the cedar, the Spanish mountain-ash, the iron-tree, and the *Laurus chloroxylon*, which is well adapted for the construction of mills. The dwellings of the settlers are shaded by orange, lemon and pomegranate trees, which fill the air with the perfume of their flowers, while their branches are loaded with fruit. The apple, the peach, and the grape ripen in the mountains. The date, the sapata and sapotilla, the mammee, several Oriental fruits, the rose-apple, the guava, the munga, and different species of spondias and annonas grow on the sultry plains.

"The winter season is the great vegetating season in the West Indian Islands. The sap then circulates with activity and energy—the trees are almost at the same instant covered with flowers and fruit, mosses and lichens cover the walls—and stramoniums of gigantic size and purple euphorbiums spring up in the unfrequented paths.

"*The Sugar-cane.*—Sugar is the great staple commodity of the West Indies. To this day, it is not exactly known what country the sugar-cane was originally imported from; but it is generally believed that it came from the East Indies. In the 12th or 13th century it was transplanted into Sicily; whence it was taken to Madeira, then recently discovered by the Portuguese. About the same time the Spaniards introduced it into the Canary Islands. Attempts were made to plant it in Provence; but they did not succeed. In the beginning of the 17th century, France had no sugar but what came from Madeira and the Canaries; but towards the end of the century, the English had monopolized this article of trade, and all the north of France was in general supplied with sugar from England. From the Canary Islands the cane was conveyed to the American Continent and islands, and afterwards to Madagascar, the coasts of Coromandel and Malabar, Ceylon and Manilla, and, at length, even to Otaheite. The sugar-cane is propagated by grains or seed. There are several varieties of this plant. One of these, which is white, with a thin bark, and knots at spaces five fingers in length, is very productive both of juice and sugar. A second species is of a reddish color; its knots lie nearer together; its bark is hard; and its produce of sugar less considerable but sweeter. In a third species, the stalk is not above an inch thick; the bark is thin, the flutings are green, the knots very distant; this last has a very sweet taste, and yields a great quantity of sugar. All the three species ripen in nine or ten months."—*English Gazetteer.*

Island of Cuba—Its Climate, etc.

This large and noble island has its east terminus at Point Maysi, in 74°8′ west longitude from Greenwich, and its west point, Cape San Antonio, in 84°59′; lying between three degrees E. and eight degrees W. from Washington. The extreme S. point of the island is in north latitude 19°50′, and the extreme north point, 23°9′. It is of an elongated narrow shape, being 648 miles in length from east to west, and from 20 to 127 in breadth from south to north; containing an estimated area of 35,750 square geographical miles, including several small islands attached to it and under the same government. The coast line exceeds 2,000 miles in extent; but such is the prevalence of reefs, rocks and sand-banks in the surrounding waters, that little more than one-third of this coast line can be considered accessible to mercantile vessels. There are, however, about fifty ports and anchorages. "Commanding the entrance to the Gulf of Mexico, and possessing one of the noblest harbors in the world, Cuba crowns by her political importance the commanding advantages of a rich soil, a varied and teeming productiveness, and a climate which enjoys the genial warmth, but escapes the fiercer heats of the tropics. The occupation of such an island must give strength and wealth to any nation."

"The climate, although tropical, is marked by an unequal distribution of heat at different periods of the year, indicating a transition to the climates of the temperate zone. When the north wind blows several weeks, ice is sometimes formed at night, at a little distance from the coast, and at an inconsiderable elevation above the sea. Yet the great lowerings of temperature which occasionally take place are of so short duration, that the palm-tree, banana and sugar-cane do not suffer from them. Snow never falls, even on the Sierra-del-Cobré, and hail so rarely that it is only observed during thunder storms, once in several years.

"The average annual temperature of Havana is 25° centigrade (76° Fahrenheit); the highest, 32° (89.6°); and the lowest, 10° (50° Fahr.) The average temperature of the warmest month was 27° (82.6°), and of the coldest 21° (69.8°). The average humidity of the atmosphere, as shown by the hair hygrometer is 85°; the maximum, in November and December, being 100°, and the minimum, in April, 66°; or 97° and 75°, without the extremes.

"The dew falls very copiously, especially during the dry season, but chiefly in December and January. Fogs also occur principally in the season of drought. The rain has so fixed and definite a period as to determine the seasons, which are divided into two, viz.: the *rainy season*, and the *dry season*, or season of '*northers*.' The first commences between May and June, and ends in November, being most active in September and October. The average fall of rain at Havana is 1,029 millimetres; the most recorded for a year is 50 inches 6 lines, and the least, 32 inches 7 lines. The most for a month (August), 11 inches, and the least (November and December), 2 lines."

In the interior of the island much more rain falls than at Havana or on the sea-board.

"The common cereal grasses are cultivated in Cuba, together with all the productions of tropical climates. The hills and savannahs are decorated with different species of palms, and the wild orange tree attains a height of from ten to fifteen feet. The mountains, in many parts, present a naked appearance; but the sweet pea, the myrtle-leafed vine and the night-blooming cereus clothe their crags in some quarters. Immense districts, especially the hilly and mountain regions, are still covered with trees, among which the magnificent olive and gigantic mahogany tree, with the red cedar, ebony, *lignum vitæ*, and a variety of other valuable woods for furniture and ship building abound. Many varieties of palms, plantains, and some beautiful hard-woods occur. Maize, manioc, cocoa, and the yaca are grown in Cuba, and numerous esculent roots and fruit are indigenous; among the latter, the delicious pine-apple. Precious woods, building timber, plants for other useful purposes, medicinal plants, and a great variety of fruits abounds in different parts of this rich and fertile island."

The chief agricultural products are sugar, molasses, tobacco, coffee, Indian corn, garden fruits, esculent vegetables, honey, wax, etc. The products of the forest are cedar, mahogany, etc., all of which articles are annually exported to a very large amount.

AGRICULTURAL PRODUCTS OF CUBA.

Sugar-cane.—The cane is a tropical plant, and although it is cultivated in the Southern States of America, it does not flower in that climate, and it is also cultivated under other disadvantages.

"There are two powerful influences under which the life of sugar-cane appears to be shortened, viz.: a low temperature in winter and poverty of soil. The cane, as it is well known, is a perennial plant, and when its stalks are cut close to the ground every year, the roots, like those of the willow, retain their vitality, and send up fresh shoots for another season. On some of the richest soils in Cuba, it has frequently been known to last for twenty-five years; and it is said that there are instances on record of its lasting forty years without being renewed.

But after the soil becomes somewhat exhausted, it is planted every five or eight years, according to the fertility of the land. On the other hand, the influence the cold winters exercise on the longevity of the cane is rather curious. The soil in the lower delta of the Mississippi is, perhaps, as fertile as any of the soils in Cuba ; yet there the canes are seldom got from the rattoons, or old roots, oftener than once, when the fields have again to be replanted. The life of the cane is thus evidently shortened by the cold winter. As we go further north, such as central parts of Alabama, where the winters are still colder, it is found necessary to plant sugar-cane annually. The effects of a rich soil in prolonging the life and vigor of the cane are well known in Louisiana, for an additional year's growth from the rattoons is obtained from crops planted on new land. But in general, but two crops only are got from one planting on the best sugar lands of Louisiana. From this circumstance alone, it may readily be conceived that the cultivation of the sugar-cane is much more expensive in Louisiana than in Cuba. And although the produce of the sugar in the former is little more than one half of that of the latter, yet strange to say, the sugar-lands of the Mississippi are of far greater value than some of the best lands of Cuba. The reason why the value of land is higher in America, notwithstanding the disadvantage of climate, is owing principally to the circumstance of more economical management," and, also might be added, the different forms of government of the two countries.

The quantity of sugar and molasses produced on the Island of Cuba in 1852, was as follows :

	SUGAR. Arrobas.	MOLASSES. Puncheons.
Western Department, . .	25,397,767	258,204
Eastern Department, . .	3,767,469	8,981
Total, ...	29,165,236	267,185

This is the great source of wealth of the island of Cuba. An acre of average sugar-cane in Cuba will yield about 3,500 pounds of syrup, that is, sugar and molasses. This is a very large quantity of saccharine matter, when it is borne in mind that it is estimated there is 25 per cent. of the whole amount of sugar left in the "begasse," or stalks, after being pressed. This is a greater weight of pure saccharine matter than the weight of the grain of two of the largest crops of Indian corn in Cuba ; while Indian corn requires a richer soil than sugar-cane.

Tobacco.—"The culture of tobacco forms the other great source of agricultural wealth to the island of Cuba. The quality of the fine Havana tobacco is the result of peculiarities of soil and climate. In regard to climate, it is worthy of observation that tobacco is only cultivated *during winter*, when there is little rain. It grows more luxuriantly in summer with the increased heat and moisture; but the leaves grown in this season are devoid of those qualities for which the weed is esteemed. The conditions of growth are less powerful in winter, when the temperature is ten degrees lower, and the fall of rain small. At the same time, there is more sunshine to impart those aromatic qualities which are so much relished by smokers of tocacco.

"But the quality of the tobacco depends as much upon the nature of the soil as of the climate. That plant, as we have already said, requires peculiarities of soil to develop certain of its qualities, and these peculiarities are such that art cannot furnish the conditions to produce them where they are naturally wanting. The sugar-cane grows chiefly on soils derived from calcareous formations; but few or none of these are fitted for tobacco, which is cultivated only on sandy loams. Both the Cuban and American planters concur in asserting that a large quantity of silicious matters in soils is essential for the growth of good tobacco. The culture of tobacco is extensively carried on in the western parts of the island, over a region of of country along the south coast, about eighty miles in length by twenty in breadth. The soils rest upon the primary formation. Even in the tobacco district, the planters know the spots in the different fields that produce the various qualities of leaf. The whole yield of tobacco on the island of Cuba in 1852, was 222,020 cargas."

Coffee.—The vermilion soils are most esteemed for the growth of the coffee plant. It rises to the height of six feet, and looks like a hardy, slow-growing shrub. It is an evergreen, and is kept closely pruned. When in blossom, the fields appear quite white. The labor of the coffee estates is light compared with what it is on the sugar estates. Coffee was formerly more extensively cultivated in Cuba than it is now; a number of coffee estates have been planted with sugar-cane, which is found to be more profitable.

Rice.—Considerable quantities of rice are sown in summer on the ordinary soils of the country; and the crops receive no more moisture than what the rains afford, but they sometimes suffer greatly when the season is dry. The Chinese laborers have an allowance of rice; for, having been accustomed to it in their native country, they still retain a preference for it over Indian corn.

Indian Corn.—"Two crops of Indian corn are raised on the

same land in one year. The winter, or 'dry crop,' as it is called, is usually sown in October, and reaped in February; and, as scarcely any rain falls during this season, its growth is almost entirely maintained by copious dews. The crop at this season yields only about thirty bushels an acre on the richest land, as it requires to be planted wide in the rows; and at the same time, the soil must be well cultivated, to promote the absorption of moisture. The summer, or 'rain crop,' is more abundant than the winter one; but, from all I could gather, Indian corn is not nearly so productive in Cuba as it is in Kentucky or Ohio. On the best soils, it seldom produces more than forty bushels to the acre; for the climate, as in the Southern States of America, has a great tendency to produce stems and leaves."

Vegetables.—The winter climate is well adapted for all kinds of vegetables; the common potato grows most luxuriantly in the vicinity of Havana; the sweet potato is also extensively cultivated as a winter crop. Vegetation in the island is very rapid. Beans are ready for use in about six weeks after being sown. Radishes are fit for the table in the course of three weeks. Lettuce, a plant of rapid growth, attains a great size in a short period. Tomatoes are also raised during winter. In summer, rain frequently falls about noon; and under a tropical sun vegetation in consequence becomes very active. Neither tomatoes, sweet or common potatoes are raised during this season owing to the great heat and moisture. They are grown only in the dry or winter season, when rain does not fall for several weeks successively. Showers are more frequent in the winter on the coast; but vegetation in the interior is chiefly sustained by the copious dews.

"Notwithstanding the fruitfulness of Cuba, all kinds of vegetables and provisions on the island are high-priced, which renders living expensive in large towns. A great deal of time and expense is incurred in taking the produce of the fields to market, as most of it is transported thither on the backs of small horses, in consequence of the roads being so bad in the interior."

Plantain, or Banana.—"The plantain is seen growing over the whole island of Cuba, affording shade and shelter to every cabin, however small or humble. Though it wants the grace and beauty of the cocoa-nut or palm, its form is peculiarly tropical—none more so. In good soil it grows to the height of twenty feet. Its trunk, or rather fleshy stem, is hollow, resembling in outward appearance that of the lily of the Nile, seen in our green-houses. It is about nine inches in diameter at the base, tapering towards the top, where it sends out long, broad leaves, and also a short stalk, bearing a heavy cluster of fruit. The plantain requires to be renewed, on good land, only once in forty years. It sends from the root a fresh shoot every nine months, and the old trunk dies as soon as the fruit becomes ripe. Little care is bestowed upon its culture, being planted in check-rows twelve feet apart. It is not unfrequently seen, however, growing in the shallow soils of the coral formation, where there is little in which to fix its roots except in the crevices of the rock.

"With a little attention, a constant supply of plantain fruit is obtained all the year round. It is largely used by all classes, and is commonly pulled when green and cooked with oil or grease. The banana is merely a smaller but less productive variety of the plantain. It is usually allowed to become ripe before it is eaten, and is then found to be a most delicious fruit."

DISEASES.—"From what has been said under the head of *Physical Climate*, it may be inferred that the temperature of the island of Cuba is mild, although humid and warm. The better to classify the diseases incident to the climate, we shall divide the year into three periods, viz.: 1. From December to May, the season of drought and of the finest weather. 2. From May to September, a period of excessive heat, rain, and of most atmospheric electricity. 3. From September to December, the season of deluging rains and of the greatest atmospheric changes. During the first period, the following complaints prevail; catarrhs, ephemeral and intermittent fevers; croup, rheumatism, and, in some years, pleurisy, inflammation of the lungs, and eruptive fevers. During the second period, the most prominent are: diarrhœa and other disorders of the digestive apparatus; yellow fever, small pox, liver complaint, Asiatic cholera and eruptive fevers; at the same time, instances offer of violent congestion, pulmonary inflammation and pleurisy, likewise neuralgia and nervous affections. The third period comprises nearly the same diseases as the second; however, the yellow fever and the cholera begin to decline, and gradually disappear; dysentery is more common during this period, also tetanus or locked jaw. Within a few years those cases of very acute consumption which sometimes destroy the patient in two months, are of very frequent occurrence. Notwithstanding that it has been stated by some writers that the climate of Cuba is unfavorable to human life, many and remarkable instances of longevity can be cited, principally of colored persons."

The climate of Cuba is exceedingly fine in winter, and is becoming a favorite resort for people from the United States, who are threatened with pulmonary diseases.

The estimated population of Cuba is 1,500,000; of which number about 900,000 are whites, mostly of Spanish descent, 200,000 free colored, and 400,000 slaves.

YELLOW FEVER IN HAVANA.—The total number of cases and deaths of Yellow Fever, or *vomito*, in the city of Havana, from 1854 to '59, were as follows:

Years.	Cases.	Deaths.
1854	5,452	1,028
1855	3,521	675
1856	5,984	1,407
1857	7,040	2,048
1858	5,326	1,401
1859	4,453	1,193
Total	31,776	7,752

Island of Porto Rico.

The Spanish island of PORTO RICO, the smallest and most easterly of the Great Antilles, is situated in the Atlantic Ocean between latitudes 17°54′ and 18°30′ north, and longitudes 65°37′ and 67°16′ west of Greenwich. Length, 96 miles; breadth, 35 miles. Area, 3,695 statute miles.

The climate is warm and moist, but salubrious, except in low and marshy places. The extremes of temperature on the plains are about 62° and 95° Fahrenheit, but it must be several degrees cooler on the highlands, especially the mountainous regions north and east. The atmosphere is very humid, but least so on the south side. The wind generally blows from the east and north-north-east, except from November to March, when northers prevail, though seldom with great violence; and during the wet season it often blows from the south-east. The land breeze is light and fitful. Whirlwinds sometimes occur during the dry season, but seldom strong enough to occasion much damage.

DISEASES.—The most prevalent during the dry season are—common catarrhal affections, epidemic influenza, rheumatism and intermittent fevers. Consumption also has its victims. During the months of hot and rainy weather—intermittent, bilious, typhus and brain fevers, small pox and erysipelas. Yellow fever visits the island only at intervals, while the Asiatic cholera has never appeared, although it has raged fiercely in some of the other West India Islands.

The population of the island is estimated at 500,000, of which number about one-half are whites; the remainder mostly free blacks, there being comparatively but few slaves, being estimated at only one-tenth of the entire number of souls.

Humboldt on the Climate of Cuba.

"The Climate of Havana is that which corresponds to the extreme limit of the torrid zone; it is a tropical climate, in which the unequal distribution of heat through the various seasons of the year presages the transition to the climates of the temperate zone.

Calcutta (N. lat. 22°34′), Canton, (N. lat. 23°8′), Macao (N. lat. 22°12′), Havana (N. lat. 23°9′), and Rio Janeiro (S. lat. 22°54′), are places whose location at the level of the ocean and near the tropics of Cancer and Capricorn, being equi-distant from the equator, makes them of the greatest importance in the study of meteorology. This science can advance only by the determination of certain *numerical elements*, which are the indispensable basis of the laws we wish to discover. As the appearance of vegetation on the confines of the torrid zone and under the equator is the same, we are accustomed vaguely to confound the climates of the zones comprised between the 0° and 10°, and 15° and 23° of latitude. The region of the palm, the banana, and the arborescent grasses, extends far beyond the tropics, but we should err in applying the result of our observations on the limit of the torrid zone, to the phenomena we may observe in the heated plains under the equator.

"It is important to establish first, in order to correct these errors, the means of temperature for the year and the months, as also the oscillations of the thermometer at different stations under the parallel of Havana; and by an exact comparison with other places equally distant from the equator, Rio Janeiro and Macao, for example, to demonstrate that the great decline of temperature which has been observed in Cuba, is owing to the descent and irruption of the masses of cold air which flow from the temperate zones toward the tropics of Cancer and Capricorn.

"The mean temperature of Havana, as shown by excellent observations made through four years, is 25° centigrade (78° Fahrenheit), being only 2° C. (3.6° F.) lower than that of the regions of America under the equator. The proximity of the sea increases the mean temperature of the coasts, but in the interior of the island, where the northern winds penetrate with equal force, and where the land has the slight elevation of 250 feet, the mean temperature does not exceed 23° C. (73° F.), which is not greater than that of Cairo and all Lower Egypt.

"The difference between the mean temperature of the hottest month and that of the coldest is 12° C. (21.6° F.) in Havana, and 8° C. (14.4° F.) in the interior, while at Cumaná, it is barely 3° C. (5.4° F.) July and August, which are the hottest months, attain in Cuba a mean temperature of 28.8° C. (83° F.), and perhaps even 29.5° C. (85° F.), as under the equator.

"The coldest months are December and January; their mean temperature is 17° C. (62° F.) in the interior of the island, and 21° C. (69° F.), in Havana, that is, from 5° C. to 8° C. (9° F.), (14° F.) less than during the same months under the equator, but yet 3° C. (5° F.) higher than that of the hottest month in Paris.

"As regards the extremes touched by the centigrade thermometer in the shade, the same fact is observed near the limits of the torrid zone that characterizes the regions nearer the equator (between 0° and 10° of north and south latitude); a thermometer which had been observed in Paris at 38.4° (101° F.), does not rise at Cumaná above 33° (91° F.); at Vera Cruz it has touched 32° (89° F.), but once in thirteen years. At Havana, during three years, (1810–12), Señor Ferrer found it to oscillate only between 16° and 30° (61° and 86° F.). Señor Robredo, in his manuscript notes, which I have in my possession, cites as a notable event that the temperature in 1801 rose to 34.4° (94° F.), while in Paris, according to the interesting investigations of Mons. Arago, the extremes of temperature between 36.7° and 38° (97° and 100° F.) have been reached four times in ten years (1793–1803.)

"The great proximity of the days on which the sun passes the zenith of those places situate near the limit of the torrid zone, makes the heat at times very intense upon the coast of Cuba, and in all those places comprised between the parallels of 20° and 23½°, not so much as regards entire months as for a term of a few days. In ordinary years the thermometer never rises in August above 28° or 30° C. (82° or 86° F.), and I have known the inhabitants complain of excessive heat when it rose to 31° C. (87.8° F.)

"It seldom happens in winter that the temperature falls to 10° or 12° C. (50° to 53° F.), but when the north wind prevails for several weeks, bringing the cold air of Canada, ice is sometimes formed at night, in the interior of the island, and in the plain near Havana. From the observations of Messrs. Wells and Wilson, we may suppose that this effect is produced by the radiation of caloric when the thermometer stands at 5° C. (41° F.), and even 9° C. (48° F.) above zero. This formation of a thick ice very near the level of the sea, is more worthy the attention of naturalists from the fact, that at Caraccas (10°31′ N. lat.), at an elevation of 300 feet, the temperature of the atmosphere has never fallen below 11° C. (41.8° F.); and that yet nearer to the equator we have to ascend 8,900 feet to see ice form. We also observe that between Havana and St.

Domingo, and between Batabanó and Jamaica, there is a difference of only 4° or 5° of latitude, and yet, in St. Domingo, Jamaica, Martinique and Guadalupe, the minimum temperature in the plains is from 18.5° to 20.5° C. (65° to 68° F.)

"It will be interesting to compare the climate of Havana with that of Macao and Rio de Janeiro, one similarly situated near the northern extreme of the torrid zone, but on the eastern shore of Asia, and the other near the southern limit of the torrid zone, on the eastern shore of America. The means of temperature at Rio Janeiro are deduced from three thousand five hundred observations made by Señor Benito Sanchez Dorta; those of Macao from twelve hundred observations which the Abbé Richenet has kindly sent me.

Mean.	Havana. N. lat. 23°9'.	Macao. N. lat. 22°12'.	Rio Janeiro. S. lat. 22°54'.
For the year, . . .	78.00° F.	73.94° F.	74.30° F.
For the hottest month, .	83.84° F.	83.12° F.	80.96° F.
For the coldest month, .	69.98° F.	61.88° F.	68.00° F.

"The climate of Havana, notwithstanding the frequent prevalence of north and north-west winds, is warmer than either that of Macao or Rio Janeiro. The first named of these places is somewhat cold, because of the west winds which prevail along the eastern shores of the great continent. The proximity of very broad stretches of land, covered with mountains and high plains, makes the distribution of heat through the months of the year more unequal at Macao and Canton, than in an island bordered by sea-shores upon the west, and on the north by the heated waters of the Gulf Stream. Thus it is that at Canton and Macao the winters are much more severe than at Havana.

"The mean temperatures of December, January, February and March, at Canton, in 1801, were between 15° and 17.3° (59° and 62° F.); at Macao, between 16.6° and 20° (61° and 68° F.); while at Havana they were generally between 21° and 24.3° (69° and 75°); yet the latitude of Macao is one degree south of that of Havana, and the latter city and Canton are on the same parallel, with a difference of one mile, a little more or less. But although the isothermal lines, or lines of equal heat, are convex toward the pole in the *system of climates of Eastern Asia*, as also in the *system of climates of Eastern America*, the cold on the same geograpical parallel is greater in Asia.* The

* The difference of climate is so great on the eastern and western shores of the old continent, that in Canton, lat. 23°8', the mean annual temperature is 22.9° (63° F.), while at Santa Cruz de Teneriffe, lat. 28°28', it is 23.8° (74° F.), according to Buch and Escolar. Canton, situate upon an eastern coast, enjoys a continental climate. Teneriffe is an island near the western coast of Africa.—H.

Abbé Richenet, who used the excellent *maximum* and *minimum* thermometer of Six, has observed it to fall even to 3.3° and 5° (38° and 41° F.), in the nine years, from 1806 to 1814.

"At Canton, the thermometer sometimes falls to 0° C. (32° F.), and from the radiation of caloric, ice is formed on the roofs of the houses. Although this excessive cold never lasts more than one day, the English merchants residing at Canton light fires during the months of November, December and January, while at Havana fires are never needed.

"Hail of large size frequently falls in the Asiatic countries round Canton and in Macao, while at Havana fifteen years will pass without a single fall of hail. In all three of these places the thermometer will sometimes stand for hours between 0° and 4° C. (32° and 39° F.); yet notwithstanding (which seems to me more strange), it has never been known to snow; and although the temperature falls so low, the banana and the palm grow as well in the neighborhoods of Canton, Macao and Havana, as in the plains immediately under the equator.

"In the present state of the world, it is an advantage to the study of meteorology, that we can gather so many numerical elements of the climates of countries situate almost immediately under the tropics. The five great cities of the commercial world—Canton, Macao, Calcutta, Havana and Rio Janeiro, are found in this position. Besides these, we have in the Northern hemisphere, Muscat, Syene, New Santander, Durango, and the Northern Sandwich Islands; in the Southern hemisphere—Bourbon, Isle of France, and the port of Cobija, between Copiapo and Arica, places much frequented by Europeans, and which present to the naturalist the same advantages of position as Rio Janeiro and Havana.

"Climatology advances slowly, because we gather by chance the results obtained at points of the globe where the civilization of man is just beginning its development. These points form small groups, separated from each other by immense spaces of lands unknown to the meteorologist. In order to attain a knowledge of the laws of nature regulating the distribution of heat in the world, we must give to observation a direction in conformity with the needs of a nascent science, and ascertain its most important numerical data. New Santander, upon the eastern coast of the Gulf of Mexico, probably has a mean temperature lower than that of the Island of Cuba, for the atmosphere there must participate, during the cold of winter, in the effects of the great continent extending towards the north-west.

"On the other hand, if we leave the *system of climates of Western America*, if we pass the lake, or, more strictly speaking, the submerged valley of the Atlantic, and fix our attention upon the coasts of Africa, we find that in the *cis-Atlantic system of climates* upon the western borders of the old continent, the isothermal lines are again raised, being convex towards the pole. The tropic of Cancer passes between Cape Bojador and Cape Blanco, near the river Ouro, upon the inhospitable confines of the Desert of Sahara, and the mean temperature of those countries is necessarily hotter than that of Havana, for the double reason of their position upon a *western* coast, and the proximity of the desert, which reflects the heat, and scatters particles of sand in the atmosphere.

"We have already seen that the great declinations of temperature in the island of Cuba are of so short duration, that neither the banana, the sugar-cane, nor the other productions of the torrid zone, suffer the slightest detriment. Every one is aware how readily plants, that have great organic vigor, sustain momentary cold, and that the orange trees in the vicinity of Genoa resist snow storms and a degree of cold not lower than 6° or 7° C. below zero (21.2° or 19.4° F. above zero).

"As the vegetation of Cuba presents an identity of character with that of regions near the equator, it is very extraordinary to find there, even in the plains, a vegetation of the colder climates, identical with that of the mountains of Southern Mexico. In other works, I have called the attention of botanists to this extraordinary phenomenon in the geography of plants. The pine (*pinus occidentalis*), is not found in the Lesser Antilles, and according to Mr. Robert Brown, not even in Jamaica (between 17¾° and 18° of latitude), notwithstanding the elevation of the Blue Mountains in that island. Further north only do we begin to find it, in the mountains of St. Domingo, and throughout the Island of Cuba, which extend from 20° to 23° of latitude. There, it attains a height of sixty or seventy feet, and and what is still more strange, the pine and the mahogany grow side by side in the plains of the Isle of Pines. The pine is also found in the south-eastern part of Cuba, on the sides of the Cobre Mountains, where the soil is arid and sandy.

"The interior plain of Mexico is covered with this same class of coniferas, if we may rely upon the comparison made by Bonpland and myself, with the specimens we brought from Acaguisotla, the snow mountain of Toluca, and the Cofre of Perote, for these do not seem to differ specifically from the *pinus occidentalis* of the Antilles, as described by Schwartz. But these pines, which we find at the level of the sea in Cuba, between the 20° and 22° of latitude, and only upon its southern side, do not descend lower than 3,200 feet above that level upon the Mexican continent, between the parallels of 17½° and 19½°. I have even observed that on the road from Perote to Jalapa, on the eastern mountains of Mexico, opposite to Cuba, the limit of the pines is 5,950 feet, while on the western mountains, between Chilpancingo and Acapulco, near Cuasiniquilapa, two degrees further south, it descends to 3,900 feet, and at some points, perhaps, even to the line of 2,860 feet.

"These anomalies of position are very rare under the torrid zone, and depend probably less on the temperature than on the soil. In the system of the migration of plants, we should suppose that the *pinus occidentalis* of Cuba had come from Yucatan, before the opening of the channel between Cape Catoche and Cape San Antonio, and not, by any means, from the United States, although the coniferous plants abound there, for the species of whose geography we are treating has not yet been found in Florida.

MEAN OF OBSERVATIONS AT HAVANA.

North Latitude, 23°9′.

Months.	1810–12. ° Fahr.	Seasons. ° Fahr.	Months.	1810–12. ° Fahr.	Seasons. ° Fahr.
March,	79.0		September,	82.6	
April,	78.6		October,	79.0	
May,	82.0	— 79	November,	75.6	— 79
June,	82.7		December,	70.0	
July,	82.9		January,	70.0	
August,	83.4	— 83	February,	72.0	— 70
Mean,				78.0° Fahr.	

Comparison between the mean temperature in the interior, and on the shore of Cuba, and at Cumaná, in South America. See following Table.

	Ubajay, int. of Cuba.	Havana coast.	Cumaná. N. lat. 10°27′.
December to February,	64.4° F.	71.2° F.	80.4° F.
March to May,	71.2 "	79.2 "	83.7 "
June to August,	81.8 "	83.0 "	82.0 "
September to November,	74.8 "	78.6 "	82.6 "
Mean,	73.2 "	78.0 "	81.7 "
Coldest months,	62.0 "	70.0 "	79.2 "
Hottest "	83.5 "	83.0 "	84.4 "

At Rome, N. lat. 41°53′—Mean temperature, 59.0° F.
" " " Hottest month, . 77.0 "
" " " Coldest " . 42.3 "

Fall of Rain in Havana.

The mean annual fall of rain in Havana for a period of five years was about 40 Inches. Average fall during the Seasons:

Spring,	7.51 inches.	Autumn,	11.48 inches.
Summer,	14.23 "	Winter,	6.54 "

"Notwithstanding the frequency of rain during the hot season, that is, during the months of July, August and September, these months do not present the greatest number of cloudy days. The rains of summer, although copious, are of short duration, and those days on which showers do not fall, are in general perfectly cloudless. It may almost be said that during these months no clouds are to be seen in the atmosphere, except while the shower is falling, while in the other months, cloudy days sometimes occur without rain. Days during which the heavens are completely clouded are extremely rare in Cuba. The following Table gives the mean for each month:

Months.	Cloudy Days.	Clear Days partly cloudy.	Months.	Cloudy Days.	Clear Days partly cloudy.
January,	5	26	August,	6	25
February,	8	20	September,	7	23
March,	7	24	October,	7	24
April,	5	25	November,	8	22
May,	8	23	December,	7	24
June,	6	24		—	—
July,	6	25	Total,	80	285

"These tables will give some idea of the beauty of the sky in these regions, and of its effect upon the life and luxuriant growth of vegetation. A high temperature moderated by great evaporation, which pours through the atmosphere a continuous torrent of watery vapors, presents the most favorable conditions for the development of an admirable vegetation; which again contributes, on its part, to maintain the humidity of the atmosphere—soul of its exuberant life. Thus it is that through all seasons of the year the fields and forests of Cuba preserve their verdure; but it is principally at the beginning of summer, during the rainy season, that all nature seems to be transformed to flowers."

With the exception of the northern Bahamas, which lie beyond the tropic of Cancer, the entire West India Archipelago is situated in the torrid zone. The heat is, consequently, very great on the lower grounds, where, however, it is tempered by the sea-breezes, which generally blow in the afternoon, when their cooling agency is most needed.

Snow is never known to fall, but slight frosts occasionally occur in the mountainous districts of Cuba. When the sun is in the southern hemisphere, the archipelago enjoys the benefit of the trade-winds, blowing from an easterly direction, and diffusing over it a refreshing coolness; but when the sun has passed the Equator, the trade-winds retire northward, and are replaced by south-eastern winds, which are warm and gentle. The long rainy season commences in July and continues till the month of November, when the rain often falls in torrents, but rarely lasts for many hours continuously. It is ushered in by violent gusts of wind accompanied by terrific thunder-storms, and during their continuance, the destructive yellow fever and other malignant diseases are prevalent.

Climate of the Bahamas.

This extensive chain of low islands, stretching in a south-east and north-west direction, lying between 20° and 28° N. latitude, have been estimated at 500, of which, however, a great proportion are mere islets and rocks, here called *cays* or *keys;* not more than 12 or 15 are inhabited. The *Gulf Stream* runs between the Great Bahama Island and the coast of Florida at the rate of five or six miles an hour; and many vessels have been wrecked in passing this dangerous strait, through which rush the heated waters of the Gulf of Mexico.

The climate of the Bahamas is salubrious. In Turk's and other salt islands, no epidemic disease has ever been known. The more northern islands, during the winter months, are rendered cool and agreeable by the north-west breeze from the continent of America. At *New Providence*, the thermometer in the shade varies from about 85° or 90° Fahr. in summer, to 60° or 65° in winter. The more southerly islands are hotter throughout the year, but they enjoy the cooling sea-breezes which blow within the tropics, though these do not extend to Abaco, the most northern of the Bahamas.

The subjoined Table exhibits an average of meteorological observations for seven years, which shows a remarkable evenness of temperature.

	Barometer.	° Fahr.	In. Rain.		Barometer.	° Fahr.	In. Rain.
Jan., .	. 30.10	73	4.72	August,	30.15	84	7.11
Feb.,	. 30.24	71	3.75	Sept.,	. 29.97	81	9.74
March,	. 30.07	72	1.06	Oct.,	. 29.99	80	7.80
April,	. 30.03	75	1.82	Nov.,	. 30.00	76	2.25
May,	. 30.00	79	2.25	Dec.,	. 30.02	72	1.70
June,	. 30.09	81	5.12				
July,	. 30.11	84	4.68	Total inches of Rain, .			51.90

Bermuda—Gulf Stream.

"The climate of the Bermuda Islands has a mean temperature between that of the West Indies and British North America, partaking neither of the extreme heat of the one, nor the excessive cold of the other. It is greatly improved by the warmth of the *Gulf Stream*, which sweeps along between Bermuda and the American Continent; the winter months resembling the early part of October in England, but without its frosts. The sweet strains of the Bard of Erin have sounded the praises of the cedar-groves and wood-nymphs of the 'Fairy Isles,' as the Bermudas have been styled by Shakspeare—

"'No: ne'er did the wave in its element steep
An island of lovelier charms;
It blooms in the giant embrace of the deep,
Like Hebe in Hercules' arms;
The blush of your bowers is light to the eye,
And their melody balm to the ear;
But the fiery planet of day is too nigh,
And the snow spirit never comes here.'

"Bermuda is not so much subject to diseases as are the more northern climates. Epidemics are of unfrequent occurrence, and deaths from all causes, as shown by the statistical tables, amount to no more than 14.5 per cent. annually.

"The climate of Bermuda would prove eminently eligible for those natives of cold countries, who, from general delicacy of constitution, are unable to undergo active continuous labor with exposure, or who otherwise suffer from a cold and variable climate. As far as the author's observation goes, the effect of residence in Bermuda, on such persons, is usually beneficial, especially on those who are predisposed to scrofula or pulmonary consumption, or who have evinced a peculiar tendency to colds and bronchial affections during the winter months. In such cases the physical energies usually undergo a rapid and marked change, resulting in permanent good health. I believe that immigrants of this description, by observing common prudence in their mode of living, might, with perfect safety, and with every prospect of improved health, engage as farmers in the islands generally.

"The effects of the *Gulf Stream* on the climate of Bermuda are very manifest. This powerful current, after rising under the tropic, and flowing from the Gulf of Mexico through the Straits of Bahama, runs in a north-easterly direction along the American coast, washing the Great Bank of Newfoundland, and, after flowing upwards of 3,000 miles, finally reaches the Azores, and even the Bay of Biscay. The temperature of the water of this current is 8° above that of the surrounding sea at the Great Bank, and 5° above the temperature of the sea at Azores. Rennel estimates the dimensions of the current and the tract that receives it at 2,000 miles in length, and 350 in breadth. Both are marked by the sea-weed, and are well known to mariners. By this cauldron of warm water the ice-bergs from the north are dissolved; the surrounding waters and superincumbent atmosphere are warmed, and the temperature of the neighboring continent elevated. A proper retreat is also afforded to the various kinds of fish after their season of spawning has passed, and while the severity of the frosts drives them from the shores. Such are some of the leading operations perceived in the economy of nature in this part of the world.

"The effects of the climate upon the agricultural produce are more favorable than in other countries under the same mean annual temperature. Besides many of the fruits of the temperate regions, the heat of summer permits those of a tropical character to flourish; hence, a greater variety may be produced than in any other part of the world. The season of vegetation is sufficiently extended to ripen a great many kinds of grain, vegetables and fruit.

"The most agreeable season at Bermuda is the winter, or cold season, which lasts from November to March; the mean temperature being 60° Fahr. The prevailing winds are then from the westward; but if from the north-west, fine, hard weather, with a clear sky, accompanies them, the thermometer varying from 50° to 56°. This weather often terminates in a very fine, bright day, with a very slight wind and partial calms; afterwards, the wind invariably changes to the south-west, and the weather becomes hazy, damp, and attended with heavy rains and gales; the thermometer rising to 66° and 70°.

"These alternate north-westerly and south-westerly winds prevail during nine months of the year, the wind remaining at no other point for any length of time. The change is shown by a difference of 14° in the temperature.

TEMPERATURE OF BERMUDA.

Range of the Barometer and Thermometer; average for four years.

	Barometer.	Thermometer.
Maximum,	30.480	85.85
Minimum,	29,236	49.00
Oscillation, or Range,	1.244	35.05

Spring commences at the end of February, and the weather usually continues mild, with refreshing showers of rain and gentle breezes from the south and west, until the end of May.

The Summer begins in June, and the weather becomes hot. Calms about this time generally replace the gentle breezes of May; the atmosphere becomes sultry and oppressive, and long droughts are common, which are usually succeeded by severe thunder-storms. The weather, in September, changes its character, and again becomes mild and agreeable.

"These islands, which are generally and properly allowed to be healthy, have only been afflicted a seventh time since their settlement—a period of above two centuries—with yellow fever.

"The productions of the soil are varied. The wheats of the south of Europe, Egypt and Africa could hardly fail in Bermuda. The American wheat has been tried with success. Excellent potatoes are easily cultivated; the sweet potato yields abundantly. Arrowroot, cassava and yams also yield abundant crops. Ginger and tobacco are easily cultivated; and vegetable oils abound. The cotton raised in Bermuda is accounted very firm and substantial, but the flax plants are the most important of all the neglected products of Bermuda.

"Drugs are here in great abundance. The fruits could be cultivated with much advantage; the strawberry, the grape, the fig, the guava, the shaddock, and many other tropical fruits, ripen in the open air without assistance from art."—"BERMUDA," *by T. L. Godet, M.D. London,* 1859.

Bermuda.

North Latitude, 32°19′; West Longitude, 64°51′.

Monthly mean Temperature of the Air—1854.

Time.	Jan.	Feb.	March.	April.	May.	June.	
9 A.M., . .	65.1	61.3	64.2	67.4	71.8	77.8	
Noon, . .	65.4	61.9	64.7	67.7	72.4	78.1	
4 P.M., . .	66.4	62.0	65.0	67.9	72.7	78.1	
	July.	Aug.	Sept.	Oct.	Nov.	Dec.	Year.
9 A.M., . .	81.1	85.8	81.2	74.6	69.4	65.2	71.9
Noon, . .	81.4	85.5	81.8	75.1	69.5	65.7	72.4
4 P.M., . .	81.5	84.4	81.4	74.3	69.3	65.4	72.3

Maximum temperature for August, . .	87° Fahr.
Minimum temperature for August, . .	81 "
Maximum temperature for February, .	66 "
Minimum temperature for February, .	56 "
Extreme range of temperature, . .	31 "

NOTE.—The year 1854 appears to have been warmer than the average Seasons of previous years.

Climate and Productions of Hayti.

This rich and beautiful island, the second in size of the West Indies, Leeward group, lies south-east of Cuba, and separated from it by the Windward Passage, 50 miles broad. Its extreme length from east to west about 400 miles; greatest breadth, 150 miles; area, 27,000 square miles. The east part of the island is occupied by the Republic of San Domingo, and the west part by the Empire of Hayti.

Climate, etc.—There are two seasons in Hayti—a wet season and a dry season. During the former heavy rains are frequent, three and even five inches at times falling in 24 hours; and in the latter, little or no rain falls, and, in some localities, years have passed over without a single heavy shower. At San Domingo the mean temperature is 78°, and the extremes 60° and 95°; while at Port-au-Prince, the range is from 63° to 104°. The minimum occurs in December, and the maximum in August and September. Land-breezes moderate the summer heats. Hurricanes are less seldom here than in the Windward or Caribbean Islands. Earthquakes, though not frequent, have been very disastrous. Nowhere is tropical vegetation seen to greater advantage than in Hayti; contributing, with the lofty, and at times, rugged mountains and deep valleys, to render the scenery of this island unsurpassed. Majestic pines, noble mahogany trees, fustic, satin-wood, lignum-vitæ clothe the mountains, and form the principal exports of the southern provinces. The roble or oak, which yields hard, durable wood; the wax-palm, divi-divi, numerous fine cabinet woods, and the richest flowering plants, abound; together with the usual tropical vegetables—plantains, bananas, yams and batatas; also fruits, including oranges, pine-apples, cherimoyas, sapodillas, with melons and grapes. The staple cultivated products are coffee, sugar, indigo, cotton, tobacco and cocoa; the quantities of which raised have fallen off, in consequence of the unsettled state of the island.

Island of Jamaica.

This is one of the Great Antilles, and the principal of the West India Islands, lying between 17°40′ and 18°30′ N. latitude.

"The mean annual temperature at Kingston is 78°; in summer, 81°; in winter, 76°. The rainy seasons are from May to August, and from October to November. Earthquakes are frequent; hurricanes less so than in the other West India Islands. The soil is naturally less productive than in many of the West India Islands; but most of staple products of tropical climates are raised, sugar being the chief. Indigo, cotton and cocoa were formerly more important staples than at present. Maize, Guinea corn and rice are now the chief grains raised.

PART XIV.

CLIMATE OF SOUTH AMERICA.

THIS rich and fertile portion of the New World, with two-thirds of its area situated between the tropics, makes the climate of South America necessarily very hot, particularly along the Atlantic coast and shores of the Caribbean Sea. Though yielding in this respect to Africa, the corresponding continent of the Old World, the temperature is, for the most part, considerably higher than that of North America ; for while the latter has its maximum breadth in the Arctic regions, the former attains its greatest width in the torrid zone near the Equator.

The highest mean temperature occurs in the parts of New Granada, Venezuela and Guiana, which is enclosed within the isothermal line of 81° Fahr., while in portions of Abyssinia, the mean temperature exceeds 85°. This great difference is, no doubt, chiefly owing to the greater humidity of South America, its vast forests, the absence of sandy deserts, the influence of the trade winds, and the freer access to its shores of the great oceans of the globe. In contradistinction to the other great divisions of the land, the western shores of this continent are considerably colder than the eastern, owing to the low temperature of the "Antarctic Drift," or Humboldt Current, which, setting out from the Antarctic Ocean, flows north-eastward against the shores of Chili, then northward along the coast of Peru to the vicinity of the Equator.

"South America is also characterized by great moisture, which attains its maximum in the extreme north, where the temperature is highest, but which is everywhere more copious on the eastern than on the western side of the Andes. Within the tropics, the wide plains on the east are deluged by the heavy periodical rains from November to May, while the narrow margin between the Cordilleras and the Pacific is almost entirely rainless. In some places, the deposition of moisture is surprisingly great; for, while in the tropical regions of the New World generally it amounts to 112 inches, on the north coast of Dutch Guiana 220 inches fall annually; and in some places on the east coast of Brazil, near the Equator, in the valley of the Amazon, no less than 270 inches have been observed. This astonishing quantity falls, moreover, in a comparatively brief period. The number of clear days, however, in many portions of the country, is much more considerable than in our temperate climates; while, during the long-continued drought that precedes the wet season, the ground is parched, the sun glares with intense radiance, and the cattle and wild animals, tormented alike by hunger and thirst, perish in great numbers.

"The climate and productions of the middle and southern portions of South America, including Bolivia, Paraguay and the Argentine Republic, drained by the Rio de la Plata, are of a different character in many particulars. The southern plain, named the *Pampas*, is a dead level, destitute of trees, but covered alternately with luxuriant pasturage, and vast crops of gigantic thistles, and interspersed with a multitude of salt lakes, some of which are of large size.

"The northern part of this region belongs to the plain of *Gran Chaco*, or Great Desert, which extends from the 18th to the 28th south parallel, and from longitude 58° to 63° west. Besides the north of La Plata, it embraces a large section of eastern Bolivia, being bounded on the east by the Paraguay, and traversed by its tributaries. It has an average elevation of from 300 to 500 feet; the northern portion is covered with grass, while the southern, consisting of an arid and desert plain, is thinly inhabited by roving Indians. The climate is characterized by great diversity, but is in general hot and very dry—the Patagonian Andes on the one side, and the mountains of Brazil on the other, intercepting the rain-bearing winds from the great oceans. At intervals of about fifteen years apart, the rains are wholly suspended in the interior of the country, the ground assumes the appearance of a dusty highway, and great suffering ensues from want of food and water. The Pampas are also subject to violent hurricanes, called *pamperos*, accompanied with terrific thunder and lightning. These carry so much dust and sand into the air as to produce darkness at noon as far south as Buenos Ayres. The mean annual temperature ranges from 58° Fahr. in the south to 72° in the north; January, from 68° to 77°; and July (the middle of winter), from 48° to 68° Fahr. In general, the heat of summer is not excessive, and the climate is more salubrious than that of other countries equally near the tropics."

The more southern portion of South America, including a part of Chili and all of Patagonia, has a cool atmosphere, ranging from 50° to 30° Fahr., mean annual temperature; while along the Straits of Magellan, in 52° south latitude, the climate becomes intensely cold during the winter months, in the southern hemisphere, of June, July and August. *Tierra del Fuego*, and the southern extremity of the continent, terminating with *Cape Horn*, in 56° south latitude, has a frigid, inhospitable climate.

Climatic Boundary of South America.

South America, like the northern portion of the continent, is washed by two great oceans, one on the east and one on the west, each exercising great and varied climatic influences. On the *east*, the great EQUATORIAL CURRENT of the Atlantic strikes Cape St. Roque with its warming influence, where it divides, a portion flowing southward along the coast of Brazil, and a larger portion northward along the Guiana coast into the Caribbean Sea, and thence into the Gulf of Mexico, forming the GULF STREAM of the North Atlantic. On the *south-west*, the ANTARCTIC DRIFT, or Humboldt Current, strikes the coast of Chili and extends northward along the shores of Peru, toward the Equator, with its cooling influence, crowding down the thermometer in its onward course; when, finally, it forms a great EQUATORIAL CURRENT, crossing the broad Pacific Ocean toward the Caroline Islands; forming another *ocean stream* off Japan, that ultimately re-crosses the North Pacific Ocean and strikes the north-west coast of North America.*

In a climatic point of view, South America, extending through sixty-eight degrees of latitude, the greater portion lying within the Tropics, may be thus divided:

1. The *Tropical* portion, lying north of the Equator (extend-

* The climatic influence of these two great Ocean Currents, one *warm* and the other *cold*, when striking the coast of South America, are of a most wonderful and different character—the one on the *east* coming laden from off the coast of Africa with warm water and a moist atmosphere brings deluging rains, which clothes the eastern shore of the continent with verdure; while the one on the *west*, with cold water and a chilling atmosphere, is unaccompanied by rain or perceptible moisture, other than what falls in copious dews along the coasts of Chili and Peru. So in regard to the continuation of this great Ocean current, which, after crossing the Pacific, strikes the American coast to the far north in Russian and British America, bringing a warm current of air and moisture like unto the *Gulf Stream* which strikes against the west coast of England.

ing to 12 N. latitude), comprises the States of New Granada, or Colombia, Venezuela, Guiana, and the northern part of Brazil. Here the atmosphere, near the level of the sea, varies from 82° to 70° mean annual temperature. The *pampas* and elevated lands of New Granada and Venezuela sustains human life at an altitude of 10,000 or 12,000 feet above the ocean, where the temperature assumes that of the frigid zone. This division, in the extensive *llanos* and low grounds of the coast forms the hottest part of the continent, being thinly inhabited by Europeans and their descendants, and native Indians.

2. The *Tropical* region lying south of the Equator, comprises the larger portion of Brazil, Ecuador, Bolivia, Peru, and part of Paraguay. This division constitutes the largest half of South America, running through 23½ degrees of latitude and 47 degrees of longitude. Here are found all degrees of temperature from the Equatorial, 82° Fahr. to the frigid temperature of perpetual winter on the elevated peaks of the Andes, rising from 20,000 to 24,000 feet in height.

3. The *Sub-Tropical* region, extending from 23½° to 32° S. latitude, embraces the southern portions of Brazil, Paraguay and Bolivia, having a mean annual temperature averaging from 60° to 70° Fahr., near the level of the ocean. This rich and fertile section of country is capable of sustaining a dense population, where is to be found a fertile soil and the most luxuriant growth of valuable trees and other vegetation, comprising a great variety of useful plants.

4. The *Temperate* division, extending from 32 to 48 S. latitude, embraces Buenos Ayres, or the Argentine Confederation, Uruguay and Chili, and the northern part of Patagonia. This most favored region produces all the grasses, cerials and fruits peculiar to the temperate zone. Here are four regular seasons of three months each, and in general is favored with a healthy climate, where man attains his highest development. Here the climate varies from 40° to 60° mean annual temperature.

While the Northern Hemisphere is blessed with a broad stretch of country lying within the temperate zone, the Southern Hemisphere is limited to a comparative small extent of country, here running through only 20 degrees of longitude, from the Atlantic to the Pacific Ocean. On the line of this zone going round the globe is embraced the large Island of New Zealand, in the South Pacific, which is one of the most favored portions of the earth's surface.

5. The *Cold* or *Frigid Zone*, embracing the southern portion of South America, extends from 48° to 56° S. latitude. It comprises the southern part of Patagonia, and the sterile peninsula, or island of Tierra del Fuego. This inhospitable region is in part sparsely inhabited by Indians in a low state of civilization, while northern Patagonia is peopled by a large and warlike race of men. Here the climate varies from **20°** to **40°** mean annual temperature, corresponding in many respects with the climate of Labrador in North America.

While Greenland on the north produces *icebergs*, this region is celebrated for its *glaciers*, where alone they are found at the present day on the continent of America. Another striking contrast is the Esquimaux of low stature on the coast of Labrador, and the stalwart Patagonian race of South America.

Climate of Costa Rica—Isthmus of Panama.

The small state of Costa Rica is bounded on the south-east by New Granada, forming the north-western extremity of the Isthmus of Panama, where runs the boundary, from sea to sea, between North and South America. In climate and vegetable productions it possesses the same general character as the whole of the Isthmus, extending through seven degrees of longitude. The physical aspect of Costa Rica is very uneven, presenting extensive valleys, table-lands and mountains; and the face of the country is at various levels above the ocean, which, according to their height, have here, as in all other parts of Central and South America, different temperatures and productions. Between the foot of the mountains and the shores of the two seas, the surface is low and flat.

The principal productions of Costa Rica are dye-woods, drugs, grain, fruits, indigo, tobacco, cocoa and coffee. The wild and white sugar-cane, and that of the species called *birota*, which, spread out, forms strong planks, are abundant. Coffee is the staple export; and when properly plucked and dried, resembles that of Mocha.

"The climate of Costa Rica is as varied as its aspect. In the principal inhabited places it may be asserted that the climate is the finest in the known world—no extremes of heat or cold. Fahrenheit's thermometer usually varies between 50° and 80° Fahr.; but the thermometer ranges through every degree of the scale, from the freezing point to 100°, in proportion to the elevation above the level of the sea. At many places a short distance from Carthage, and in other parts, the cold is so intense that it frequently happens that running waters are found frozen in the morning; and the inhabitants of Carthage and San José enjoy the luxury of ice; so that the territory of Costa Rica can produce all the fruits and productions of every climate in the world." (*For Fall of Rain and Temperature,* see pages 251, 253.)

Climate and Physical Features of New Granada.

The Republic of New Granada, or Colombia, embracing the northern portion of South America, including the Isthmus of Panama, is bounded on the north and east by the Caribbean Sea, and on the west by the Pacific Ocean; extending from the Equator to 12°20′ N. latitude. This country is the most equally diversified in soil and climate of all the South American States. Neither plain or mountain can be said to predominate; the sea-coasts are ample and commodious, and, owing to the wide ramification of the Andes, there is a great extent of country at an elevation of from 5,000 to 10,000 feet, which, in such a latitude, is most favorable to industry and the progress of civilization. Yet the insalubrity of the zone surrounding this highly favored region has hitherto counterbalanced its apparent advantages, and prevented the development of its varied and abundant resources.

The climate of New Granada presents the most remarkable contrasts of almost any portion of the New World. At Honda, nearly 1,000 feet above the sea level, it is intensely hot and unhealthy. The yellow fever is endemic at Cartagena and on the west coasts. But in the elevated country, the air is perfectly salubrious and the temperature (from 56° to 70° Fahr.) seems that of perpetual spring. Here the rains in the wet season darken the sky only a few hours daily in the afternoon. At Mompox, the day is always cloudy, the night clear. The summits of the Cordilleras are often shrouded in mists; torrents of rain fall almost unceasingly in the forests of Darien; the Gulf of Chocho is perpetually vexed with violent storms; but these excesses of the elements are all unknown in the middle regions or Templadas, and, excepting the earthquakes, which have left here, as elsewhere in the Andes, deep traces of their destructive visitations, there is nothing which detracts from the general benignityof nature. Even up to the limits of perpetual congelation the climate continues healthy, though it may cease to be agreeable.

The remarkable equality of the climate in this part of the

world, where the seasons differ little from each other, seems unfavorable to the multiplication of vegetable species. Each kind seizes on some locality or region wherein it predominates to the almost total exclusion of others. On the plains of Bogota, in the region of perpetual spring, though vegetation is most luxuriant, the species are not numerous. Yet the woods, imperfectly explored, teem with valuable productions. The wax palm, 200 feet high, clothes the sides of Tolima to an elevation of 8,000 feet. The forests of Popayan yield china or cinchona (the cascarilla or Jesuit's bark of commerce) in abundance. Cotton, rice, tobacco, cocoa, sugar-cane, with all tropical fruits, are among the productions of the coast; while the elevated plains yield maize, wheat, and all the cereals and fruits of Europe. With nature so bountiful, the wants of the population so few, and the demands of commerce very moderate, the cultivation of the soil is carried on, as might be expected, very remissly, and the reclaimed land bears but a small proportion to the whole. In the llanos, towards the Orinoco, the people are occupied wholly with the rearing of cattle and horses.

Climate, Productions and Physical Features of Venezuela.

This Republic, occupying the north-east portion of South America, lies between N. latitude 1°8′ and 12°16′, is bounded on the north by the Caribbean Sea, and forms the hottest portion of the American Continent. The mountains hold a secondary importance, and occupy but a third of the whole territory. The plains in the vicinity of Lake Maracaibo, have but a moderate elevation, rarely exceeding 4,000 feet, and are nowhere cultivated to any considerable extent. Thick forests which cover the whole territory shelter numerous independent Indian tribes. The *paramos*, or summit-plains, have generally an elevation of 10,000 or 12,000 feet. Where cultivation has obtained a footing on the slopes of the mountains, it succeeds to a height of 8,000 or 9,000 feet, the line which separates the cereal crops of temperate climates, wheat, barley, etc.; from tropical productions, maize, cocoa, coffee, the yuca, etc., being at an elevation of about 4,000 feet.

"The *Llanos*, or plains of Venezuela, are of vast extent, having an area of over 100,000 square miles. They are generally destitute of trees, which, in the most favored spots, occur only in small clusters. In the dry season, the greater part of the llanos presents to the view a bare sunburnt desert with intense heat. But no sooner does the rain fall—and it pours down with the violence peculiar to the tropics—then the scene changes totally; vegetation springs forth and spreads itself abroad with surprising rapidity; the arid waste becomes a rich garden, the moistened earth seems to heave and open, and forth comes the crocodile and boa-constrictor, shaking off their lethargy, and releasing themselves from their temporary imprisonment; the streams and rivers being quickly flooded.

"The great river of Venezuela is the ORINOCO, flowing into the Atlantic, which holds the third rank among the great rivers of South America. The exuberantly fertile valley of this noble river, into which flow above 300 other rivers reputed navigable, watering a territory of 150,000 square miles, offers to advancing civilization all the natural conditions of an opulent and populous state.

"The climate of Venezuela exhibits in the highest degree the equatorial character. The change of seasons is scarcely perceptible, and vegetation goes on perpetually. On the coast, the thermometer ranges from 80° to 85° Fahrenheit the year round. But, notwithstanding the continuous heat that prevails along the coast, epidemic diseases are rare, and the climate is comparatively healthy. To those unacclimated, however, a due amount of care is necessary, as a too great exposure and inattention to diet are often followed by violent fevers. The table-land bordering the coast has an almost uniform range of temperature throughout the year, the thermometer varying only about ten degrees, from 70° to 82°. In the llanos, especially those portions subject to inundation, the climate is not very salubrious."

Climate and Surface of Guiana.

This portion of South America, lying between the parallels of 1° and 9°20′ N. latitude, is known as British Guiana, Dutch Guiana and French Guiana, comprising altogether an area of 142,000 square miles, with a population of about 400,000. The maritime region is low and level, but very fertile and hot. The country rises in successive terraces to the Sierra of Acaria, which separates it from Brazil, where the temperature becomes less heated.

The climate is tropical, but more genial than that of most places in the torrid zone, owing to the trade winds from the Atlantic, the sea and land breezes, and the frequent rains. It has two wet and dry seasons on the coasts, each continuing for three months; but in the interior, there is only one rainy season, from April to the middle of August. The mean temperature of the year is 81° Fahr. Violent thunder-storms occur at the change of the seasons; but hurricanes so destructive in the West Indies are unknown. Yellow fever and other malignant diseases occur periodically, being very fatal to the white population.

APPENDIX.

Culture of various Agricultural Products in North America, as influenced by Climate, &c.

The *Sugar-cane*, one of the most desirable tropical plants, grows in Central America, Mexico, the West India Islands, and as far north, in Louisiana, as 31° north latitude, where is found a mean annual temperature varying from 70° to 80° Fahr.

Coffee is also produced within the tropics, Costa Rica being celebrated for its culture, as well as other portions of Central America; it can be raised to advantage in Jamaica, Hayti, and Cuba, although sugar-cane is found to be more profitable.

Cotton is cultivated in portions of Central America and the West India Islands, but still more extensively in the southern portions of the United States, between the 28th and 34th parallels of north latitude, where the mean annual temperature ranges from 62° to 72° Fahr.

Rice requires a peculiar soil and situation, contiguous to water, where it can be overflowed. The Carolinas, Georgia, and Louisiana produce large quantities; while it is raised in the West India Islands and Central America.

Indian corn or Maize is produced over a wide extent of country, extending from the Equator, at different altitudes, to 45° north latitude. Its greatest yield may be said to be between the 38th and 41st parallels, where the mean annual temperature varies from 50° to 56° Fahr.; extending from the Atlantic Ocean to the base of the Rocky Mountains.

Tobacco is grown within the tropics and temperate region, extending as far north as the southern portion of Canada in 43° north latitude. Within the United States, it is raised in the greatest quantities in the Border States, including Maryland, Virginia, West Virginia, Kentucky and Missouri, where a mean annual temperature of from 52° to 60° Fahr. prevails. On the Island of Cuba the finest tobacco is raised during the winter season.

Flax and *Hemp* are raised in the United States under about the same climatic influence as tobacco ; the greatest yield being in the Middle and Border States.

LIMIT OF THE WHEAT REGION.—The limits of the culture of *Wheat* and the common cerealia are not so well defined in the United States, and Canada and other portions of British America, owing to the want of correct meteorological observations in the different parts of this extensive and unexplored region. It is safe, however, to say, that in Canada it extends north as far as the 48th parallel of latitude, from the Bay of Chaleurs to near the mouth of the Saguenay River, and from thence to the Lake St. John, 48° 30′ north, including the valley of Lake Temiscaming and all the head sources of the Ottawa River, extending to Michicopoten Bay, situated on the north shore of Lake Superior, 47° 56′ N. lat., having a mean *summer* temperature of 59° Fahr.

To the west of Lake Superior it embraces the valley of the Lake of the Woods, on the 49th parallel, running northward and embracing the whole of the valley of Lake Winnipeg, elevated 700 feet above the ocean; and the great valley of the Saskatchewan River, extending still further northward to the 60th parallel of north latitude, in the valley of Mackenzie's River. To the west of the Rocky Mountains, in the northern part of British Columbia, and on the Island of Sitka, 57° north latitude, the culture of wheat and other cereals is prevented, owing to the low summer temperature which exists along the North-west coast of America.

On the south, wheat can be raised profitably in the western portion of Texas and Arkansas, commencing at about the 30th parallel of latitude, excluding the Gulf Coast, where cotton flourishes to great perfection. Thus it appears evident that wheat can be raised to advantage from Texas to the British Possessions on Mackenzie's River, running through about one-third of the distance from the Equator to the North Pole, and from the Atlantic to the Pacific Ocean.

The harvesting of wheat through this extensive belt may be said to commence in the latter part of May, and continue until the latter part of August. "It is said that the ripening of the 'staff of life' will move steadily northward about twelve or fifteen miles per day, like a wave, until it sweeps up to the northern margin of the great wheat belt. A marching regiment in Texas, starting for the north, could barely keep before the ripening wave, and if they halted a day to rest, it would pass them. This wave stretches east and west across the Union, from the Atlantic to the confines of Kansas, and as it moves north it will grow longer and denser." Minnesota, extending northward to the 49th parallel of latitude, is one of the finest wheat-growing regions on the Continent. Indian corn also flourishes in the valley of the Red River of the North, which empties into Lake Winnipeg in about 50° north latitude.

The northern limit of wheat on the American Continent may be said to be on the line of the isotheral or mean *summer temperature* of 58° Fahr., where is found a fertile soil; while Indian corn requires a mean summer temperature of 66° Fahr. and upwards, embracing a still larger area of the earth's surface for its growth than that of wheat.

Meteorological Table,

SHOWING THE TEMPERATURE ALONG THE NORTHERN BELT OF THE WHEAT GROWING REGION.

Stations.	Latitude.	Longitude.	Altitude.	Y'rly M'n T'mp	Four Seasons.			
					Spring.	Summer.	Autumn.	Winter.
				°Fahr.	°Fahr.	°Fahr.	°Fahr.	°Fahr.
St. John, N. F.	47°33′	52°43′	140	39.00	33.00	54.00	44 00	23.00
Fort Kent, Maine	47°15′	68°35′	575	37.00	35.22	61.68	39.88	11.36
Quebec, Canada	46°50′	71°16′	100	40.50	38.00	65 56	43.00	13.68
Fort Temiscaming, Canada	47°20′	80°00′	612	39.50	37 60	65 20	40.00	15.11
Michipicoten, "	47°56′	85°06′	620	38 00	37.00	59.00	41.00	16.00
Fort Gary, H. B. Com.	50°15′	97°00′	750	36.00	36.00	68.00	40.00	8.00
Norway House, "	54°00′	98°00′	650	30.00	23.00	60.00	30.00	-2.00
Fort Chepewyan, "	58°43′	111°48′	700	32.00	24.00	59.00	34.00	-5.00
Fort Liard, " "	59°30′	121°00′	500	30.00	22 00	60.00	30.00	-6.00
Sitka, N. W. Ter.	57°00′	135°18′	50	42 00	40.00	54.00	44.00	32.00

NOTE.—The singular climatic influence operating in Newfoundland, 47° north latitude, and on the Island of Sitka, 57° north, determines the unfitness of either locality for producing the cereals; while inland on about the same belt of mean annual temperature, wheat can be grown in abundance, owing to a higher summer temperature prevailing.

In Europe, on the coast of Norway, and in Finland, wheat is raised as far north as 61°, in favored spots; while the hardier cerealia, rye, oats, and barley, are cultivated as high as 68° north latitude.

The growth of *Grass* or *Hay* as an article of commerce is less limited than wheat or the other cereals. It may be said to flourish from the 38th to the 45th parallels of latitude, although its limits in perfection are much less extensive. The belt included within the parallels of 39 and 43 north, within the United States, having a mean annual temperature from 47° to 53° Fahr. is its most favorite region, where are produced the largest quantities, and the best quality of butter and cheese. South of 39° north latitude, except in elevated regions, grass is of an inferior quality, and not much cultivated. In importance, as regards its value as an article of commerce, it vies with the product of either wheat, Indian corn, or cotton.

The *Potato*, the most valuable of the vegetable species, is produced in different varieties, in all climes, from the Equator to the northern limit of the Temperate Zone, varying, in North America, from 47° to 57° north latitude, where a mean annual temperature of 40° Fahr. exists; and even higher, in British America, west of Lake Winnipeg, where a Siberian summer prevails. On the south the sweet potato is most exclusively raised, while to the north the common or Irish potato flourishes in great perfection, forming nutritious food for man and beast.

AGRICULTURAL PRODUCTS OF THE DOMINION OF CANADA, ACCORDING TO THE CENSUS OF 1861.

PROVINCES.	Wheat.	Oats.	Barley.	Indian corn.	Potatoes.	Hay.
	bushels.	bushels.	bushels.	bushels.	bushels.	tons.
Ontario	24,620,425	21,220,874	2,821,962	2,256,290	15,325,920	861,844
Quebec........	2,654,354	17,551,296	2,281,674	334,861	12,770,471	689,977
New Brunswick	279,775	2,656,883	94,679	17,420	4,041,339	324,166
Nova Scotia...	312,081	1,978,137	269,578	15,529	3,824,864	334,287
Total.......	27,866,635	43,407,210	5,467,893	2,624,100	35,962,594	2,210,268

The vegetable kingdom is of the most magnificent description. Among cultivated plants the sugar-cane holds the highest rank, its cultivation having largely superseded the cotton and coffee formerly grown. The fauna resembles that of the valley of the Amazon.

Climate, Topography and Productions of Brazil.

This extensive empire comprehends the eastern and a large part of the central portion of South America, extending from 4° North, to that of 33°50′ South latitude. The name of BRAZIL, —which was for a long time restricted to a narrow though long extended portion of the American coast, extending from the mouth of the Amazon, under the Equator, nearly to that of the La Plata, is now used to designate all the former possessions of the Portuguese in this quarter of the globe, comprehending most of the valley of the Amazon, including the vast region between the sea and the mountains—the greater part of the interior country being formerly called Amazonia—and the extensive territory to the north of the Marañon, called Portuguese Guiana. Estimated Area, 2,500,000 square miles.

When first discovered by Cabral, it was denominated by him *Terra da Santa Cruz*, or "the Land of the Holy Cross." But this appellation was soon superseded by its present name, derived from *Braza*, a valuable species of wood with which this country abounds. Other useful varieties of hard wood grow in the Amazonian forests, numbering upwards of two hundred different kinds, being suitable for building purposes, ship building and fancy work; medicinal plants, flowering plants and fruit-bearing trees are also abundant.

VALLEY OF THE AMAZON.—The excessive hot temperature of the Valley of the Amazon, ranging from 70° to 100° Fahr., with at times deluging rains and alternate dry seasons, in the interior of the country on the *llanos* or plains, renders this portion of Brazil, lying near the Equator, say from 10° North to 10° South, altogether too hot for the North American or European race of men used to the temperate climate. This objection to the Equatorial region, however, is greatly modified on ascending to the head-waters of the Amazon, and its tributaries, to the foot of the Andes, where the country rises gradually to an elevation of several thousand feet, affording, at different altitudes, all the climates of the sub-tropical, temperate and cold zones.

RIVERS.—Brazil is watered by a profusion of great rivers. The chief of these is the mighty and majestic Marañon or Amazon. On the side of Guiana, the Amazon is a Brazilian river for 1,200 miles direct distance from Cape North to the mouth of the Yapuro ; on the south side, from Belem to Tapatinga, a distance of 1,600 miles direct, it flows through the Brazilian territory. The immense tributary streams which on both sides fall into the Amazon, and intersect the interior of Brazil in every direction, when opened up by steam navigation, will give to a great part of the interior of Brazil all the advantages of a maritime shore. The lower part of the Rio Negro, and the whole course of its great affluent, the Parana, belong to Brazil; also all the streams which join the Amazon, on the left bank, from the mouth of the Rio Negro downwards. On the right or south bank of the Amazon, the Yatay, the Tefe, the lower part of the Purus, the Madeira from about the parallel of 10° north, and its affluent the Guapore, and the whole water-system of the Topayos, Xingu and Tocantins, belong to Brazil. Proceeding southwards along the coast, from the mouth of the Para, or estuary of the Tocantins, we have the Gurupy, the Maracassame, the Turiassu, the Maranham, the Parnahyba, the Camucim, the Jajuaribe, the Capibaribe, the Unna, the great Rio San Francisco, the Peruaguassu, the Rio Contas, the Ilheos, the Rio-Grande-do-Belmonte, the Rio Doce, the Parahyba, and a multitude of minor streams, and affluents, flowing into the Atlantic. To the south of the parallel of 20° south, the rivers of Brazil mostly belong to the water-system of the Parana, which is wholly a Brazilian river to within 1° of the Stropu ; and throughout a large portion of the remainder of its course, form the common boundary of Brazil and Paraguay. The headstreams of the Paraguay, descending from the Serra Parecis, likewise belong to Brazil; and the tributary stream of the Cuiaba, or Cuyaba, a large river, almost equal in size to the Paraguay, which it joins in 17°57′ south. The sources of the Paraguay approach within a few miles of those of the Xingu and Araguaya ; and in many places, owing to the configuration of the ground, the tributary rivers of the Amazon and the La Plata seem as if their respective head-streams inosculated.

CLIMATE.—In such an extensive region as Brazil, both the climate and soil must necessarily vary greatly according to the locality. The climate may, however, be generally characterized as mild and regular. In the vicinity of the Amazon, and in the northern parts, great tropical heats prevail; but these are tempered by the excessive humidity of the atmosphere, and the copious dews. The great alluvial plains in the north-west and west, being inundated for several months in the year, are exceedingly unhealthy. The following is a summary of thermometrical observations made in the capital and the four northern cities of Brazil:

	S. Latitude.	W. Long.	Mean Temp.	Average Max.	Average Min.
Rio, . . .	22° 0′	42°50′	74°	82°	67°
Bahia, . .	13° 0′	38°32′	80°	86°	74°
Pernambuco, .	8° 6′	35°01′	80°	86°	70°
Maranham, .	2°31′	44°16′	81°	86°	76°
Para, . .	1°21′	48°28′	82°	93°	75°

In the southern parts, the climate is more mild and temperate, and frequently even cold, Fahrenheit's thermometer sometimes falling below 40°. This takes place, especially in ascending towards the sources of the great rivers, where the elevation of the ground modifies the temperature; and within the lofty plains which spread out into the interior, fertile valleys occur which are both salubrious and temperate, and in which all the fruits of Europe grow to maturity, along with the native productions of America. Of this climate are the inland provinces of Minas-Geraes, Villa-Rica, San-Paulo, Goyaz and Mato-Grosso. The mean temperature of the central table-land of Brazil is from 8° to 10° lower than that of the low districts on the coast. The west wind, passing over vast marshy forests, is frequently found to be unhealthy in the interior parts. These unhealthy blasts, however, are corrected by the influence of the atmospheric plants which abound in the woods, and which fill the air with a fragrance perceived at several leagues from shore when the wind blows from the land. Over all Brazil, December, January and February are the hottest months; June, July and August the coolest. The rains commence in March and continue until May, with intervals. During part of June and July a cessation of wet weather frequently takes place, and is called *veronica,* the short summer. The rains resume in August, and continue, with short intervals, until September. During the hot months, there is almost constant dry weather; and under the influence of the dry and parching blasts, vegetation languishes, and on the higher and more exposed parts appears burnt up and withered. In the northern provinces of Ceara, Pernambuco, and neighborhood, sometimes no rain falls for two or three years together, when the consequences are most disastrous. A famine ensues; cattle die of thirst; and the wretched inhabitants rush to the sea-coast, dying in hundreds by the way. The sea-breeze, which ushers in the rainy season, refreshes the atmosphere, and reanimates vegetation. The south-east trade-winds sweep the whole coast, and arrive tolerably cooled down by their passage from the burning coast of Africa on the opposite side of the Atlantic. This tendency to east winds receives, however, very regular modifications from the sun's progress in the ecliptic; a monsoon setting down the coast from September to April, and in the contrary direction the other half of the year. The heaviness of the rains can only be imagined by those who have been in such latitudes.

VEGETATION.—The interior of Brazil, with the exception of the Campos Parecis, and table-land already mentioned, forms a vast and impenetrable forest, the trees of which are closely interwoven with brushwood, and with innumerable shrubs and creeping plants, which cling round them to their summits, and being generally adorned with the most beautiful flowers, give a peculiarly rich appearance to the scenery. These plants, after encircling the tree to the top, frequently grow downwards, and taking root in the ground remount anew; so that the whole forest becomes laced together, and is rendered quite impenetrable. Luccock describes the various tints of a Brazilian forest as extending from a light yellow-green to one bordering on blue; and these mingled again with red, brown, and a gradation of deeper shades almost to black. The forests of Brazil abound in varieties of useful and ornamental wood. One species, called the *sippipira,* resembles the teak of India. The *peroba, oraubu* and *louro* resemble the oak and the larch. The *vinhatico,* "amarello venatico," yields large broad planks for flooring and cabinet-work, like mahogany. There are, besides, many lighter species of wood, similar to fir, besides log-wood, mahogany, and an infinity of ornamental and dyeing woods. Of the palm tree, nearly a hundred species are known and described as natives of Brazil; and amongst them that celebrated species, the long serrated lancet-formed leaves of which are composed of innumerable fibres, which rival silk in strength and fineness; and are used for fishing-lines, and sometimes converted into bridles. The Brazilian cocoa tree is thicker and more elevated than that found in the West Indies. The Brazilian myrtle is distinguished by the shining of its bark. The *ibiripitanga,* or Brazil-wood tree—called in Pernambuco, the *pao da rainha* or "Queen's wood," on account of its being a government monopoly—is now more rarely to be seen on the coast, owing to the improvident manner in which it has been cut down. It grows chiefly in the northern provinces. It is not a lofty tree; at a short distance from the ground, innumerable branches spring forth in every direction, in a straggling, irregular manner; the leaves are small and not luxuriant; the wood is very hard and heavy, takes a high polish, and sinks in water; the only valuable portion of it is the heart, as the outward coat of wood has not any peculiarity. The name of this wood is derived from *brasas* or *brazas*, a "glowing fire," or "coal;" its botanical name is *Cæsalpinia Brasiletto*. The leaves are pinnated; the flowers are white, and papilionaceous, growing in a pyramidal spike. One species has flowers variegated with red. "Almost every one of these sovereigns of the forest," says Von Spix, "is distinguished, in the total effect of the picture, from its neighbor. While the silk-cotton tree (*Bombax pentandrum*), partly armed with strong thorns, begins at a considerable height from the ground to spread out its thick arms, and its digitated leaves are grouped in light and airy masses, the luxuriant lecythis and the Brazilian anda shoot out at a less height many branches profusely covered with leaves, which unite to form a verdant arcade. The jaracanda (rose-wood tree) attracts the eye by the lightness of its double-feathered leaves: the large gold-colored flower of this tree and the ipe (*Bignonia chrysantha*), dazzle by their splendor, contrasted with the dark green of the foliage. The spondias (*S. myrobalanus*), arches its pennated leaves into light oblong forms. A very peculiar and most striking effect in the picture is produced by the trumpet tree (*Secropia peltata*) among the other lofty forms of the forest: the smooth ash-grey stems rise slightly bending to a considerable height, and spread out at the top into verticillate branches, which have at the extremities large tufts of deeply lobated white leaves. The flowering cæsalpina; the airy laurel; the lofty geoffrea; the soap trees with their shining leaves; the slender Barbadoes cedar; the ormosia with its pennated leaves; the tapia or garlic pear tree, so called from the strong smell of its bark; the maina; and a thousand not yet described trees are mingled confusedly together, forming groups agreeably contrasted by the diversity of their forms and tints. Here and there, the dark crown of a Chilian fir (*Araucaria imbricata*), among the lighter green, appears like a stranger amid the natives of the tropics; while the towering stems of the palms with their waving crowns are an incomparable ornament of the forests, the beauty and majesty of which no language can describe. If the eye turns to the more humble and lower which clothe the ground with a rich verdure, it is delighted with the splendor and gay variety of the flowers. The purple blossoms of the rhexia; profuse clusters of the melastoma, myrtles and the eugenia ; the delicate foliage of many rubiaceæ and ardisiæ, their pretty flowers blended with the singularly formed leaves of the theoprasta ; the concocarpus ; the reed-like dwarf palms ; the brilliant spadix of the costus ; the ragged hedges of the maranta, from which a squamous fern rises ; the magnificent stiftia, thorny solana, large flowering gardenias and coutereas, enlivened with garlands of mikonia and bignonia ; the far-spreading shoots of the mellifluous paullinias, delechampias, and the bauhinea with its strangely lobated leaves ; strings of the leafless milky *liānes* (bind-weed), which descend from the highest summits of the trees, or closely twine round the strongest trunks, and gradually kill them ; lastly, those parasitical plants by which old trees are invested with the garment of youth, the grotesque species of the pothos, and the arum, the superb flowers of the orchideæ, the bromelias which catch the rain water, the tillandsia, hanging down like *Lichen pulmonarius*, and a multiplicity of strangely formed ferns : all these admirable productions combine to form a scene which alternately fills the European naturalist with delight and astonishment." Among the products peculiar to the Amazonian forests is the caoutchouc tree, *Siphonia elastica*, which grows in general to the height of forty or fifty feet without branches ; then branching, runs up fifteen feet higher, with a thick and glossy foliage. The leaf is about six inches long, thin, and shaped like that of a peach tree. The juice of the caoutchouc is sometimes used as milk, and the negroes and Indians who work with it, are said to be fond of drinking it. The aboriginal name of this substance was *cahuchu*, the pronunciation of which is nearly preserved in the word caoutchouc. At Para it is now generally called *borracha*. On the slightest incision the gum exudes, having at first the appearance of thick yellow cream. The trees are generally tapped in the morning, and about a gill of the fluid is collected from one incision in the course of the day. It is caught in small cups of clay, moulded for the purpose with the hand. These are emptied, when full, into a jar. No sooner is this gum collected, than it is ready for immediate use. Forms of various kinds, representing shoes, bottles, toys, etc., are in readiness, made of clay. When shoes are manufactured, it is a matter of economy to have wooden lasts. These are first coated with clay, so as to be easily withdrawn. A handle is affixed to the last for the convenience of working. The fluid is poured over the form, and a thin coating immediately adheres to the clay. The next movement is to expose the gum to the action of smoke. The substance ignited for this purpose is the fruit of the *wassou* palm. This fumigation serves the double purpose of drying the gum, and of giving it a darker color. When one coating is sufficiently hardened, another is added, and smoked in turn. Thus any thickness can be produced. It is seldom that a shoe receives more than a dozen coats. The work when formed, is exposed to the sun. For a day or two it remains soft enough to receive permanent impressions. During this time the shoes are figured according to the fancy of the operatives, by the use of a style or pointed stick. They retain their yellowish color for some time after the lasts are taken out and they are considered ready for market. Several other trees, most of them belonging to the tribe euphorbiaciæ, produce a similar gum, but none of them is likely to enter into competition with the India rubber tree of Para. Another tree, not uncommon in the province, called the massarandúba, yields in profusion a white secretion, which so resembles milk that it is much prized for an aliment. It forms when coagulated a species of plaster, which is deemed valuable. The Brazil nut, "*Castanha do Maranham*," or Maranham chestnut, which grows upon the lofty branches of a giant tree, *Bertholletia excelsa*, is only produced in the neighborhood of the Amazon River, in the forests of which it grows spontaneously in great abundance. It would, however, be impossible to enumerate all the products of this wonderful region. Amongst the products general over the empire are vanilla, sarsaparilla, ipecacuanha, copal, cinnamon, cloves, tamarinds and cinchons. The most useful fruit cultivated in Brazil is the banana, which forms a principal part of the food of the Indians, and in its season of the free black population, whose locations, in the low, warm, thickly wooded spots, are favorable to the culture of this plant. The fruit is from ten to twelve inches in length, and about two in diameter. Several varieties of the orange, which comes to perfection in most of the provinces of Brazil are cultivated. The pine-apple is abundant; but the necessity of cutting this fruit the moment it gives out its odor, as it is then immediately attacked by the ants, is prejudicial to its flavor. The *maracuja*, or fruit of the passion flower, is highly esteemed. The mango is uncertain in its produce. Among other fruits known are the *fruta do conde* or custard-apple, the guava, the cashew, the *jamba* or rose-apple, melons, and *melonçias* or water melons.

Agricultural Productions.—As no country is blessed with a more genial climate than Brazil, so no country exceeds it in natural fertility. Its vast extent, its diversified surface, and its varied soil, enable it to produce all the fruits of tropical climates, and perhaps in favorable situations some kinds of European grain. In no country perhaps would agriculture yield equal returns to the industrious cultivator, but unhappily, in few countries is it more generally neglected. It is estimated that not more than one acre in 150 of the whole cultivable area of Brazil is under any kind of culture; probably not one acre in 200. The articles of food raised in the maritime provinces of Brazil in fact fall short of the consumption; and wheat is imported from the United States, and occasionally from Europe, owing to the industrial strength of these districts being devoted to the preparation of products for exportation to Europe and the United States. Maize, beans, rice and cassava root are very generally cultivated, and in some places, wheat and other European grain is reared. The flour of the cassava root, *farinha de mandioca*, is the staple article of farinaceous food for all the less wealthy classes, and is so especially of the Indians and slaves. The common garden pea has been sown and gathered in the neighborhood of Rio within twenty-one days. Coffee is the great staple mercantile product in the provinces of and around Rio-de-Janeiro, and is the most valuable in amount of all the exports of Brazil. At the commencement of the present century, the quantity grown was trifling. Its increase may be dated from 1810. The construction of a highway to Minas Geraes added greatly to the cultivation in the interior. Some of the coffee estates near Rio-de-Janeiro are extensive, and occupy 800 to 1,000 slaves in the culture and preparation; on the other hand, many of the smaller *lavradores* have not more land under it than their own family and two or three slaves can manage. A superior quality is grown by a colony of Germans at Caravellas, in the province of Bahia, but most of it finds its way to, and is disposed of in, the market of Rio.

The cultivation of sugar is extensive in Brazil, but is confined to the sea-board, and margins of rivers and streams having a convenient outlet to a port for exportation. As sugar cannot be grown with advantage except on the richest soils—and these extend in the respective provinces only where alluvial deposits have been formed—the quantity grown has not increased during the present century, nor is likely to do so. In the middle of the last century it formed the principal riches of the country. In the course of 150 leagues along the coast, from 25 leagues beyond Pernambuco to 25 leagues beyond the bay of All Saints, Perard counted above 400 sugar mills, each of which manufactured annually about 100,000 arrobas, or 2,500,000 pounds of sugar. While the Dutch were in possession of Northern Brazil, 250,000 chests of sugar were annually remitted to Holland. Although the cultivation is now spread over a wider space, it is chiefly confined to the same districts as in these epochs. In the interior, and where it would not bear the expense of sending to the ports of export, sugar is made into cakes called

Rapadoura, and consumed by the natives. Tobacco is cultivated, but not to a great extent. The tobacco is put up in rolls of from 200 to 300 pounds each, prepared with a syrup of sugar, and is exported to Europe and to Guinea. The cultivation of cotton is pursued to a considerable extent in the northern provinces, as Pará, Maranham, Pernambuco and Bahia; and were the colonists enterprising and industrious, more might be raised, and of a superior quality. The cultivation is chiefly followed in the table-land or interior elevated plateau of the north, on account of the dryness of the climate. The plantations, therefore, lie generally at a distance from the coast. The culture is rude and primitive. Little or no capital is embarked in it, or is likely to be, so long as the expense of transit—on mules, over jungle and wild wastes for hundreds of miles—absorbs a heavy percentage of the price. The cotton is gathered in small quantities, and collected by local dealers until a quantity large enough to transmit to next town has been got. It passes through many hands before it reaches the port of embarkation, and chiefly in barter, or in payment of debts. The cacao region of Brazil covers several hundred square miles, along the banks of the Amazon. The cacao trees are low, not rising above fifteen or twenty feet; and are distinguishable by the yellowish green of their leaves. They are planted at intervals of about twelve feet; and, at first, are protected from the sun's fierceness by banana palms. Three years after planting the trees yield. The tree tops are suffered to mat together until the whole becomes dense as thatchwork, and the ground below is constantly wet. The trunk of the tree grows irregularly. The leaf is thin and smooth-edged. The flower is very small, and the cone-shaped fruit grows direct from the trunk or branches. It is eight inches in length, and five in diameter. Within the cone is a white acid pulp, and embedded in this are from thirty to forty seeds, an inch in length, narrow and flat. These seeds are the cacao of commerce. The cacao tree yields two crops annually.—*Edwards' Voyage up the Amazon.*

The Amazon River, Climate, &c.

Prof. AGASSIZ, in his Lecture before the "*New York Association for the Advancement of Science and Art,*" delivered in the Great Hall of the Cooper Institute, Feb. 11, 1867, remarks:

"The Amazon flows nearly parallel to the Equator in a west-easterly direction, the main trunk not deviating from the Equator more than two or three degrees, while its southern tributaries rise from twelve to fifteen degrees south, and its northern, from six to seven degrees north; so that the width of the valley at some points is nearly as great as its total length. The fact that this main portion of the Amazon flows in one and the same latitude, brings a result very different, with reference to the climate, from that which we observe along the banks of other large rivers which flow in a north-southerly direction, or in a south-northerly direction. Our Mississippi begins its course in very cold regions, and ends it almost in the tropics. The Nile begins under the Equator, and further south, and terminates in the Mediteranean, where the climate is always temperate. You see, therefore, that those rivers are, as they flow on, under very changing climatic influences. Not so with the Amazon, which occupies a belt under the Equator, and retains the same climatic conditions for its whole length, and would present a great monotony were it not for the peculiar character of its tributaries, and for the peculiar economy of the waters which fill its basin. Extending its trunk across the whole continent, and sending its branches north and south over such a wide area, the basin of the Amazon establishes communication with all the adjoining Republics of South America. And this is a point of great importance with reference to the fact that the Amazon is this year to be opened to the commerce of the world; for, in consequence of the natural physical relation of the Amazon, its tributaries, and the areas drained by these tributaries, the opening of the Amazon does not only bring the internal commerce of Brazil into immediate contact with the commerce of the world, but also that of those Republics, the surface of which is mainly drained by the tributaries of the Amazon. Mark how extensive this communication is. Here we have the Guianas—French, Dutch, and English Guiana—then the Republic of Venezuela, through which flows the Orinoco, and which is connected directly with the Rio Negro through the Casiquiare. Here we have the Republic of New Granada, the eastern rivers of which all empty into the Amazon, several into the Rio Negro, and others, such as the Japura and the Iça, empty into the Amazon. Then we have the Republic of Ecuador, the principal rivers of which also empty into the Amazon. Then we have Peru, the three great rivers of which empty into the Amazon. Then the Republic of Bolivia, the great rivers of which flow also into the Amazon. And, finally, we have the rivers which come down from the table-lands of Brazil, which drain two of the most fertile provinces of Brazil itself; the Province of Matto Grosso, through which the Tapajos and Xingu flow, and the provinces of Goyaz, through which the Araguay and the Tocantins flow to meet the Amazon. So that those countries, which we are in the habit of considering only from their maritime side, have an extensive area which slopes toward the Amazon.

"When thinking of Venezuela, we generally remember Caraccas; when we think of New Granada, it is to Panama that our thoughts fly; and when of Ecuador, it is in Guayaquil that we stop; when we think of Peru, it is at Lima; when we think of Bolivia, it is at Mansilla, or along the sea-coast, or among the high mountains. You see the whole space of Bolivia, the whole space of Peru, the whole space of Ecuador, the whole space of Granada, the whole area of Venezuela, and even some parts of the Guianas, all slope toward the Amazon; so that he who has a foot upon the mouth of these rivers has also the key to that internal trade with these provinces. You see, therefore, what an extensive prospect is open to the enterprise of seafaring nations by the mere fact that the navigation of the Amazon will be free as the sea itself to the mercantile shipping of all nations. You may realize what it is for us by considering for a moment what it would be to the shipping of England or of France, if the United States were at once to declare that the Mississippi be open to their navigation; if the flags of Europe could float at Cincinnati, or St. Louis, and all the tributaries of the Mississippi, as well as the main river itself (stopping at intermediate ports), should be allowed to be navigated by foreign ships. It is that step which the Emperor of Brazil has taken. It is thus that he opens his country to the enterprise of the world; and no nation is more likely to be as greatly benefited by it as the United States.

"Now, you may ask, 'How can it be that a nation throws away its wealth in that manner into the hands of foreigners?' Very serious considerations must have weighed in the scale to induce the Government to divest itself to that extent of its internal property. The case is simple. The whole valley of the Amazon has not yet been peopled. The whole tract of this country, which is as large as many Empires of the first rank in the Old World—the whole of that country drained by the Amazon does not nourish at this moment 250,000 individuals, including the Indians; and no doubt the Government of Brazil has thought that the only way of settling that rich country was to offer its treasures to all nations. Let me, therefore, say a few words of the character of that country, and the facilities which are offered there for settlement, for commerce, and for travel. In the first place, when we speak of the valley of the Amazon, we ought to at once divest ourselves of the ordinary idea which we combine with the word 'valley.' There is not a bottom, with walls or banks rising on both sides, and forming an inclosure to the water that runs in the bottom of the valley. Here the basin of the Amazon is an extensive plain. It is so flat that the slope is hardly more than a foot in ten miles; and over the whole of this extent of 2,500 miles the slope is not more than 210 feet. It is only 45 feet from Obydos to the sea-shore, and it is only 200 feet from Tabatinga to the sea-shore, and yet the distance is, in a straight line, over 2,000 miles; so that really the slope is hardly a foot in ten miles. The impression to the eye is that of an absolute plain, and the flow of water is so gentle, generally, that in many parts it hardly seems to flow. It makes the impression of a fresh water ocean far more than a river, and the width of this basin compares favorably to its extraordinary length. There is not one channel through which the bulk of the water flows, but a multitudinous number of channels, connected with one another in the most various ways, so that instead of travelling in a straight course, you may ascend the Amazon in any number of parallel channels, and pass from one to another by any number of intersecting communications. And this net-work of rivers spreads over an area which is sometimes 50, 100, or upwards of 150 miles wide. In that region you have to travel for about 250 miles from the mouth of the Madeira before you come to the rising land over which the water falls in cascades. At the Tocantins, and Tapajos, and Xingu, you have to ascend 150 or 180 miles before you come to those higher grounds which determine rapids into water courses, and on the north side it is equally at a considerable distance from the centre of the basin that the land rises into gates. Over the whole expanse of the valley, for hundreds of miles on either side of the main channel, the bottom of the valley is as flat at the side as it is in a longitudinal direction; so that it is an inundated plain rather than a river; and if there are channels in which water flows more constantly than others, these channels are so frequently overflowed by the rise of the water, you have frequently, for a month at a time, the water covering this wide expanse without break. But there is a great difference in the character of these water courses, and a great difference also in the nature of the water itself.

"Before, however, I enter into details concerning the river, let me say a few words concerning the climate. The valley of the Amazon has a rather temperate climate. Though under the Equator, it is not among the hottest parts of the globe. The hottest point of the earth's temperature extends to the north of the valley of the Amazon, along the northern shore of Guiana, Venezuela, and the more northern part of South America. The valley of the Amazon is of milder temperature, owing to two circumstances; the extent of submerged land, with the constant evaporation, and the regular flow of the trade winds, which are constantly blowing in the face of the Amazon, and sending an air cooled by the amount of moisture received over the whole of its surface. The trade winds blow in the mouth of the Amazon and over the whole valley, so that there is an unceasing cool breeze from the Atlantic to the base of the Andes, reducing markedly the average temperature of the valley. Indeed, the average temperature of the valley is only 82°. The maximum temperature is from 90° to 92°; the minimum about 72° to 74°. It is only about Manaos, the junction of the Rio Negro, that the temperature rises to 95°. The temperature between day and night is always perceptible, and toward morning the nights are always remarkably cool. Under these circumstances, you see that, far from sharing the intensity of heat characteristic of tropical regions, the valley of the Amazon is favored to a degree which will make it a pleasant habitation for the people of our race. During nearly a year of residence there, I do not feel that the climate had the slightest unpleasant influence. My companions enjoyed it as well as I did; and, in fact, we found it was as agreeable a residence as we could wish, preferable to the intense heat of the dog-days, and so uniform as to save the inhabitants from those sudden changes of our climate so injurious to health. If the bracing air of our northern climate has a more stimulating influence upon the energies of man, we know how many it kills. It is the strong and healthy that survive; and many diseases which are the result of our northern climate are only cured by a residence at the south, while the south is saved from all these inconveniences, if it has some of its own. I would sum up my description of the valley of the Amazon as a healthy country, which will prove genial to the white race as much as any other part of the world having a similar temperature.

"I know that my statements are contrary to generally received notions, and that the valley of the Amazon, in particular, has a very bad reputation. But it does not deserve it; and the origin of this reputation is one which ought to be explained. It is owing to the reports of the officers of the Government of

Brazil, sent there to administer the affairs of the country, who, desirous of being relieved from a kind of exile in an unsettled land, and wishing to return to society, to the capital, or to the luxury they may have enjoyed, in order to accomplish that end, represent the country as injurious to health, and their residence there as a great sacrifice, deserving an advance in their social position. And that this is a true explanation, I have by the acknowledgment of some gentlemen who had been there, and who had themselves played that game.

"Now, let us look at the river and its banks. The average temperature of the water is 81°; the maximum temperature is 84°; the lowest temperature of the river is 77° to 78°; so that the waters are constantly tepid. It is only the streams which flow through the forest that are temperate and cool. Where there is dense vegetation (and nearly the whole surface of the land is covered with it), you may find at all times cool water. The whole of this extensive area is covered by vegetation. This plain is not, like other plains under the tropics, partly desert and partly covered with vegetation. The whole is covered with the most luxuriant vegetation—a vegetation sometimes so dense that it is almost impenetrable. Of its character I shall give some account at the close of this lecture.

"And now let me point out to you the fact that the River Amazon has three different regions, which present different aspects. At the lower turn of its course two great tributaries join it (one of them comes from the table lands of Guiana and the low lands between the Andes and Guiana); the Rio Negro on its northern shores, and the Rio Madeira on its southern shore, which comes from the mountains of Bolivia and the Andes at that latitude. These two tributaries are so large that, from their junction with the main stream, the stream assumes a greater dimension. The whole basin is full of water after it has received these two tributaries, and it is that part alone which generally goes by the name of the Amazonas. The Rio Amazonas begins at the junction of the Rio Negro, and extends to the Atlantic. Above the junction of the Rio Negro, and through the territory of Peru, it is called the Rio Solimões; and it receives, from Peru to Rio Negro, two very important tributaries on its northern side. One is the Iça, and the other the Japura. On the southern side it receives other tributaries, and then comes the Rio Madeira, to which I have already alluded. That part of the Amazon which occupies the middle track of the continent, goes by the name of the Solemões. The upper part, which is in Peru, and which comes down from the Andes, is called Mirañon, and it receives tributaries on the northern side and on the southern. The southern shore of the

Amazon, below the Madeira, receives the Xingu, the Tapajos, and Tocantins, all three of which flow from the northern slope of the table lands of Brazil. To the east of the Rio Negro there are a number of rivers, which are hardly known among us by their names, and which are yet very important and remarkable for their peculiar character; very broad, but not long. One of them opens into the Amazon with a mouth of over thirty miles. The Tocantins presents a front of sixty miles, and it is only one of the smaller tributaries of the Amazon; so that, as you reach its mouth, it seems as if a broad ocean were spreading before you, and you were passing from the river into an open sea, instead of meeting an affluent of the river on which you navigate. So also the Xingu. It is not over forty miles in length, and has a width of over twenty miles; so that these rivers are remarkable for their width. They are comparatively shallow, and their current is very light. The natural consequence is that they carry little material in suspension. Their waters are therefore clear, transparent, and somewhat tinged by vegetable substance. The waters of the Tapajos are greenish, those of the Xingu are a tinge of gray, those of the Tocantins about the same, a little yellowish; but all three are clear waters, as are also the rivers I have named before. There are other rivers of the same character which are tributary to the Madeira, but I need not name them particularly. Now the Madeira is of a totally different character. It is a very deep river, which flows rapidly, and carries with it a large amount of loose material, mud, and whitish clay; so that its waters are turbid, of a milky color; and hence the Madeira is called the 'White Water River.' The same characteristic is shared by all the rivers that flow into the Amazon to the west of the Madeira. The Purus, which is one of the largest of the tributaries, the Jurua, and the Japura, are all three white water rivers. They differ, however, from the Madeira in one respect—that though they are very deep, they are very tortuous; their course is not straight, but they are meandering. They are literally destitute of islands, while the Madeira has numerous islands. The consequence is that these rivers admit of navigation to a great distance.

"An officer in the Brazilian army (Major Continho), who was my companion during the whole journey, and who is as familiar with the Amazon as our pilots are with the Mississippi, told me that he explored the principal course of the Purus for over 500 miles, and found the river navigable for vessels drawing fifteen to eighteen feet. The Rio Negro, again, presents a very different aspect. It is a river which is very wide, but it is deep, and has a very slow course. It is dark and transparent; dark, owing to the immense amount of vegetable matter held in solution in the water; somewhat like those amber-colored streams we find in our woodlands, which are dark and yet transparent. Such is the Rio Negro, only that instead of dark amber, its tint is so rich that, seen from above, the whole river appears as black as ink. There are other rivers which have the same character, but not so dark. You see, therefore, that not only in width and depth and bulk of water, but also in the character of the water, every region of the Amazon has its peculiarities.

"The Amazon is the main stream; it is a white water river; it is the widest of all; it is that which occupies the widest area, the ramifications of which go over the largest surface, and which flows most evenly along its whole course; and as it flows from the mountains, and its main tributaries come from the mountains, it is for its whole course a white water river. The large amount of water which the Rio Negro throws into the Amazon hardly tinges it at all, and you may trace the yellowish white tint into the ocean about 50 miles before you see land. The front of the Amazon itself, as it enters the Atlantic, is 150 miles; so that it is, as you may see, the largest and most voluminous of all the rivers known. The banks of the river are everywhere clear, and they rise gradually above the level of the water. The rise is hardly over 20, 30, or 40 feet in the middle course. You have frequently banks of about 50, 60, and sometimes 100 feet. In one region only do the banks rise to a greater height, and back of the river are hills of 700 or 800 feet, which, owing to the even appearance of the whole country, give the impression to the eyes of a lofty range. These really low hills (the highest not being over 1,000 feet) appear, by contrast with the flat country, like our Alpine mountains. Of these hills I shall have occasion to say something more. They are characteristic of that region. Their structure is very remarkable, as they are thoroughly stratified, and indicate what deposits formerly occupied the valley with more precision than any other feature of the valley itself.

"Now as to the change of level of this immense stream, it varies within limits which are really astonishing. The river may be at times 30, 40, or 50 feet higher than at other times. You may conceive what an amount of water must be condensed from the atmosphere, in order to fill a plain so extensive with an amount of water sufficient to raise the level of the main current to such an extraordinary amount. But this does not take place simultaneously over the whole valley; so that there is the most extraordinary distribution of freshets over the whole basin.

"The rains begin on the southern side of the valley in the months of September and October, and from the table-land of Brazil and the mountains of Bolivia the southern tributaries of the Amazon first begin to swell at such a rate that through December they reach with their new flood the valley of the Amazon; the greatest rise in the Amazon being in the month of March, when in the region below the Madeira the rise may be as much as a foot in twenty-four hours during the whole month of March. The rise continues on until the end of June, when the river is most full; so that it takes from October to June for the rivers on the southern side of the Amazon to fill and discharge their water into the main stream. At a somewhat earlier period the Andes send down their contribution to the main river in consequence of the melting of the snow on the summit of the mountains in the months of August and September.

"The great freshet resulting from this melting of the snow in Equador is felt in the valley in October and November; it is felt in November as low as Manaos, so that in connection with the waters coming down from the Andes and the waters coming from the table-land of Brazil and the mountains of Bolivia, the Amazon is filled in its centre and on its southern side, and flows over to its northern side, the whole river extending northward in consequence of this swelling—for during three months all the rivers which come to the Amazon on its northern side are at their lowest stand as empty as they ever are. In turn, they will swell to a similar height; but, in the month of December, the northern rivers are at their lowest ebb. The southern rivers flow into them; they push the waters of the main basin to a more northern latitude than during any other season. It rains in the main valley during the months of January, February, and partly during March also; but in March the rains extend chiefly over the table-land of Guiana and the northern part of the Andes, and during April and May the northern rivers begin to swell, and in June they have reached their maximum, so that by the end of June, when the southern rivers have begun to empty, the northern rivers flowing into the Amazon rise to the same great level. The Rio Negro at Manaos rises generally to more than forty-five feet above its low level, and that mass of water now pressing against the waters which occupy the centre of the valley, pushes them southward, and these rivers are now moving in another direction. So that the whole flow is, as it were, thus the main flow from west to east on that gentle plain which has such a slight slope, aided by the interflow from the south and the north at opposite seasons. The natural consequence is that, while the whole flows westward, it flows westward in its northernmost reach during our winter months, and it flows westward in its most southernmost reach during the months of our summer, and in that manner the bottom of the valley is constantly shifting to and fro. The natural consequence is that there are extraordinary water communications between these rivers.

"You may travel up the Rio Negro, and perhaps sixty miles distant from its mouth you will find a white water river tributary from the Amazon flowing into one of its own tributaries—a branch from the main river flowing into one of its tributaries sixty miles above its mouth, moving with it and meeting the main river afterward; or you may find that from the Purus there is a communication extending across the lower portion of its course with the Madeira, or that from the Madeira there is a communication with the Tapajos in such a manner that, without ever travelling in the main course of the river, it is possible to pass from the basin of one of its tributaries into the basin of the other. When the country is more settled, these channels will be of immense advantage for intercommunication, for they are limited to the lower course of the river. About 400 miles above its mouth the Purus sends a branch which goes into the Madeira, the Madeira sends branches which go into the Tapajos, the Tapajos sends branches which go into the Tocantins, and this occurs to a most extraordinary extent.

"If I had before me a detailed map representing the two arms of the Amazon, you would be surprised to see how a hundred branches intercommunicate between the northern and southern divisions of the river and establish innumerable passes from one part of the country to the other. In fact, all these passages between the rivers are natural highways, which will forever remain the principal means of communication from one part of the country to the other. The whole land is too much under the power of water to ever be susceptible of sustaining inland travel over any great extent. The patches of land which rise above the river are limited in extent, though they are sufficiently high and extensive to afford the most exquisite sites for settlements. But the main communication throughout the river country must forever be a water communication, and the whole country must be administered in order to be well administered, not as land, but as a cluster of islands, between which the communication is necessarily by water. That idea must be the prevalent idea with those who have any intention of settling in that country. The idea of travel by horse and wagon, by stage, or by railroad, is an idea that must ever be foreign to the future civilization of the Valley of the Amazon. The boat is the natural means of conveyance over the whole land, and there is something charming in the character of this water communication, covered with such luxuriant vegetation, so varied and yet so continuous that nothing can give an idea of what such a submerged country covered by forests and interlocked by plants of all kinds is. It must be seen to form an idea of its true appearance. I will try, however, to convey some idea by comparison rather than by direct description.

"The whole land is covered with vegetation and forests. There are here and there small spaces which are occupied by water, but even those are encroached upon by the vegetation, and there is no knowing where the land ends and the water begins. The aquatic vegetation is so dense that it extends over the land into the water, concealing the limits of the one and the beginning of the other. Wherever there are extensive lakes their margins are covered with this aquatic vegetation, which extends sometimes very far from the shore, and here there are extensive tracts covered with water, which appear, nevertheless, as if they were land, owing to the dense growth of all sorts of plants sufficiently high to conceal entirely the surface of the water. I have navigated for miles and miles among meadows which have presented a variety of flowers as great as our prairies in the most favorable season of the year, and over these large meadows covered in this way with aquatic vegetation the animal creation is as varied, the water-birds especially being so numerous that the scene is one of the most varied that can be conceived of.

"The forest itself has a character of its own, entirely different from the forest of other parts of the world. With us in the temperate zone, in the more northern latitudes, all the forests consist of a few kinds of trees, and these trees are clustered together, a large number of individuals of the same kind occupying exclusively a considerable tract of land. Not so with the tropical forests. Plants the most varied, the most diversified from one another, are mixed together in the most profuse manner, so that you rarely see several stems of the same tree side by side, but a mixture of the most diversified kind are crowded together, and form as dense forests as our densest. And then between them there are a variety of smaller plants, and of parasites growing upon the trees, and of vines climbing from one tree to another; and it is difficult, sometimes, to determine to which plant, vine or tree the flowers or fruit you see belong. The variety is the more astonishing as at all seasons there are some of these plants in flower. Though there are somewhat marked seasons, yet there is never a period when the trees are destitute of leaves. The forests are evergreen, and only a few kinds of trees, at particular seasons, drop their leaves; but they are so few in number that they only create the impression of a few dead trees in a thick growing forest. These forests are rich in all kinds of natural products, and it is in these products that consists the wealth of the continent. The valley of the Amazon, as a country, is not rich in mineral productions. It is only in the higher land of Goyaz and Matto Grosso that there are gold and diamond mines, and it is only in the lower parts of the Andes that you find valuable mineral productions. Throughout this extensive valley, as I have stated before, the mineral kingdom is represented only by sands, clays and loams, to which I shall allude more in detail in a future lecture, but there are no rocks except where the country begins to rise; for instance, on the Rio Negro, above its junction with the Amazon, and on the Tapajos, the Tocantins, the Xingu, and the Madeira, above the waterfalls. There the solid rock begins, and there is land in which valuable mineral productions may be obtained; but for the whole extent of this plain the chief wealth of the country consists in timber, in textile fibres, in various fruits, and all the various productions of the vegetable kingdom.

"In the first place, let me allude to the timber. The variety is incredible. I have seen at Pará, at a public exhibition, a collection of Brazilian timber, choice and varied, and susceptible of furnishing material for the most beautiful cabinet work, of 117 different kinds, collected over a piece of land half a mile square. We have not in the United States one-half of this number of different kinds of timber worth anything for building purposes, or for manufacturing; yet there the variety is so great that from a small area of half a square mile 117 different kinds could be collected. I have brought home from this short expedition of ten months' survey, in which the study of plants was only an accessory part of my examination, specimens of 300 different kinds of valuable timber, remarkable for the beauty of their grain, for their hardness, the variety of their tints, and their durability, which, if introduced into the commerce of the world, would change the art for which wood is supplied. And that wood is not yet used in any way. It is allowed to float down the river; and the only impediment to navigation that I have perceived at any time was the quantity of floating timber. So little have the inhabitants made use of it that they have no saw-mills; and when they want timber for any purpose, they cut down a tree of sufficient length, and then cut it the size they wish with a hatchet. This waste is practiced in reference to timber. With reference to textile fibres, there is an endless variety, and we would be greatly benefited, so far as regards our shipping alone, if we would make use of those tissues which are so peculiarly adapted for making cables, ropes, and the like. There are, in particular, several kinds of palm leaves which have a very resistant and strong fibre. These may be obtained in any quantity on the banks of the Rio Negro, and already the English have begun to export that *piasaba*, but I am not aware that the Americans have yet begun to make use of it. The fibre is so light that the cables may float when made.

Among other articles which are most useful, and which are produced in the largest amount, is a variety of fruit, most delicious, of which the greatest variety of preserves are made, and of which we have hardly any idea. It is curious to see how, all the world over, the plants which produce fruit belong to particular families. If we compare, a moment, the fruit trees and fruits of the tropical regions with ours, there is the most striking contrast. Most of our fruits belong to one and the same natural family of the vegetable kingdom—the rose family. Cherries, peaches, plums, apricots, apples, and pears, in fact, the choicest of our fruits belong to that family. It is only a few other kinds of native fruits that belong to other families, such as the walnut, and then the grape vines, of which we have a great variety among the native, while in the old world there is one kind only. Now in the valley of the Amazon the principal fruits belong to the myrtle family. There is as great a variety of fruits belonging to that family as we have in the rose family. The *Goyaba* (Spanish *Guava*), which you may know from the presence of that name which you get from Cuba, is one of the most common trees all over that region; but they have, also, numerous fruits similar to ours. Plums grow in immense quantities on the banks of all this network of rivers throughout the valley of the Amazon. And then other families produce fruits. You are familiar with the magnolia, and know that it produces a dry fruit that has no taste. Now, there is a family akin to that in Brazil which produces a great variety of luscious fruits. There are several kinds of fruit produced by another family which are most delicious; but I will only entertain you at intervals with these, for there are other articles which are of more importance to the commerce of the world.

"In the valley of the Amazon there is grown an immense amount of coffee. Its culture extends over the northern provinces of Brazil, and also over Ceará; and the production of this plant is so great in that country that probably its yield is greater there than anywhere else. The chocolate we derive from a plant grown there in immense quantities. It is the cocoa plant (*cacáo*), which grows in all these forests, and produces a fruit somewhat like a cucumber, but larger, in which the great seed are now growing. These seed are taken out when ripe, dried and prepared, and it is from these seed that the various preparations of cocoa are made. Then there is another fruit very extensively cultivated there, which produces a cooling beverage, of which the Brazilians are very fond. It is something like chocolate. Its cultivation covers extensive areas between the Madeira and the Tapajos. But the great staples of that country are the dyestuffs and a variety of medicinal drugs; the sarsaparilla, the ipecac, and the *chincona*, which is so extensively used in the manufacture of quinine, sugar, and the most valuable of all the productions is india-rubber. The india-rubber is obtained from a tree which grows in the submerged lands. The india-rubber of South America is principally the product of a euphorbiaceous plant—*Siphonia elastica*. We have hardly a plant of that family to compare with it which is at all similar in aspect. It is perhaps more like the mulberry, and may be compared to it, though it grows taller, and does not spread so much. A wound is made in the bark of the tree by cutting it, and the sap which flows from it is collected into a number of cups, made of the leaves of trees, and is then poured into a larger vessel, dried, smoked and prepared in the way in which you see it in commerce. Thus far india-rubber has only been collected accidentally. Nowhere is it cultivated; and it is one of the miseries of the country that all the natural productions are still in their wild condition, and have nowhere received the culture which their importance would necessarily command. The consequence of this mode of collection is the extraordinary methods connected with it. The laborers go into the forest to collect it, and tap the trees in the most irregular manner, and when they collect the sap there are many trees which they pass unnoticed. In that way an immense quantity of the material is wasted. I have been told that any one who would follow in the ordinary track of these collectors of india-rubber would find as rich a harvest as those who made the first collection. And so it is with all the natural productions of the country. The Brazilian nuts (fruit of *Bertholletia excelsa*), which are so extensively used in the West, and of which the Brazilians manufacture very excellent oil—these nuts, which are produced from fruit about the size of your two fists, fall to the ground, and are also collected at random; and I have been told by one of the most intelligent men connected with the trade of the Amazon, that he was satisfied that annually about $10,000,000 worth of the natural products of the Amazon remained rotting on the ground from want of hands to collect them, when, if there was an industrious population, all this wealth would be saved. And if the population was sufficiently extensive to cultivate regularly those valuable productions, you see at once what mines of wealth would flow into the commerce of the world.

"But the first requisite is that there shall be a settled population, and a population living regularly. The population which now occupies that valley is indolent, is irregular in its mode of life, and, in consequence of that, liable to diseases which are ascribed to the unhealthy climate. For what could be expected of a population which goes into the wood with an indifferent supply of food, and that of a poor quality, and will remain in the wet, will allow themselves to be rained upon without taking any precaution for change, but that, after a while, they would get fever or rheumatism, and all other diseases which carelessness and poor food bring on, and which are universally charged to the climate? The first step toward improving Brazil should be regular settlements—settlements on these neat banks which rise regularly above the level of the water, and which are so inviting, not only on account of the variety of vegetation, but on account of the picturesque manner in which the rivers intersect these infinite forests. There is one feature which is particularly charming. It is the narrow channels of water which cut through the forests, sometimes so narrow that the branches meet together and form a closed arch over the water, sometimes so close that the smallest boats find it difficult to follow their courses. All these constitute one of the great charms of that region, to which you may add the interest arising from the immense variety of animals of all classes which mingle in this luxuriant vegetation."

Climate and Physical Features of Ecuador.

The great feature of this country, extending from 2° north to 6° south latitude, is the stupendous chain of the Andes, which traverses it in a double chain, running north and northeast, in a direction nearly parallel to the coast, and at an average distance from it, on the west chain, of 90 miles. The two ridges are distant from each other generally from 20 to 24 miles, sometimes receding and sometimes approximating, but always preserving nearly the same direction. The elevated plain between them is from five to six leagues in breadth; and within its narrow bounds is concentrated the population of the province of Quito. From the paramo of Assuay, which, rising from 14,764 feet to 15,749 feet, unites, like an enormous dike, the East and West Andes, under the parallel of 2°30′ south, 37 leagues to the south of Quito, the Andes, as we proceed north to Quito, present the appearance of a longitudinal valley, lined with a constant succession of soaring summits on the east and west. What is called the valley or plain of Quito is actually an Andean ridge of an absolute height of from 8,860 to 9,515 feet. The great mountains, though appearing only as so many isolated tops of this summit, when viewed from the distant plains, yet seem to the inhabitants of the central vale of Quito as so many distinct mountains rising from a plain unclothed by forests; and are so arranged that, viewed from the central plain, they appear in their natural shape, as if projected in the azure vault of the equatorial sky. After the long rains of winter, when the transparency of the air has suddenly increased, Chimborazo (altitude, 21,500 feet) presents a most magnifient spectacle, appearing, from the shores of the Pacific, like a white cloud on the edge of the horizon, detaching itself from the neighboring summits, and soaring with commanding majesty over the whole chain of the Andes. Between the Andes and the Pacific the surface occasionally rises into mountains, but presents no continuous ridge.

The principal rivers descending from the west slope of the Andes to the Pacific are, in their order from north to south, the Patia, and its affluent the Telembi, the Mira, the Santiago, the Rio Verde, the Rio Esmeraldas, the Chones, the Guayaquil, and its great affluent the Daule, the Navanjea, Jubones, and Tumbez. The country to the east of the Andes, is, in great part, a vast desert, over which yet roam only wild hordes of Indians. It is intersected by several vast streams, the upper courses of still mightier rivers, all pursuing a direction prevailingly to the southwest to join the mighty Orellana, or Amazon, on its left bank. These streams are the Pulumayo, or Iça, with all its head branches; the Rio Napo, with its great head streams the Ahuaricu and the Curaray, and its hundred minor affluents.

Climate, &c.—"Although this country lies under the Equator, yet the great elevation of its central valley, and of the western table-lands, renders the climate of those sections mild and temperate. In the low country, along the coast, the heat is excessive, and the climate dangerous to foreigners. Under tropics, what are usually termed winter and summer, mean only the wet and dry seasons; and the former is often superior in warmth. The dry season may be regarded as the coldest and the most healthy. At Guayaquil, the rainy season continues from January to June; and the dry, from June to December. The inundations at this period are so great that the coast at Guayaquil is often one sheet of water up to the base of the Andes, to which the inhabitants retire with their herds. Fevers, diarrhœas, dysenteries, vomiting, and spasms, then prevail, and the mortality is often very great. The temperature of the air at Guayaquil is so uniformly between 96° and 104° that the people complain of cold when the thermometer suddenly falls to 80° or 84°. At Popayan, in the interior of New Granada, the driest months are June, July, and August, when the south winds blow from the snowy mountains and paramo of Purasi. On the table-land from Quito to Popayan it may be said to be an eternal spring, the temperature being uniform during the whole year, notwithstanding that violent storms of thunder and lightning frequently occur. On the declivity of the Andes, from 3,000 to 5,000 feet in height, a soft spring temperature perpetually reigns, never varying more than 7° or 8° Fahr. The extremes of heat and cold are unknown, the mean heat of the whole year being here from 68° to 70°. The climate is an eternal spring, at once benign and equal; and even during the four rainy months the mornings and evenings are clear and beautiful. Vegetation never ceases in the 'evergreen Quito.' The inhabitants of our wintry climes see with astonishment the plough and the sickle at once in activity; herbs of the same species here fading with age, there just beginning to bud; one flower drooping, and its sister unfolding its beauties to the sun. Standing on an eminence, the spectator here beholds the tints of spring, summer and autumn blended, while above these verdant hills and flowery vales rise the lofty cones of the Andes, clad in eternal snows, or frowning with naked rocks. Under the equator, it has been calculated that heat near the terrestrial surface diminishes one degree of Fahrenheit's scale for every 333 feet of perpendicular elevation. At 10,000 feet of elevation, one degree of heat is lost for every 297 feet; and at the height of 20,000 feet, one degree for 318 feet.* The mean temperature of the table-land of South America, at different points, is the following: At Quito, 59°; Bogota, 60°; Loxa, 66°; Popayan, 65°; whilst at Caraccas it is 70°, and at Valencia, 78°. On the plains of the Orinoco, elevated 500 feet, though the high temperature is 115°, yet the medium temperature is 78°. The mean heat of the Pacific coast is 80°, and that of the Atlantic coast, 82°. The mean heat of the interior of South America is 80°, that of the plain of Venezuela being 85° Fahr. We have thus three climates—that of the coasts, the interior, and the high table-lands."

Vegetation.—"In the region of palms, the natives here cultivate the banana, jatropha, maize, and cocoa; and Europeans have introduced the sugar-cane and indigo plant. After passing

* This estimate disagrees with recent observations made in temperate climates, where the decrement is found to be about 1° Fahr. for every 400 feet ascent.

the altitude of 3,100 feet all these plants become rare, and only prosper in particular situations. Thus, the sugar-cane has been grown even at the height of 7,500 feet; and coffee and cotton extend across both these regions. The cultivation of wheat commences at 3,000 feet, but its growth is not completely established lower than 1,500 feet above this line. Barley is the most vigorous of the cerealia cultivated in these regions, and flourishes at an altitude of 6,000 feet, one year with another producing twenty-five or thirty fold. Above 5,400 feet the fruit of the banana does not easily ripen; but the plant is met with, although in a feeble condition, 2,400 feet higher. The region comprehended between 4,920 and 5,160 feet is the one which principally abounds with the cocoa; a few leaves of which, mixed with quick-lime, support the Indian in his longest journeys across the Cordillera. It is at the elevation of 6,000 and 9,000 feet that the *chenopodium quinoa* and the various grains of Europe are principally cultivated, a circumstance greatly favored by the extensive plains that exist at this altitude, the soil of which requires little labor, resembling the bottom of ancient lakes. At a height of 9,600 or 10,200 feet frost and hail often destroy the wheat. Indian corn is seldom cultivated above the elevation of 7,200 feet. 1,000 feet higher the potato is produced, but it ceases at 12,600 feet. At about 10,200 feet barley no longer grows; rye only is sown; although even this grain suffers from a want of heat. Above 11,040 feet all culture and gardening cease, and man dwells in the midst of flocks of lamas, sheep, and oxen, which, wandering from each other, are often lost in the region of perpetual snow."

Climate of Bolivia.

This elevated portion of South America, sometimes called UPPER PERU, lies between the parallels of 10° and 25°40′ south latitude. "The western half of Bolivia is occupied by prodigious mountain ranges, and elevated table-lands, which stretch towards the centre of the country. On the extreme north, and all along the east and south frontier, we meet with vast plains—here bare grassy steppes, there covered with majestic primeval forests—through which flow the head waters of the Amazon and the La Plata, themselves, in many instances, large rivers. The *Atacama*, or coast district, is a mere desert, sterile and featureless to the roots of the Andes as the Sahara itself.

"The mountains of Bolivia are a portion of the system of the Andes, here rising to a great height. The majestic peaks of the Sorata (altitude, 22,400 feet) and the Illimani (21,250 feet) belong to the Andes of Bolivia. The great plateau, on which the lake of Titicaca reposes its vast expanse of water, at an altitude of 12,795 feet above the level of the Pacific, may be regarded as belonging to the mountain system of the Andes, and as forming one of the most remarkable features of the country."

Climate and Productions.—"Rain seldom falls on the Atacama coast, which seems to be so scantily supplied with water that it can never exchange its present desert state for the clothing and verdure of cultivation. The great table-land of Titicaca has a mean temperature of 45° Fahr., and from November to April enjoys delightfully refreshing showers. Here the pasture is rich, and numerous herds of cattle and sheep are reared; but the tropical fruits are unknown.

"The temperature of the great plains on the northeast and southern frontiers of Bolivia is most oppressive; and being combined with great humidity, these districts are often highly insalubrious. Here cotton, indigo, rice, cocoa, oranges, pineapples, and all the tropical fruits grow freely. Around Potosi, elevated 13,300 feet, the temperature varies greatly throughout the day. Early in the morning it is cold and piercing. The forenoon is pleasant; and from noon till 3 P. M. it is generally very hot in the sun; while the evenings are usually serene and mild."

Peru—Its Climate and Surface.

PERU, extending along the Pacific coast from 3°30′ to 22° south latitude, is traversed throughout its entire length by the lofty chain of the Andes, running from northwest to southeast, and forming two grand ridges, which divide the country into three widely different physical regions, viz., the Coast, the Central, and the Eastern regions.

"The Western or Coast Region, between the Andes and the ocean, is rarely more than 60 miles wide. It consists of an arid, rainless, and barren district, covered with sand, and intersected by chains of hillocks that cross it from east to west. In some places of this district no rain has fallen in the memory of man; but above the level of 500 feet slight showers occasionally occur. The scanty vegetation is sustained by dews and dense fogs, or by artificial irrigation. The climate is sultry and unhealthy. There is no navigable stream except the Piura; and the few towns are generally situated close to the coast. The Central Region, or Montaña, consists of a lofty plateau, of about 12,000 feet of average elevation, which, though difficult of access from the coast, contains numerous cities and villages, owing to the coolness and humidity of the climate at low levels. Amongst the mountains are many favored spots, with a fertile soil, amid the most magnificent scenery on the earth's surface, enjoying a temperate and delightful climate. The Eastern Region consists of immense fertile plains, traversed by the head waters of the Amazon, and covered with gigantic forests, which extend up the mountain sides to upwards of 5,000 feet. The climate here is very humid, the crests of the Andes intercepting the equatorial winds, which come laden with moisture from the distant Atlantic, causing, in some places, almost incessant rain. This region of country is very imperfectly known to foreigners; but the opening of the Amazon river to the commerce of the world will induce adventurers and emigrants to visit this fertile and romantic portion of South America, where, at high elevations, are to be found rich mines of gold and silver."

DISTANCES FROM PARA, BRAZIL, TO JAEN, PERU.

From PARA (1°21′ South latitude), to—

Obydos, ...	600 miles.
MANAOS, ...	1,000 "
Aga, or Teffe, ...	1,600 "
Tabatinja, ...	2,200 "
JAEN, Peru (5° South latitude), ...	2,500 "

Climate and Productions of Paraguay.

The Republic of Paraguay is included mostly between 19° and 27°30′ south latitude, and nearly enclosed by the Paraná and Paraguay rivers. On the north it has the Brazilian province of Matto Grosso. The above large rivers, interlocking with tributaries of the Amazon, form the *Rio de la Plata*, the second river in magnitude in South America. The territory of Paraguay is about 470 miles in length, and 200 miles in breadth; area estimated at 84,000 square miles, containing a population of about 1,000,000 souls. The inhabitants are a warlike people, chiefly the descendants of Europeans from the north of Spain, with native Indians and negroes.

A mountain range of considerable elevation stretches nearly through the centre of the country from north to south, between the Paraná and Paraguay, sending the drainage in opposite directions. From the mountain regions the surface first presents a succession of finely diversified lower heights, and then stretches out into rich alluvial plains of great fertility. ASUNCION, the capital of Paraguay, is situated on a height on the left bank of the Paraguay river, in south latitude 25°18′, being 650 miles north of Buenos Ayres. Although mostly surrounded by Brazil, this country can only be reached by a hostile foe, to advantage, by the Rio de la Plata, which is navigable for a large class of vessels.

Climate, &c.—The climate, for the most part tropical, has its heat greatly modified by the inequalities of the surface. In July and August occasionally frosts occur. The whole country is remarkable for its salubrity. The soil is of great fertility, and vegetation almost unrivaled in its luxuriance. In the forests are found about 100 different kinds of trees, furnishing timber, dyewoods, gums, drugs, perfumes, oils, fruit, &c. A principal product is the *yerba mate*, or Paraguay tea, an evergreen, the leaf of which is nearly as much used for infusion, in this and the neighboring countries of South America, as the Chinese tea in the United States. The plant grows to the height of about a foot and a half, and has slender branches, with leaves resembling those of senna. The objects of agriculture include the greater part of the most valuable products both of tropical and temperate zones. On all the alluvial tracts where cultivation is attempted, sugar-cane, cotton, tobacco of superior quality, rice, maize, and culinary vegetables yield a rich return. The large plains feed immense herds of cattle, which are slaughtered chiefly for their tallow, hides, and horns, as articles of export.

Animal Kingdom.—The wild animals of Paraguay include most all the species peculiar to South America (except the *peixe boi*, or sea-cow of the Amazon valley), of which Prof. Agassiz remarks . "As a whole, they are far inferior to the wild animals of Asia and Africa." The most prominent are the jaguar, or tiger, of which there are great numbers; the puma, or cougar, called, also, the American lion; the black bear, and ant-eater, the tapir, the capibara or water-pig, river cavies, and various other amphibious animals. Alligators are numerous in the river Paraguay, and have been seen 30 feet in length. The wild boar, deer, and other species of animals less known, inhabit the forests. The boa-constrictor is found in most places adjoining the rivers. Among the feathered tribe are the cassowary, or American ostrich, the peacock, parrots of various species, paroquets, goldfinches, nightingales, and several species of the humming-bird. Wild geese and ducks abound in the rivers and lakes; and there is, also, a bird called the toucan, resembling the crow, but having a very long beak, which is most beautifully variegated with streaks of red, yellow, and black.

A remarkable circumstance in regard to the Animal and Vegetable Kingdom is the fact that they are both most wonderfully influenced by climate, in all its phases, as you proceed from the Equator to either of the Earth's poles.

The Pampas and Llanos of South America.

Next to the Andes of South America, the pampas and llanos, which extend through the centre of the continent from Venezuela to Buenos Ayres, are of the greatest interest. They lie mainly to the east of the Andes, along the head sources of the Orinoco, Amazon, and Rio de la Plata. The immense plain of the latter stream, in connection with the Paraguay river, extends from the mountains of Brazil on the north, and the Andes on the west, to near the Atlantic ocean on the south-east, including a great part of Bolivia and the Argentine Confederation. A large portion of this region is known by the name of the *pampas*. These are plains which present one uniform expanse of waving grass, uninterrupted either by forest or eminence. They commence fifty to one hundred miles east of the Rio de la Plata, and are in some places parched and barren, in others fertile, and mostly uninhabited. They are the abode of innumerable herds of wild cattle, horses, ostriches, and other animals, which, under the shade of the grass, find protection from the intolerable heat of the sun during the warm season. They extend westward to the Chilian frontier. Over these pampas, as in the Russian and Tartarian steppes, there are no landmarks for many hundred miles, so that travellers are obliged to pursue their route by the compass. The winds which often sweep over these extensive plains with great violence, are called *pampeiros*.

The *llanos*, which are more elevated plains, are thus described by a learned and eloquent traveller: "There is something awful, but sad and gloomy in the uniform aspect of these steppes. Everything there seems motionless. Seldom does a small cloud, as it crosses the zenith, and announces the approach of the rainy season, cast its shadow on the savannah. I know not whether the first aspect of the llanos excites less astonishment than that of the Andes. Mountainous countries, whatever may be the absolute elevation of the highest summit, have an analogous physiogomy; but we accustom ourselves with difficulty to the view of the llanos of Venezuela and Casanare, or the pampas of Buenos Ayres and Chaco, which recall to mind continually, during journeys of twenty or thirty days, the smooth surface of the ocean. Owing to the unequal mass of vapors diffused through the atmosphere, and the various temperatures of the different strata of air, the horizon was in some parts close and distinct; in others, undulating, sinuous, and as if striped—the heaven was there confounded with the sky. The *llanos* and *pampas* of South America are real steppes. They display a beautiful verdure in the rainy season; but in continued drought, assume the aspect of a desert."

PART XV.

TEMPERATE AND COLD ZONES OF SOUTH AMERICA.

Buenos Ayres, or the Argentine Confederation.

THIS is one of the largest and most important federative States in South America, being very favorably situated in a climatic and commercial point of view. The characteristic feature of the country is that of an immense level plain of fertile soil, and diversified by only a few slight elevations. These plains present one uniform expanse of waving grass, uninterrupted by either wood or eminence, somewhat resembling the prairies of North America. They are the abode of innumerable animals, which, under the shade of the grass, find protection from the intolerable heat of the sun during the summer months of December, January and February. In winter, during the months of June, July and August, it is reckoned cold when the thermometer falls to 45° Fahr.; but in some seasons it has fallen as low as 30°. A southwest and southeast wind always cool the air, while a north wind invariably brings heat. East and north winds are the most common. The southwest wind is always bracing and healthy, while the north wind produces languor and headache. During the summer rains are frequent, and are commonly accompanied by thunder and lightning. Long continued droughts occasionally occur, at intervals of several years, followed by excessive and long continued rains.

The most striking feature in the scenery, and the greatest disadvantage under which this region labors, is the almost entire want of trees. There are no forests in this part of South America, and no considerable growth of wood. It is not easy to account for this absence of timber; for the supply of moisture is greater than in many regions where woodlands abound. Darwin, however, says "that the limit of the forests coincides, in South America, with that of the region over which the damp winds travel; *that where they go laden with moisture from the Atlantic and Pacific oceans, the country is thickly covered with wood.*"

May not this theory satisfactorily account for the prairies and want of wood in portions of North America, and the desert regions in other parts of the world? The *pampas* of South America lie near the centre of the continent, being flanked by the Andes, as are the prairies of the United States by the Rocky Mountains; both lying under similar climatic influences as regards elevation, moisture, temperature, and winds.

BUENOS AYRES, the chief city of the Republic, derives its name from the peculiar salubrity of its climate. It is situated in 34°36′ south latitude, and 58°23′ west longitude, on the south side of the Rio de la Plata, and enjoys a foreign and inland trade of growing importance. The La Plata, which rises in the centre of Brazil, and interlocks with the Amazon, is a large and noble stream, comparing favorably with the St. Lawrence river of North America. Its mouth is upwards of 100 miles in width, and furnishes navigation from the Atlantic to Asuncion, the capital of Paraguay, a distance of about 1,500 miles; its whole length being about 2,000 miles.

The following TABLE exhibits the mean elevation of the Barometer and Thermometer for one year:

MONTHS.	MEAN OF BAROMETER.	MAX. TEMP. ° Fahr.	MEAN TEMP. ° Fahr.
January,	29.50	92	72
February,	29.58	89	73
March,	29.61	82	71
April,	29.73	78	62
May,	29.76	68	58
June,	29.77	66	54
July,	29.65	67	53
August,	29.84	66	52
September,	29.74	72	55
October,	29.67	81	59
November,	29.61	88	68
December,	29.45	86	71

The mean temperature for the year was 62° Fahr.

It is only between 24° and 48° south latitude, on the Atlantic side of South America, running through the sub-tropical and temperate zones, ranging from **40°** to **70°** mean annual temperature, that the white race can with safety, in regard to health and life, emigrate to, and take up their permanent abode. This region of country, however, is in many respects dissimilar to the same latitudes in the northern hemisphere, as the *pampas* are far inferior to the prairies of the United States, being better adapted to pasturage and the raising of cattle than the cereals. It embraces Southern Brazil, Paraguay, Uruguay, and Buenos Ayres, or the Argentine Confederation. Chili, on the Pacific side, has also, for the most part, a temperate climate. Here health can be enjoyed, and life prolonged; while the Rio de la Plata and its numerous tributaries afford uninterrupted navigation during the whole year, draining an immense and fertile region of country, abounding in large forests, containing many kinds of valuable wood, and rich vegetable productions.

Climate and Productions of Chili.

The Republic of Chili, lying on the Pacific side of South America, runs through eighteen degrees of latitude, extending from 25° to 43° south latitude, embracing the Chilian Archipelago. Its breadth is various, being determined by the greater or less distance of the summit of the Andes from the ocean. In the north part of Chili the country rises in a series of successive terraces from the coast to the foot of the Andes.

Mountains.—The grand belt of the Andes separates Chili from the provinces of the La Plata; and its western declivities occupy a considerable portion of the surface. Three small ranges likewise extend in nearly parallel lines between the Andes and the ocean. Of these parallel lines the Peuquenes ridge is considerably higher than the others, attaining an elevation, where the road crosses it from Santiago to Mendoza, by the pass of the Portello, of 13,210 feet. The highest mountains of Chili are: Manfla, in south latitude 28°45′; Aconcagua, in 32°38′ (altitude, 23,910 feet); Tupongato, in 33°20′; Descabesado, in 35°; Blanquillo, in 35°4′; Longavi, in 35°30′; Chilian, in 36°; Antuco, in 36°50′; and Corcovado, in 43°11′. Molina had not an opportunity of taking the altitude of the above mountains, but the Chilenos suppose them to rise upwards of 20,000 feet above the sea. There are no fewer than 14 volcanoes in a state of perpetual combustion in Chili, and all of them belong to the main ridge of the Andes.

Climate and Seasons.—The climate of Chili is delightful and salubrious. The four seasons occur here as regularly as in Europe, though in inverse order, being in the southern hemisphere. Spring commences on the 21st of September, summer on the 21st of December, autumn on the 21st of March, and winter at our summer solstice, or 21st of June. From the commencement of spring to the middle of autumn, between 24°

and 36° south latitude, the sky is always serene, it being rare that rain falls during that period. The rains begin in the middle of April, and continue, with greater or less intervals, till the end of August. In the northern provinces of Copiapo and Coquimbo, little rain falls; but in the middle provinces, three or four days' rain alternates with fifteen or twenty dry days; and in the southern provinces, the rain sometimes continues nine or ten days uninterruptedly. In Copiapo and Coquimbo, the comparative want of rain is compensated by very copious dews. The transitions from heat to cold, and *vice versa*, are moderate, and their extremes are equally unknown. The air is so much cooled by sea-breezes on the one hand, and by the winds from the snowy Andes on the other, that the thermometer in the shade seldom exceeds 76°. In winter it rarely sinks to the freezing point, but a perceptible cold is generally felt till noon. Snow, except on the Andes, is very uncommon. It is entirely unknown on the coast; and though it sometimes falls in the middle districts, it often melts ere it reaches the ground, and is seldom known to lie above one day. On the Andes, however, from April to November, which is the rainy season on the plains, snow falls so abundantly as to render the passes wholly impracticable for the greater part of the year. Thunder is unknown except amid the Andes. The winds, in Chili, are considered by the inhabitants as nearly infallible indications of the weather, and serve as barometers. The south winds, coming directly from the Antarctic pole, are cold, and attended with fair weather. The north winds, on the contrary, are hot and humid, and, on the east of the Andes, are more suffocating than the sirocco. The south wind prevails while the sun is in the southern hemisphere. It relaxes about noon, and is then supplanted for two or three hours by a fresh breeze from the sea, which, from its returning regularly, is called the meridian-breeze, and the clock of the peasants. In the afternoon the south wind returns; and, at midnight, it is once more succeeded by the before-mentioned breeze. "At first, it appears rather surprising that the trade-wind, along the northern parts of Chili, and on the coast of Peru, should blow in so very southerly a direction as it does; but when we reflect that the Cordillera, running in a north and south line, intercepts, like a great wall, the entire depth of the lower atmospheric current, we can easily see that the trade-wind must be drawn northward, following the line of mountains, toward the equatorial regions, and thus lose part of that easterly movement which it otherwise would have gained from the earth's rotation" (Darwin). The east wind is seldom felt in Chili. The meteorological history records only one hurricane.

The gales of wind which have been at times so destructive to shipping on the coast of Chili, come from the northwest, and are common in winter. The nights are magnificent, from the clearness of the atmosphere, and the brilliancy of the heavenly bodies. Fiery meteors are frequent, proceeding from the Andes to the sea. The *aurora australis* seldom appears.

In respect of productions, Chili appears to be divided by nature into three sections. That to the north of the 32d parallel is barren, but abounds in copper and silver. The central section is composed of rich valleys, and corn is here raised in abundance; but there is little wood. The south portion is also fertile, and abounds in good timber, some of which attains a large size. The soil of Chili, in the northern provinces, is sandy and saline; but it improves as we advance from the coast to the Andes, and likewise as we proceed south. The valleys of the Andes are superior, in this respect, to the middle districts; and these latter excel the maritime tract. The soil of the latter often resembles the fat land of Bologna, being of a reddish brown, friable, mixed with a little clay or marl, and sometimes presenting white or brown pebbles, arsenical and martial pyrites, with shells, madrepores, and other marine productions. That of the midland and Andine vales is of a yellowish black color, porous, friable, flints, and decomposed marine bodies. In other quarters the soil is a stiff clay, abounding in water-worn pebbles. Agriculture in this happy climate requires little attention. Many of the cereals and plants raised are the same as those of Europe; while the herbage, especially in the valleys of the Andes, is tall and luxuriant, sustaining large numbers of cattle, horses, and sheep. In fruit trees Chili is greatly inferior to the tropical countries of America.

Mountain System of the Andes.

The Andes Proper may be subdivided into four sections. The first, forming the southern section of the system, and extending from the southern extremity of the continent to the 44th southern parallel, may be distinguished as the Patagonian Andes, sometimes called Sierra Nevada de los Andes, and is that portion of the system which is least known to geographers. The second section, extending from the 44th to the 20th southern parallel, is the Andes of Chili and Potosi. The third section is the Peruvian Andes, extending from the 20th parallel to the plateau of Almaguer, under north latitude 1°50′, and sometimes called the Royal Cordillera, or Grand Cordillera of Peru. The fourth section is the Cordilleras of New Granada.

The western of these ridges runs parallel to the shores of the Pacific, and is called the Cordillera of the Coast. The eastern, or that of the interior, is called the Cordillera Real. The intermediate plain is the basin of the celebrated Lake Titicaca, the physical features of which are scarcely less extraordinary than its history is interesting. Generally speaking, the western Cordillera is the most elevated, attaining, at many points, an absolute height of from 22,000 to 24,000 feet; while the eastern Cordillera, between the latitudes of 19° and 16°40′ south, nowhere exceeds 17,000 feet. In the latter parallel, however, the gigantic Illimani springs to the height of 24,200 feet; and north of it, several other elevated points even surpass the height of the western ridge. The most elevated is the Nevardo de Sorata, in south latitude 16°10′, the height of which is 25,250 feet. In general shape and character the two also differ. The heights in the western are chiefly dome or bell shaped; those in the eastern are peaked, giving the range generally a serrated form. The descent of both, east and west, is rapid; but that of the western Cordillera, into the basin of Titicaca, is less so than that of the eastern. The breadth of the former is about 100 miles; that of the latter it is less easy to determine, in consequence of its throwing out many lateral chains on its eastern side, the length of which may be considered portions of the breadth of the main ridge. Excluding these, however, this may be estimated at from 35 miles where narrowest (17°58′ south), to above 70 miles where widest (16°50′ south). The entire width of the two ridges, including that also of the basin of Titicaca, varies from 200 to 300 miles, exclusive of the projecting chains; including them, it approaches 500 miles; and the length of this portion of the Andean chain, bounded by the 14th and 20th southern parallels, is nearly 400 miles.

According to Humboldt, all the great elevations of the New World stand connected with that prodigious chain, which, following the direction of the western coast, stretches, under different names, and with considerable interruptions, from the northern extremity of the continent to the southern, over a space of 10,000 miles in length. This mountain system vastly exceeds its only rival, the Himalaya, in length, and never, like it, loses its mountain character in the wide level expanse of an elevated plateau, or is interrupted by broad valleys and the intersection of mighty streams. Unlike its Asiatic rival, moreover, it rises at once from near sea-level on both sides of the range. It may be considered as formed of three sections: *First*, the Andes, properly so called, extending from Cape Horn to the isthmus of Panama; *second*, the Central American system, extending from Panama to the peninsula of Tehuantepec; *third*, the North American chain, extending from the Mexican plateau to near Behring's Straits.

Cruise through the Straits of Magellan.

EXTRACT OF A LETTER FROM ON BOARD THE U. S. STEAMER POWHATTAN, GIVING DESCRIPTIONS OF THE STRAITS OF MAGELLAN.

"UNITED STATES FRIGATE POWHATTAN,
"*Off Cape Virgins, Eastern Entrance to Straits of Magellan* (52°30′ *S. lat.*),
"*February* 5, 1866.

"We are just entering this far-famed, much abused, and dreaded strait. The weather is fine, but we, just from the tropics, are muffled up in our great coats and furs. The next two weeks will probably be the most eventful, as they must be the most interesting part of our voyage. The navigation of a crooked and dangerous passage, but slightly known, in the latitudes of almost constant gales, is quite a different thing from a pleasure trip across the ocean in warmer latitudes. We have daylight to cheer us from two o'clock, A. M., to half-past ten at night; but I dread the cold and snow which we must expect. I cannot pretend to give you the faintest idea of the intense blackness and loneliness of this coast. A large portion of the banks of the straits are perfectly treeless, shrubless, and grassless. The sky is seldom cloudless, the wind almost continually blowing a gale, and the sea has a cold leaden color, which adds to the general dreariness of the aspect. We have not as yet seen any traces of the Indians, either on the Patagonian or Terra del Fuegian shore, but llamas, guanacos, ostriches, and other wild animals and birds, besides seals and sea lions, have been frequently seen on the shores.

"OFF CAPE GREGORY (53° *S. lat.*), *February* 8.

"I have just returned from a run on the shore, where we were lucky enough to fall in with a band of Indians. They are strange looking fellows, and always on horseback. They are hugely built, perhaps six feet four or five inches tall, with splendid chests, brawny arms, but small, ill-shaped legs. I do not wonder that Drake and other early voyagers described these natives as giants. We had a large party, well armed, and compelled them to take us to their village, much against their wishes. I can assure you the long tramp we had was amply repaid by the novelty of the scene we beheld. At first the whole camp took flight; but on our offering them some copper coin, we soon established a good feeling. They live in the rudest kind of huts, made of skins, banked up with earth. The faces and bodies of the young women and girls were besmeared with red, yellow and black mud, giving them a most grotesque appearance. Near by their village we saw one of their "toldos," or tombs, which consists of a mound of earth, flanked on either side by the effigies of their horses, rudely cut out of wood, and bits of skin, cut like pennants, stuck on poles over and about the mound. While we were at the village, a party of hunters returned, bringing guanacoes, foxes, and several large ostriches.

"ENTERING THE PACIFIC, *February* 19.

"We cleared the straits on the 15th, and, after a boisterous passage of three days, arrived at this out-of-the-way refuge. Human eye never rested on grander scenery than that by which we are now surrounded. The mountains rise from the water's edge to the height of 15,000 feet, covered with eternal snow. We have three huge glaciers in sight, one of which runs down to the water, and is two miles broad at its base! These waters have never been surveyed to their limits, and it is supposed they communicate inland to the Isle of Chiloe, in Chili. There are no natives near, nor has the gulf been visited since the time of Fitz Roy, except by one or two sealers."

The Straits of Magellan, or Magalhaens, forms a navigable channel of about 300 miles in length, and varying in breadth from 2 to 40 miles, between the Atlantic and Pacific oceans; the most southern portion lying in about 54° south latitude. This channel was discovered in 1519 by Fernando Magalhaens, who sailed through it. The western shore of Patagonia here presents numerous sinuosities, studded with islands of various dimensions, forming the Adelaide, or Patagonian Archipelago. Through these islands there is a ship channel to the Gulf of Peneas, in 47°30′ south latitude.

The *Andes of Patagonia,* extending in one range from latitude 42° south, to Cape Horn, are of comparatively moderate elevation, varying from 8,000 feet downwards. The snow line descends here to 3,000 feet, and glaciers make their appearance, though unknown in the rest of the Andes. *Mount Yanteles,* south latitude 43°30′, elevated 8,000 feet, is an active volcano; Mount Stokes, south latitude 50°, 6,400 feet; Mount Darwin, Tierra del Fuego, 6,800 feet; Cape Horn, the southern extremity, 300 feet.

The Patagonian Indians are a tall, muscular race of men, averaging about six feet in height, leading a nomadic life, and subsisting on the produce of the chase and fishing. They are represented as a warlike people, and good horsemen; but the natives of the mountain region, and of the Fuegian Archipelago, are a stunted race, sunk in the deepest ignorance and degradation.

Ascent of the Peak of Orizaba.

In order to give an idea of the impassable barrier to the highest *Mountain Peaks*, as well as the impossibility of reaching the higher latitudes toward the *Poles*, we add the following graphic account of a late attempt (January, 1867) to ascend the snowy peak of ORIZABA, according to Humboldt the highest peak in Mexico, and the handsomest in the world. The party had resolved on treading where a white man's foot never ventured before, and anticipated remaining upon the mountain, above the clouds, two nights at least. The correspondent of the New Orleans *Picayune* describes the incidents of the first day's ascent, and continues:

"At sunset we reached the stone chapel and tower, 13,000 feet above the level of the sea. One after the other our company filed into the gateway for the night. Of its origin there is no history; of its age, generations long since dumb knew naught. The frost that night was sharp and heavy. The bare hard earth was white, and the morning light revealed the neighboring stream iced over as it slept almost on the narrow level above the abysses. The air drifted down from the snow region, and man and horse shivered in the blast. Thermometers were low, and so were the spirits of the party. Complaints were many, and enthusiasm had flagged alarmingly. A vote then would have disclosed two-thirds of the party in favor of retreating. From this point there was no horse-path. Here ended all signs of human or animal travel, and upward, for four or five thousand feet, in interminable layers of rock and cinders, and above these, and on the pyramidal ridges, the ruined tower gleamed in the sun, until one's head became dizzy at the sight of close-packed snow and thousands of ice-pinnacles.

"Over these we were to clamber before reaching the summit. Some of the party breathed heavily even at this height, and were averse to proceeding further. Horses were picketed within the walled yard of the ruined chapel; artists packed their apparatus, the engineers their instruments, and with a plenteous supply of brandy in each man's pocket, the guides were directed to commence the ascent. Then followed promiscuously Americans, Englishmen, Mexicans, one after the other, singing, whistling, jesting as we went. Not long did these noisy demonstrations last, for the breath came hard, and the hands and feet and senses were required for the hazardous journey. As we advanced, new difficulties arose; ledges were precipitous and barely passable; rocks of round sandstone came rolling and sliding downward by us; drifts of snow from the topmost ridges glided swiftly from their places as the sun rose in the heavens; and huge flat ice-blocks at times came whirling by us like cannon balls.

"When on a rise of 14,000 feet the party separated, some taking the high snow ridges, others the gorges or gulleys. There is a mean difference in height between the two, often 800 feet. The gulleys run up to the summit, with occasional breaks, parallel with the ridges; and the surface is composed of *débris*—a collection of centuries—a spongy, black earth, through which we sank to the knees, and where no snow or ice lay, but through which, at noon-day, ran the drippings that trickled from the high snow cliffs. The leaping fountains, winding through the gorges, increased as they descended, and uniting, sometimes, below, swept in a thundering torrent down the mountain side. These streams had worn beds a hundred feet deeper yet than the general level of the gulleys. The formation of rocks and earth differed in no wise from that further down. Here and there lay huge piles of gray limestone and sandstone, and specks of quartz, promiscuously intermingled, angled, some flat and edged, others with regular layers of lime and sandstone. Rents, fearfully deep, in the mountain side, disclosed curiously disposed strata of the upper and lower and intermediate sections of geologic formation. Great, gaping mouths in the rocky sides send out sulphuric fumes; and in one mammoth opening lay heaps of sulphur, and, further back, pillars of purplish stone (the result of drippings), thirty feet high. There are no evidences of recent eruption (perhaps none for two hundred years), but the fact that the summit is bare and black, and that occasional whirls of smoke are emitted from the crater, indicates the smouldering condition only of the volcano at present.

"The ascent was continued in an almost direct line toward the top. Up to within two thousand feet of the summit level, the whole company were in motion, but scattered at great distances from each other, some almost out of sight on the conical cliffs, some toiling abreast up the dark gulleys. At this time some began to fail and fall by the way, blood began to pass from the nose and ears, and faces were swollen so that old friends knew each other only by the dress. A few continued the journey a thousand feet higher, lay down, slept on the snow or black dust, gasped for breath, and awoke. Some dropped every few minutes (it was impossible to keep awake all the time), but started up again as soon, catching the breath.

"The sun was by this time in mid-heaven, and beat down

fiercely, blinding us, and starting a thousand little rills from the exposed ledges, that seemed, in the sunbeams, silver veins, as they slid down noiselessly from the tall ice-pillars, and ran along the rocky sides, clear as crystal, till they went foaming and leaping into the surging stream below. No sign of tree, or shrub, or grass blade, or hardy flower—all silence, and snow, and black desolation; rifted rocks, weird, unseemly piles of frozen earth and ice, upward; mist and cloud below; the sun and sky, deep blue, overhead; beneath, the cloud-field and the abyss. Snow banks would start off themselves from their places, and, with a sharp, cutting sound, drop into the abyss, and be seen no more. Shafts of ice thirty feet long, loosened by the falling boulders and snow slides, slipped from their moorings, fell upon the sandstone cliff below, ground into fragments, and, bounding onward between sun and cloud, sparkled like diamonds as they fell. The winds were sharp and cold, but not high. Sometimes, in the deep hollow, it struck the sharp crag, and shrieked like the night tempest on a rocky reef. Once, and once only, it chopped round, and swept the mass of cloud away eastward, and then distant landmarks, and cities, and plains, were visible. Popocatapetl and all the Mexican volcanoes were distinguishable, and, with a good telescope, we looked out, over the Chiquite Mountains, into the placid waters of the Gulf. In a few minutes the wind shifted, and cloud and mist trooped back again, and hung with a sort of affectionate embrace around the mountain top and sides.

"Sound at this height was very distinct, although it appeared distant when actually near. Amid the silence that reigned, the snapping ice shafts, and snow slides, and falling rocks, and even the little waterfalls, fell painfully upon the ear. The crashing noises one experiences in caverns when a stone strikes the floor, or a rill plays upon the rock, resemble very nearly the sensation; and when a boulder broke upon the lower ledge, the sound quivered with a vibratory motion for a long time before it died away. The sense of isolation is acute, existence is a dream, the senses half benumbed, memory in a mist, and thought lost in a maze of uncertainty. Were it not, indeed, for the continuous struggle to retain vitality, the sensation of losing breath, and the constant loss of blood, one might easily be induced to dream on in a seeming sleep on a sunny snow ledge or cinder gorge.

"We were now nearly 16,000 feet above sea level. Distinctly, as if at our elbow, the sound of the guide's feet striking the solid drift, 1,000 feet away, fell upon the ear. Evidently the Indian pilots, who did not count upon our advancing so far, became alarmed, and indicated a wish to return.

"Two-thirds of our party were out of sight, down the slope. Three alone, beside the affrighted guides, held their way. Blood oozed from ears and nostrils and mouth, and veins stood out on the forehead like great black lines. Our footing became more and more uncertain, the ascent more abrupt, the stones constantly turning and crumbling away, and, betimes, huge masses of earth and boulders and scoria, loosened by the melting snow, came thundering and hissing from above, fairly flying past our heads on to the next projecting ledge; and great snow drifts, broken and crumbled by the colliding rocks, avalanched down upon our heads a perfect storm of snow, and icicles, and black earth, and lava dust. One of the guides, smitten by a passing drift, rolled, half dead, three hundred feet down the slope, and was buried for awhile in the *débris* of snow and earth.

"The miniature cascades disappeared. Even the drippings disappeared from the rocks; for we had passed the line of thaw. Snow was beaten down hard and compact, and glistened like ice as the sun fell upon it. But an abundance of loose rocks lay on the surface, poised for motion at the slightest touch. The guide started more than one as he picked his way some distance in front. We heard by the footfalls that the courageous S. was pushing on. He was within 500 *feet of the top*, turning into a shallow gulley to avoid the falling boulders, when a sliding, tumbling noise was heard, then a heavy dull click, then a fall, and in a moment a heavy boulder came whizzing by on its downward course. Some one called out, 'S. has fallen!' The rock struck him on the shoulder, breaking it, and hurled him a hundred feet down the steep gulley. The guide reached him soon after, and we bore him slowly down the steep slope, abandoning, for the time, our enterprise.

"Arrived at the tower, the mountain streams, swollen by the melting snow, went foaming and roaring down their rocky beds. Our horses were picketed as we left them in the morning. We passed another night within the roofless chapel, and with all the quaint stories and goblin fables associated with it, slept soundly till 'rosy fingers' of morn streaked the eastern sky; and down again, with our wounded comrade, into the soft warm winds and pine groves, we picked our way; and yet further on, to the balmier air of the lowlands, where cool streams from the hills and peaks danced merrily through maguey fields, and in the orange shade, through broad pampas, to the Rio Blanco, where reigns perpetual spring or summer."

The Antarctic Ocean and Continent.

The expanse of water surrounding the South or Antarctic Pole, called the *Antarctic Ocean*, may be strictly regarded as extending from the Pole to the Antarctic Circle, or 66°30′ south latitude. This portion of the globe's surface has been hitherto very imperfectly explored, and is even less familiar to navigators than corresponding latitudes in the opposite polar region, within the Arctic Circle, which has been partially described in the beginning of this work. The *Antartic Continent* and adjacent islands, discovered and explored by English, French, and American navigators, is ascertained, however, to be much less habitable than the Arctic regions of North America, since it is limited in extent, and the vast space within the Antarctic Circle is mostly occupied either with sea or ice, in latitudes corresponding to parts of the northern hemisphere far within the limits of man's occupation. While the absolute limits of vegetable life have not yet been attained in the frigid regions of the north, not the minutest trace of a moss or an algæ was discovered in the vast Antarctic Continent, traced by Sir James Ross; nor is it possible that any tribes of the human family can exist in these high southern latitudes; but the animal kingdom has numerous representatives. The penguin and the blue petrel are everywhere to be seen, and the hump-backed and finned-backed whale abound; also, the sea-elephant, and different species of seals, which yield a valuable article of fur. The furthest point in these southern latitudes hitherto attained (78°10′) was reached by Capt. James Ross, in 1842. He says:

"Still steering to the southward, along the coast, a mountain, of 12,400 feet above the level of the sea, was seen emitting flame and smoke in splendid profusion. This magnificent volcano received the name of *Mount Erebus*. It is in south latitude 77°32′, and east longitude 167°. An extinct crater to the eastward, of a somewhat less elevation, was called *Mount Terror*. In other parts of the coast are lofty mountain peaks of from 9,000 to 12,000 feet in height, perfectly covered with eternal snow. The glaciers that descended from near the mountain summits projected many miles into the ocean, and presented a perpendicular face of lofty cliffs."

The United States Exploring Expedition, under the command of Lieut. Charles Wilkes, in 1840–'41, penetrated to 67° S.,

159° E. (magnetic pole 90° S., 140° E.) The Antarctic Continent was reached January 16, 1840, the middle of the summer in this hemisphere; temperature of air 32°, and water 31° Fahr. Island, berg, and field ice surrounded the land. Only about 1,900 miles of the coast of this inhospitable region has been explored by the different navigators.

Thus is exhibited the wonderful phenomena of nature as you approach either of the poles of our globe—the Arctic or the Antarctic. So in regard to every parallel of latitude that is passed in succession, although less marked, in going through the frigid, cold, temperate, sub-tropical, and tropical zones, until you reach the Equator, as well as on ascending to higher altitudes, the Animal and Vegetable Kingdom are found to change; so much so as to place an eternal barrier between the two extremes—one abounding in animated life, and the other in the solitude of the desert.

The Temperate and Sub-tropical Zones, which lie about equidistant from the poles and the equator, are the only highly favored portions of the Earth's surface, as are abundantly proved by the facts exhibited in this compilation. An eminent writer remarks, when speaking of the different races of men, as influenced by climate: "History bears out this theory when it sums up what the nations of the extreme north and south have done for civilization. Were they stricken from the earth, it would feel it no more than the steamship does the wave which sends a shower of spray over its bows, without checking the revolution of the wheels."

In conclusion, it may be said that one-third of the earth's surface is given up to *heat*, too intense and enervating for the advance of the human species, while rank vegetation and animals of an inferior order abound. About one-third is also given up to *cold*, too intense for advancement, MAN being compelled to toil incessantly in order to gain food and clothing sufficient to sustain life. It is only within the remaining third of the earth's surface, or the *Temperate Climates*, ranging from 40° to 70° mean annual temperature, where the seasons are about equally divided, averaging three months for Spring, Summer, Autumn, and Winter, that the human race really thrives, and advances in moral and intellectual culture, and where science and the arts are encouraged.

INDEX.

A

B